W9-CSP-034

Process Dynamics, Control and Protection

By

Brian Roffel

John E. Rijnsdorp

230 Collingwood, P.O. Box 1425, Ann Arbor, Michigan 48106

Library of Congress Card Catalog Number 81-68029
ISBN 0-250-40483-4

Manufactured in the United States of America

Butterworths, Ltd., Borough Green, Sevenoaks, Kent TN15 8PH, England

PREFACE

In various fields of science and technology, actual systems are represented by models. These models can be of different types; they can express relationships in terms of energy, material or information flows, or in terms of cause and effect. The latter models are useful in many respects, including:

1. understanding the actual system behavior;
2. assessing its sensitivity to outside influences;
3. analyzing its stability;
4. predicting its future behavior;
5. investigating its controllability;
6. designing control systems; and
7. designing protection systems.

Most textbooks treat cause-effect models and their applications in terms of specialized mathematical techniques, such as Laplace transform, frequency domain, etc. This is very much to the point for control engineers or students in control engineering. However, most chemical engineers and chemical engineering students have only a secondary interest in control theory; hence, they would benefit from an approach based on more generally known and used mathematical approaches.

This book tries to fill this gap. It uses a straightforward operational notation, which on the one hand is strongly linked to linear differential equations and, on the other hand, to the convenient handling of transfer functions. Much attention is paid to the physical interpretation of mathematical results, which, by the way, can also be of value for process control engineers. In addition, various realistic examples are given and worked out in detail.

Chapters 1 and 2 can be seen as an introduction. Chapters 3 to 8 give an approach to process modeling. Chapters 9 to 13 deal with distributed systems, and in Chapters 14 and 15 the approximation of the solution of partial differential equations is introduced. Chapter 16 deals with control quality, and Chapters 17 and 18 with the development of control systems.

Finally, Chapter 19 gives an introduction to process protection. An introductory course in process dynamics might include Chapters 1 to 8 and an introduction to process control and protection, Chapters 16 to 19. An advanced course in process dynamics could include Chapters 9 to 15.

Brian Roffel John E. Rijnsdorp

ACKNOWLEDGMENTS

The authors would like to express their gratitude to *Polytechnisch Tijdschrift*, for permission to use articles published in that journal as a basis for a number of chapters in this book; to Mr. Arends, for preparing the drawings; and to Miss Jolanda Grob, for typing the manuscript.

Brian Roffel is a Computer Applications Engineer at ESSO Chemical Canada. He holds a MSc in chemical engineering and a PhD in control engineering.

Previously, Dr. Roffel was Chief Senior Research Officer at the Twente University of Technology. During 1976, Dr. Roffel was with the Council for Scientific and Industrial Research in South Africa.

Dr. Roffel has published numerous articles on process dynamics and optimal control.

John E. Rijnsdorp is a Professor of Process Dynamics at Twente University of Technology. He holds a MSc in electrical engineering. He worked as a process control engineer with the Royal Dutch Shell Group, and was appointed as a visiting professor at the Polytechnic Institute of Brooklyn during 1967–1968. He has written many publications on distillation dynamics and control, and human factors in automation.

CONTENTS

CHAPTER 1

INTRODUCTION

Cause-effect relationships play an important role in understanding process behavior. In process technology, relationships are usually expressed in terms of material and energy streams, or exchange between phases. This is, for example, apparent in process flow diagrams, where arrows indicate directions of flow. Process dynamics, however, are related to process operation–more specifically, to changes in process conditions. "What happens if . . . ?" is the major question. As a result, relationships studies are of the cause-effect type, and time-dependent behavior receives much attention. This is evident for batch processes, although process dynamics also finds a field of application in continuous process operation. The changes in continuous process operation can generally be divided into the following categories:

1. dynamic operations,
2. internal changes, and
3. external changes.

DYNAMIC OPERATIONS

Dynamic operations take place when the process is not operated under constant conditions, for example, during startup, shutdown or switchover to other conditions (other product specifications). These operations are usually accompanied by a certain loss of product. Mathematical techniques (dynamic programming [1], maximum principle [2]) are available that can be used to find an operation pattern for minimizing loss of product. However, most modern chemical plants are so complicated that these techniques are difficult to apply, except after gross simplifications have been made.

It also can happen that a small part of a process is operated dynamically, for example, regeneration of catalyst beds or cleaning of burners. Evidently, this causes disturbances in the operation of other parts of the process.

Finally, there is a type of dynamic operation where, during normal operation, the continuous process is varied intentionally, often in periodic fashion. An example is the pulsation of extraction columns, where a regularly repeated pulse promotes the mixing of the two liquid phases [3]. Research has been done on the application of this method to other processes, such as distillation and chemical reactors [4,5]. In the following cases, processes are supposed to operate under constant conditions. Internal or external changes can interfere with this desired state.

INTERNAL CHANGES

Drastic internal changes are caused by failures in process equipment, control instrumentation and the like. A more subtle case occurs when a process runs away from the desired constant state without malfunctioning. An example is a chemical reactor in which an exothermic reaction takes place. Here a small increase in reaction temperature causes a larger heat production, which cannot always be met by increased heat discharge. Another case of process instability is the phenomenon of spontaneous periodic fluctuations. In both cases the process is unstable. A borderline case is the slow change of conditions of a continuous process by fouling, poisoning of catalyst, coke depositing in furnace tubes, etc.

EXTERNAL CHANGES

For this type of change, continuous operation is disturbed by external influences. In many cases it is a matter of disturbances that enter the process by feed, heating, catalyst flows, etc. Here the term "disturbances" is used with a special meaning: it means relatively small changes in process conditions without malfunctioning of process apparatus and control instrumentation. The response of process variables to simple disturbances will be treated in more detail at a later stage. Because all changes from internal and external causes are undesirable at constant operation, the process has to be corrected. Corrections can be manual, but the development of process control has led to a large-scale changeover to automatic correction.

Process behavior, however, imposes restrictions on the reduction of changes. This will be dealt with in a later chapter, where control quality will be analyzed quantitatively.

INFORMATION FLOW DIAGRAMS

Cause-effect relationships can be visualized in information flow diagrams. Figure 1.1 shows a process and information flow diagram of a simple level process. The difference is remarkable. In the process flow diagram the arrows have the same direction as the process flow; in other words, the inlet flow ϕ_{in} is an input variable and the discharge flow ϕ_{out} is an output variable. In the information flow diagram, the arrows are arranged according to cause-effect relationships. Therefore, ϕ_{in} and ϕ_{out} are both input variables, because a change in one or both flows will cause a change in the level L. If the input variable of the block is independent of the system, the input variable is marked by a double arrow. In Figure 1.1, such an independent input variable is given: the inlet flow ϕ_{in}, which can be seen as a disturbance for the process. These information flow diagrams will be developed further in the chapters dealing with process dynamics.

MATHEMATICAL MODELING

In practice, an engineer is often confronted with a vague problem description. For example, there may be a difficulty with an existing system or process, or an analysis must be made of certain aspects of a new design. Hence, he has to know how to translate the vague physical picture into a relevant mathematical description. In process dynamics, this usually results in a dynamic model, which can be interpreted as a generalization of a static model.

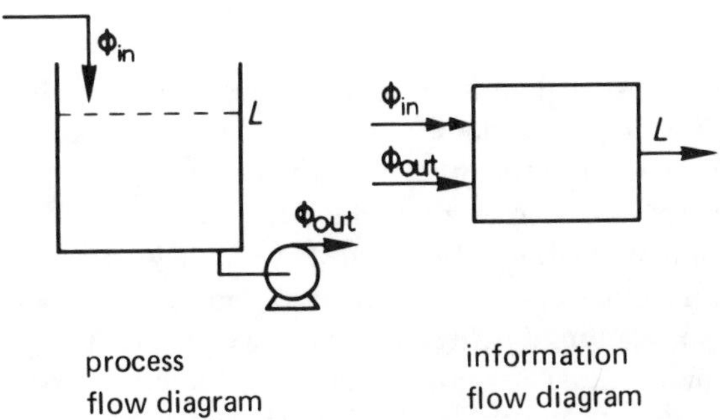

Figure 1.1 Process and information flow diagram for a level process.

It should be stressed that a mathematical model is not a fixed, invariable matter. In fact, different models can be made for the same physical or chemical system. The selection of a model will depend strongly on the purpose for which the model will be used. Purpose definition is thus the first step in model building. After this is done, the part of the process which has to be modeled is demarcated with respect to its environment. This leads to the choice of relevant inputs and outputs which can be represented in an information flow diagram.

After this qualitative phase, quantitative relationships must be found as building blocks for the mathematical model. Usually it is convenient to start with the laws of conservation of mass, momentum and energy, which take the form of differential equations for the dynamic case. In these equations, the derivatives represent the rate of change of mass, momentum or energy content of the system.

It may happen that variables are not only a function of time but also depend on one or more geometric dimensions. In that case, the system is called a distributed parameter system. An example of a distributed parameter system is a tubular reactor where the temperature and concentration of reactants are a function of time and of distance along the reactor. Examples will be given in Chapters 9 to 13 of distributed systems, and the way they can be analyzed.

In the mathematical model, derivatives can be approximated by differences:

$$\frac{\partial x}{\partial z} = \lim_{\Delta z \to 0} \frac{\Delta x}{\Delta z} \tag{1.1}$$

where x = a process variable
z = a geometric dimension

In this way, the partial differential equation is replaced by one or more ordinary differential equations containing only derivatives with respect to time. This approach is followed in Chapters 2 to 8.

When the laws of conservation do not yield a sufficient number of equations, other equations are needed describing fundamental physical or chemical relationships, such as material properties, equations of state or equilibrium relationships, reaction rates and so on. These equations are usually algebraic. Several equations may be empirical. In uncomplicated cases it is possible to substitute the algebraic equations into the differential equations, and to write these equations explicitly in terms of derivatives. The general expression is:

$$\frac{dx_1(t)}{dt} = f_1[x_1(t), x_2(t), \ldots x_n(t); u_1(t) \ldots u_r(t)] \tag{1.2}$$

$$\frac{dx_n(t)}{dt} = f_n[x_1(t), x_2(t), \ldots x_n(t); u_1(t) \ldots u_r(t)] \tag{1.3}$$

In these equations, $u_1(t) \ldots u_r(t)$ are process input variables, and $x_1(t) \ldots x_n(t)$ are the state variables. This indicates that they characterize the state of the system at any time. When the values of the state variables are known at a particular time, past behavior is irrelevant because the past is represented in the value of the state variables.

STABILITY ANALYSIS

One of the first aspects to study is stability. This is important for both uncontrolled and controlled processes. In practice, large instabilities should be avoided, and a reasonable safety margin to this situation should be maintained. Only after a solution has been found for the stability problem can sensitivity to disturbances and control quality be analyzed. A simple but incomplete approach is to analyze stability using linearized versions of the equations. This is based on the assumption that the process remains close to the average conditions. Then every state variable x_i can be replaced by the sum of an average value and a small deviation

$$x_i = \bar{x}_i + \delta x_i \tag{1.4}$$

Similarly, input variables u_j are replaced by:

$$u_j = \bar{u}_j + \delta u_j \tag{1.5}$$

Substitution of Equations 1.4 and 1.5 in Equations 1.2 and 1.3 together with:

$$\overline{\frac{dx_i}{dt}} = 0 \ , \quad i = 1 \ldots n \tag{1.6}$$

and therefore,

$$f_i[\bar{x}_1(t), \bar{x}_2(t), \ldots \bar{x}_n(t); \bar{u}_1(t) \ldots \bar{u}_r(t)] = 0 \ , \quad i = 1 \ldots n \tag{1.7}$$

results in a set of simultaneous ordinary linear differential equations, which can be written in the form:

$$\left.\begin{aligned} \frac{d}{dt}(\delta x_i) + \sum_{i=1}^{n} a_{1i}\delta x_i &= \sum_{j=1}^{r} b_{1j}\delta u_j \\ \frac{d}{dt}(\delta x_n) + \sum_{i=1}^{n} a_{ni}\delta x_i &= \sum_{j=1}^{r} b_{nj}\delta u_j \end{aligned}\right\} \tag{1.8}$$

The coefficients $a_{11} \ldots a_{nn}$ and $b_{11} \ldots b_{nr}$ depend on the average values $\bar{x}_1 \ldots \bar{x}_n$, $\bar{u}_1 \ldots \bar{u}_r$. For stability analysis, the right sides of these equations are set equal to zero as the stability corresponds to system behavior with constant inputs. General solutions are of the following type:

$$Ce^{\alpha t}, Cte^{\alpha t} \ldots Ce^{\alpha t}\sin(\omega t + \phi), Cte^{\alpha t}\sin(\omega t + \phi) \tag{1.9}$$

These terms determine the behavior of the system when it is uncontrolled. The coefficients α are of great importance: if α is negative in all terms, they converge to zero, which means that the system moves to an equilibrium; in other words the system is stable. If one α is equal to zero with all others negative, the system is on the limit of stability. If one or more α are positive, the deviation will increase without limit; hence, the system is unstable. Mathematically spoken, the deviations go to infinity. This will not happen in practice, because linearization of the original equations is not allowed when deviations become too large. This indicates a danger, which in non-linear mechanics is known as "hard oscillations": the system is stable for small deviations, but it may be unstable for large deviations. To find hard oscillations, one has to study the original mathematical model in terms of its state space representation of the state variables.

RESPONSES TO DISTURBANCES

After the stability of the system has been analyzed, we are in a position to investigate the influence of the independent variables. First we shall look at disturbances. As the disturbance pattern can seldom be predicted, a standard form is often chosen. Figure 1.2 shows the most frequently used possibilities (solid line) with the response of the uncontrolled process (dotted line). We will not go into further details of the use, advantages and

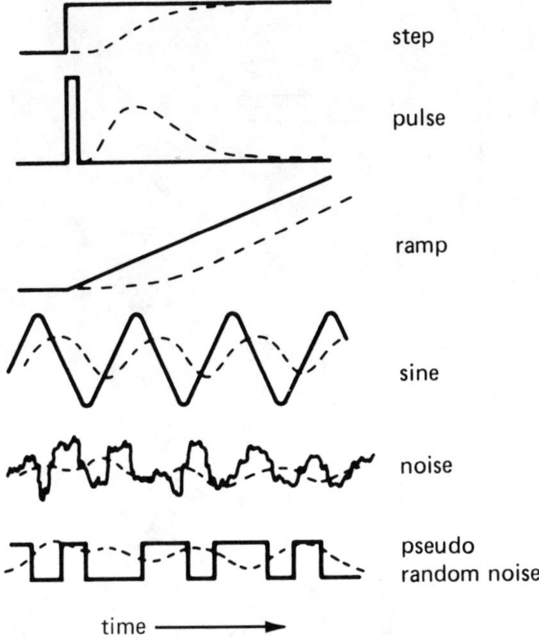

Figure 1.2 Survey of the most frequently used disturbance signals. Solid line denotes standard disturbance signal, dotted line indicates response.

disadvantages of the different disturbance patterns for experimental determination of process dynamics. In the following chapters the assumption is made that the disturbance has the form of a step. This type of disturbance often happens in the process industry, e.g., in case of failure of apparatus. When a control system can cope with a severe disturbance like a step, it may be expected that it can also cope with minor disturbances.

THE AUTOMATIC CONTROL LOOP

The information flow diagram of an automatic control loop is shown in Figure 1.3. The measured value of the controlled variable is compared with a set value. The difference between these two values, the deviation, results in an output signal of the controller, acting on the correcting unit, which should eventually reduce or, preferably, eliminate the deviation. When analyzing the stability of different controlled processes, it will first be assumed that control is achieved by means of the most simple type of

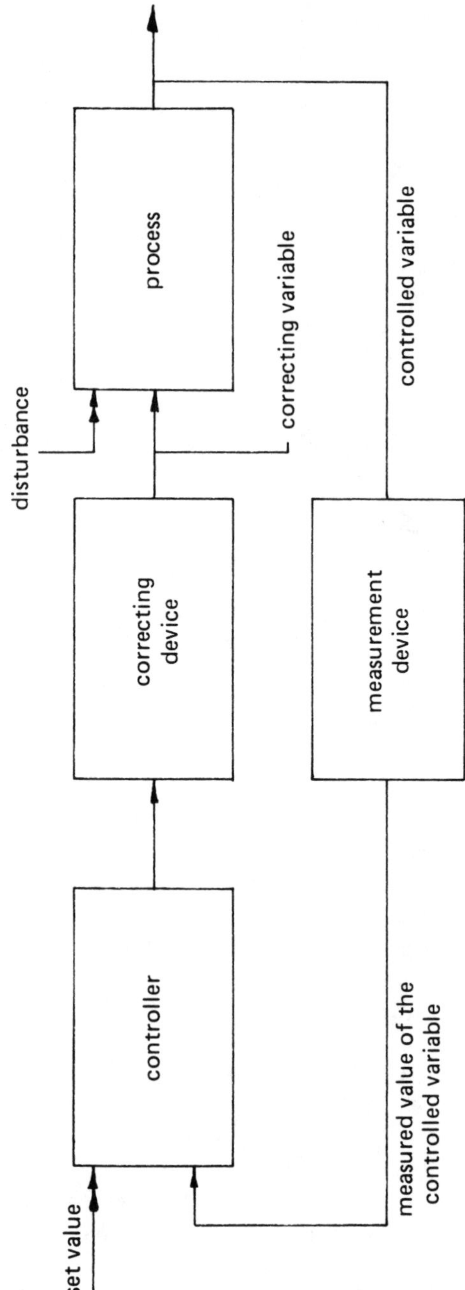

Figure 1.3 Information flow diagram of the control loop.

controller: the proportional controller. The mathematical expression for this controller is:

$$\delta u = K_c \epsilon \tag{1.10}$$

where δu = the change in output signal of the controller (the correction)
K_c = the controller gain
ϵ = the deviation (the difference between measured value and set value of the controlled variable).

When the controller gain K_c is small, controller action is small, and the deviation ϵ must be large to produce a reasonable correction δu. It would therefore be attractive if the controller gain could be set at a large value: a small deviation would produce a large correction.

However, if one tries to increase the controller gain in actual applications, one will discover that the control loop starts to oscillate above a certain value of the controller gain, and that at higher values, the system is even unstable. In practice, an unstable control system is unacceptable. Figure 1.4 shows these differing responses to a step in set value.

In practice, of course, an unstable system is unacceptable, but a low controller gain is not satisfactory either. In Chapter 16, other control actions will be discussed to cope with this problem and to improve control quality.

CONTROL QUALITY

For a given dynamic process behavior, a certain control quality can be expected for correct adjustment of controller parameters. The control quality can be judged on the basis of the control speed and power. The speed of control is related to the period of oscillation of the control loop at the limit of stability (Figure 1.4). The shorter this period, the higher the speed of control. The power of control is related to the range over which control is effective. In the case of a control valve, the most that can be done is to open or close it entirely (apart from leakage). If the disturbances are so large that these maximum corrections are not sufficient to correct deviations, the power of control is insufficient. It is evident that one should try to avoid this situation as much as possible. One possibility is to increase the capacity of the control valve. However, this possibility cannot always be used, because even without a valve, flow through a pipe is limited. Another disadvantage of an increase of the control valve capacity is the increased leakage in closed position; thus, the minimum flow is increased. Finally, there is a third difficulty: when the size of the control valve is not in accordance with

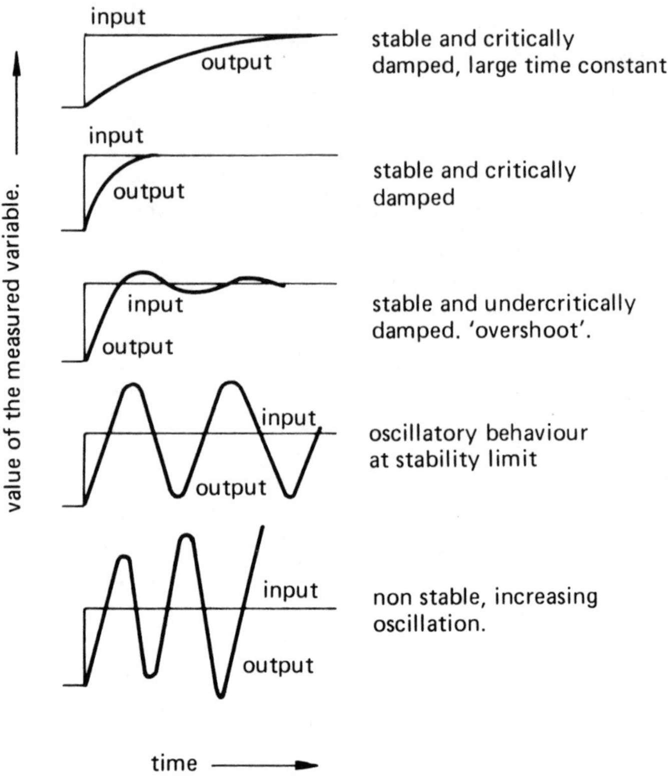

Figure 1.4 Possible responses of a controlled system on a step in the input.

other flow restrictions in the circuit, the relationship between valve opening and flow becomes strongly nonlinear. Then a control loop, which is stable at one valve opening, can become unstable at another valve opening.

CHOICE OF CONTROL SCHEME

The analysis of process and control loop dynamics is applied in the synthesis of control schemes. First, the correcting variables must be selected from the input variables, which indicates where correcting units should be installed in the process. This selection cannot be made independent of the selection of the controlled variables from the output variables of the process. Here the control objective plays a large role. On the basis of both selections, a table can be made with columns for the controlled variables and rows for the correcting variables. In every field of this table, data can be filled in for the control speed and power belonging to that combination of correcting and controlled variables.

With the aid of this table the best connections between controlled and correcting variables can be selected. Evidently the correcting variable with highest speed of control and largest power of control is preferred. It may happen, however, that one correcting variable has a high speed of control and another a high power of control. Then all circumstances should be examined carefully before making a decision.

The control scheme thus constructed can be called basic control scheme, as it contains the minimally required number of controllers. However, to reduce the effects of disturbances still further the control scheme may be extended, for example by "master-slave" control or "feedforward" control. Fortunately, the control engineer has a large range of possibilities for solving the control problem in a flexible and adequate way.

PROTECTIONS

An important function of instrumentation in a chemical plant is to increase availability and safety. The conventional control loops can contribute to safety by keeping process variables close to desired values, thus avoiding abnormal values. However, this is not sufficient. Control loops can be deficient in different ways, for example, as a result of insufficient power of control, or they may simply be put on "manual" by the operator. If the operators do not notice lack of control action, a further "line of defense" is formed by the alarm system (auditory and visual) to warn them. Finally, if they do not return the process to normal operation, an automatic safety device can come into action, bringing the process back to a safe situation (which may be a shutdown). A difficulty of a too-extensive alarm and protection system is the probability of spurious trips. If this happens too often, operators will take no more notice of alarm signals and disable the protection devices. With special techniques (high-integrity alarms and protections) the probability of spurious trips can be decreased.

In the following chapters the various subjects that were introduced in this chapter will be discussed and analyzed in more detail.

CHAPTER 2

DYNAMICS AND CONTROL OF A LEVEL PROCESS

In this chapter, various aspects of process dynamics and control will be demonstrated for a simple case: the level in a buffer vessel (Figure 2.1). Simplifying assumptions will be made that eliminate some of the finer points, but still keep the main features intact. After the process description, the simplifying assumptions will be formulated. Then the dynamic model will be analyzed, the dynamics of the level measurement will be modeled, and a synthesis of the control loop will be made.

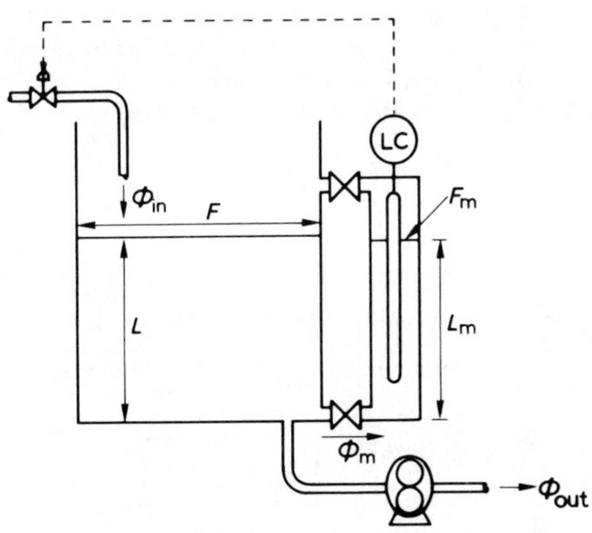

Figure 2.1 Level control.

PROCESS DESCRIPTION

Buffer vessels are used to attenuate flow variations that travel from one process section to another. The holdup in the vessel allows the operator some time to take action, e.g., in case of malfunctioning in the upstream process. The level in the buffer vessel provides net positive suction head for the pump, to avoid cavitation.

The pump shown in Figure 2.1 withdraws a constant flowrate ϕ_{out} from the buffer vessel, which ensures a constant feed rate to the downstream process section. This means that the flowrate withdrawn from the upstream section (ϕ_{in}) must be adjusted for keeping the level in the buffer vessel within bounds. This is ensured by an automatic level controller, acting on a control valve in the inlet flow. The result is a type of control which acts countercurrently to the direction of material flow. In practice, it is more common to apply a co-current relationship, where the level is controlled by the outgoing flow. The choice between these alternatives depends on different factors (see Chapter 17).

ASSUMPTIONS AND MODELING APPROACH

The characteristic of the control valve and the correction dynamics will be dealt with in the next chapter. It is assumed here that the inlet flow follows the output of the controller according a linear relationship without any time lag. The vessel and measurement tube show a certain holdup effect. The flow of the liquid to the measurement tube encounters friction in a valve. The combination of holdup and friction leads to a characteristic time lag. The information flow diagram of the control loop is shown in Figure 2.2. For each

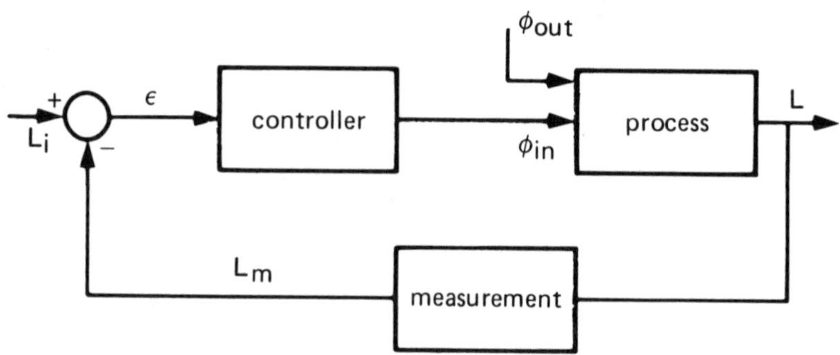

Figure 2.2 Information flow diagram of level control.

block in this figure, the transfer function will be determined. The transfer function is nothing more than a generalized mathematical expression for the relation between input and output. To find the transfer function, the differential operator s will be introduced:

$$sx = \frac{dx}{dt} \tag{2.1}$$

This operator can only be used in the case of linear differential equations. For linear time-independent models, initially at equilibrium, this operator is equivalent to the Laplace operator s (see Appendix).

For the process, a mass balance can be set up: change of mass holdup per unit of time is equal to mass flow rate in minus mass flowrate out:

$$\frac{d}{dt} M = \phi_{in} - \phi_{out} \tag{2.2}$$

The mass of the liquid holdup in the vessel M is equal to the product of the density ρ, the area F of the vessel and the liquid height L. If ρ and F are constant, Equation 2.2 may be written as:

$$\frac{dL}{dt} = \frac{1}{\rho F}(\phi_{in} - \phi_{out}) \tag{2.3}$$

Equation 2.3 is already linear; hence, it need not be linearized. However, to be in line with the notation which will be introduced in the next sections, we shall isolate the influence that a small change in the inlet flowrate has on the level:

$$\frac{d\delta L}{dt} = \frac{1}{\rho F} \delta\phi_{in} \tag{2.4}$$

After replacing d/dt by s, this equation can be written as:

$$s\delta L = \frac{1}{\rho F} \delta\phi_{in} \tag{2.5}$$

which leads to the transfer function (ratio of output and input):

$$\frac{\delta L}{\delta \phi_{in}} = \frac{1}{\rho F s} \tag{2.6}$$

In Chapter 1 the term capacitative effect was introduced. This is actually not the same as capacity. The capacity of the vessel indicates the maximum amount of liquid the vessel can contain. The capacitative effect indicates the change in quantity per unit change in an intensity variable. For the amount of liquid in the vessel, the capacitative effect C_p is defined as the change in mass per unit change in liquid height:

$$C_p = \frac{\delta M}{\delta L} \tag{2.7}$$

Now

$$\delta M = \rho F \delta L \tag{2.8}$$

hence

$$C_p = \rho F \tag{2.9}$$

and Equation 2.6 may be written as:

$$\frac{\delta L}{\delta \phi_{in}} = \frac{1}{C_p s} \tag{2.10}$$

INHERENT REGULATION

When the pump that withdraws a constant flow from the vessel is replaced by a restriction, e.g., a valve, the liquid can still flow out of the vessel, but not with a constant flowrate. An increase in liquid level will obviously lead to an increase in outlet flow. This effect tends to restore equilibrium and is called inherent regulation. If the flow through the restriction is turbulent, the following pressure drop equation holds:

$$\frac{\Delta p}{\rho g} = L = \left(\frac{\phi_{out}}{c_1}\right)^2 \tag{2.11}$$

where L = height of the level above the restriction
c_1 = a parameter depending on the valve opening and physical properties of the liquid
g = the gravity acceleration

For small variations around average conditions, Equation 2.11 becomes

$$\bar{L} + \delta L = \left(\frac{\bar{\phi}_{out} + \delta\phi_{out}}{c_1}\right)^2 \tag{2.12}$$

where $\bar{L}$ and $\bar{\phi}_{out}$ refer to average conditions, δ means a variation. Average terms in Equation 2.12 cancel out, and linearization means ignoring higher-order terms, $\theta(\delta^2)$. Consequently, only the first-order terms in δ remain:

$$\delta L = 2\frac{\bar{\phi}_{out}}{c_1^2}\delta\phi_{out} = R_p \cdot \delta\phi_{out} \tag{2.13}$$

where R_p is the resistance coefficient for flow through the valve. This resistance coefficient may generally be determined from:

$$R_p = \frac{1}{\rho g}\frac{\delta(\Delta p)}{\delta\phi} \tag{2.14}$$

Combination of Equation 2.3 for small variations together with Equation 2.13 results in:

$$\frac{\delta L}{\delta\phi_{in}} = \frac{R_p}{\tau_p s + 1} \tag{2.15}$$

where

$$\tau_p = R_p C_p \tag{2.16}$$

the process time constant.

When $\delta\phi_{in}$ has the form of a unit step (going from 0 to 1), the response of the level can easily be calculated. For the case without inherent regulation (Equation 2.10) the response is:

$$\delta L = \frac{t}{C_p} \tag{2.17}$$

and for the case with inherent regulation (Equation 2.15):

$$\delta L = R_p \left(1 - e^{-t/\tau_p}\right) \tag{2.18}$$

Both responses are shown in Fig. 2.3. The transfer function of Equation 2.15 can be represented by an electrical system consisting of a resistance and a capacity. This electrical system is shown in Figure 2.4. The following equations hold:

Ohm's law:

$$U_R = i \cdot R \tag{2.19}$$

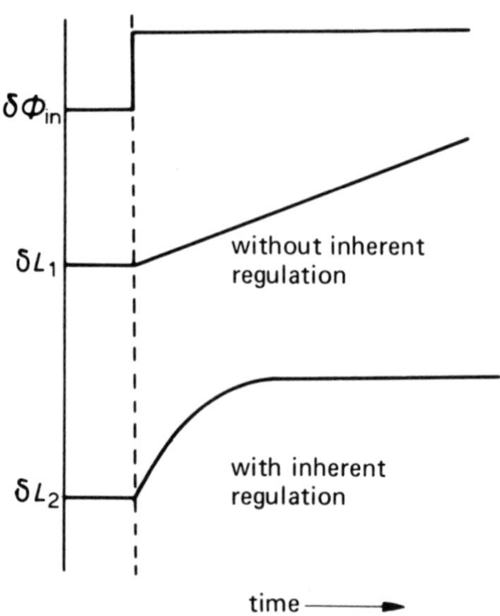

Figure 2.3 Response of the uncontrolled level process to a step in the inlet liquid flow.

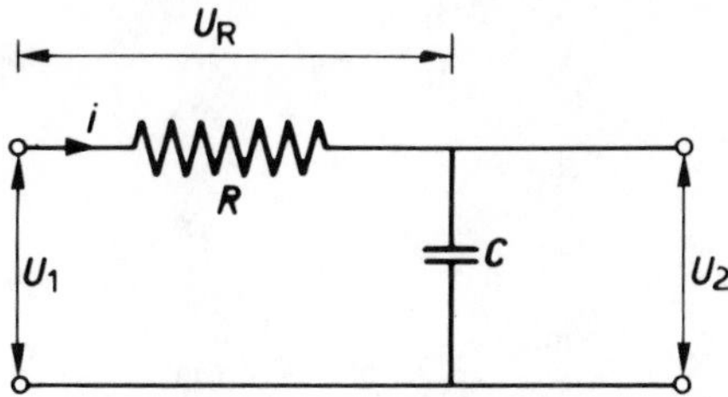

Figure 2.4 Electrical network.

$$i = C \cdot \frac{dU_2}{dt} \tag{2.20}$$

$$U_1 = U_R + U_2 \tag{2.21}$$

thus,

$$\frac{U_2}{U_1} = \frac{1}{RCs + 1} \tag{2.22}$$

where i = current
U = voltage

DYNAMICS OF THE LEVEL MEASUREMENT

When the liquid level in the vessel L is higher than the liquid level in the measurement tube L_m, liquid will flow from the vessel to the measurement tube. The difference $L - L_m$ is used to overcome the resistance due to flow and for inertia opposing acceleration of the liquid in the connecting tube and measurement device. When the valve in the connecting tube is throttled, the effect of friction is much stronger than that of inertia. Then, with good approximation:

$$\frac{\Delta p_m}{\rho g} = L - L_m = \left(\frac{\phi_m}{c_2}\right)^2 \tag{2.23}$$

The mass balance for the measurement tube is:

$$\frac{dM_m}{dt} = \phi_m \tag{2.24}$$

or

$$\frac{dL_m}{dt} = \frac{1}{\rho F_m}\phi_m = \frac{1}{C_m}\phi_m \tag{2.25}$$

Linearization of Equations 2.23 and 2.25 results in:

$$\frac{\delta L_m}{\delta L} = \frac{1}{\tau_m s + 1} \tag{2.26}$$

where

$$\tau_m = R_m C_m \tag{2.27}$$

the measurement time constant, with:

$$R_m = 2\,\frac{\overline{\phi}_m}{c_2^2} \tag{2.28}$$

In practice, flow will not always go from vessel to measurement tube, but will reverse from time to time. Then, with some approximation, $\overline{\phi}_m$ has to be taken as the amplitude of the fluctuation.

ANALYSIS OF THE CONTROLLED PROCESS

When as in the case of Figure 2.2, the controller has only proportional action, its behavior is described by the relation:

$$\phi_{in} = \overline{\phi}_{in} + K_c\epsilon \tag{2.29}$$

with

$$\epsilon = L_i - L_m \tag{2.30}$$

It can be seen that K_c in Equation 2.29 is not dimensionless. L_i is the set value of the level and $\bar{\phi}_{in}$ is the average value of the inlet flow at ϵ equal to zero. Notice that the controller provides negative feedback (see also Figure 2.2): a larger value of L results in a larger value of L_m and a smaller value of ϵ thus resulting in a decrease in ϕ_{in} and hence in L. Writing Equations 2.29 and 2.30 for small variations leads to:

$$\delta\phi_{in} = -K_c \delta L_m \tag{2.31}$$

when L_i is constant.

The information flow diagram of Figure 2.2 can be quantified to that of Figure 2.5. Combination of Equations 2.10, 2.26 and 2.31 results in the following equation for the control loop of Figure 2.5:

$$\delta L = -\frac{1}{\tau_m s + 1} \frac{K_c}{C_p s} \delta L \tag{2.32}$$

which may be rewritten as:

$$\delta L\,(\tau_L \tau_m s^2 + \tau_L s + 1) = 0 \tag{2.33}$$

with

$$\tau_L = \frac{C_p}{K_c} \tag{2.34}$$

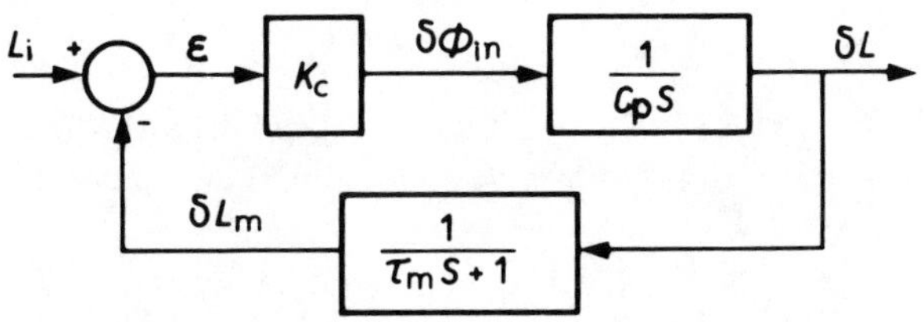

Figure 2.5 Transfer functions in the controlled level process.

Equation 2.33 is called the characteristic equation, and determines the stability of the process. When τ_L is greater than $4\tau_m$, Equation 2.33 has two real roots, and the response is overcritically damped. Here the characteristic equation may be written as:

$$(\tau_1 s + 1)(\tau_2 s + 1) = 0 \tag{2.35}$$

Equation 2.33 together with 2.35 result in:

$$\tau_{1,2} = \frac{\tau_L \pm \sqrt{\tau_L^2 - 4\tau_L\tau_m}}{2} \tag{2.36}$$

and the general solution takes the form:

$$\delta L = A_1 e^{-t/\tau_1} + A_2 e^{-t/\tau_2} \tag{2.37}$$

where the constants A_1 and A_2 may be determined from the initial conditions. When τ_L is equal to $4\tau_m$, Equation 2.33 has two equal roots, and the response is critically damped. The characteristic equation is then written as:

$$(\tau s + 1)^2 = 0 \tag{2.38}$$

with

$$\tau = \frac{\tau_L}{2} = 2\tau_m \tag{2.39}$$

The general solution is now:

$$\delta L = A_1 e^{-t/\tau} + A_2 t e^{-t/\tau} \tag{2.40}$$

where again A_1 and A_2 follow from the initial conditions. When τ_L is smaller than $4\tau_m$, Equation 2.33 is usually written in the standard notation:

$$\left(\frac{s}{\omega_n}\right)^2 + 2\zeta\left(\frac{s}{\omega_n}\right) + 1 = 0 \tag{2.41}$$

where ω_n is the oscillation frequency without damping, here given by:

$$\omega_n = \frac{1}{\sqrt{\tau_L \tau_m}} \tag{2.42}$$

and ζ the damping coefficient:

$$\zeta = \frac{1}{2}\sqrt{\tau_L/\tau_m} \tag{2.43}$$

The response is undercritically damped and the general solution of Equation 2.41 is damped oscillation:

$$\delta L = Ae^{-\zeta\omega_n t}\cos(\omega_n\sqrt{1-\zeta^2}\,t + \varphi) \tag{2.44}$$

where A and φ follow from the initial conditions. When analyzing the dynamics of the uncontrolled process, it was assumed that the discharge flow ϕ_{out} was constant. However if ϕ_{out} fluctuates, Equation 2.3 becomes, for small variations:

$$\delta L = \frac{1}{C_p s}(\delta\phi_{in} - \delta\phi_{out}) \tag{2.45}$$

combination with Equations 2.26, 2.31 and 2.34 results in:

$$\delta L\left(\frac{\rho F}{\tau_m}\right)[\tau_L\tau_m s^2 + \tau_L s + 1] = \frac{\tau_L}{\tau_m}(\tau_m s + 1)\delta\phi_{out} \tag{2.46}$$

where the part between square brackets is the characteristic equation again. In Figure 2.6 some responses on a unit step in ϕ_{out} are shown for different values of the ratio between τ_L and τ_m, thus for different values of the damping coefficient ζ. For large values of the time, when equilibrium is being approached (thus $s \rightarrow 0$), Equation 2.46 gives for the deviation in the level:

$$\delta L\left(\frac{\rho F}{\tau_m}\right) = \frac{\tau_L}{\tau_m} \tag{2.47}$$

When the value of K_c is small, τ_L and δL are large, resulting in a large sustained control deviation (offset). When the controller gain is large, the value

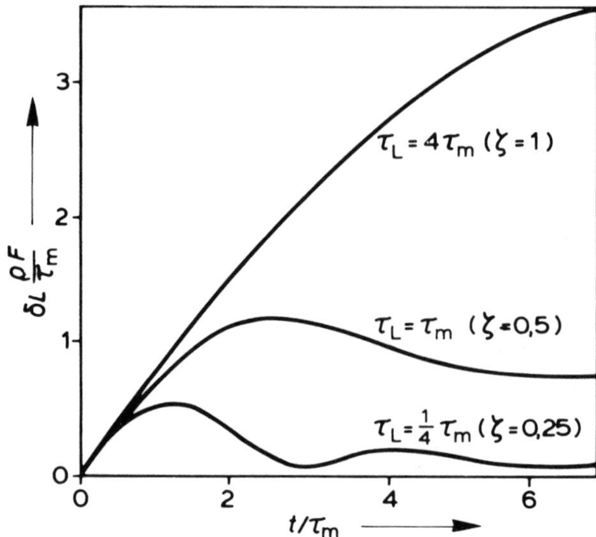

Figure 2.6 Responses of the controlled level process to a unit step in the inlet liquid flow.

of τ_L is small; hence, δL is small, which means that the level returns almost entirely back to its initial value after a unit step in the inlet flow.

However, when τ_L approaches zero, ζ approaches zero. Then the response will show increasingly oscillatory behavior. The limit of stability is theoretically reached for ζ equal to zero or K_c equal to infinity, in other words, in a second-order control loop the controller gain has to be made infinite before the system becomes unstable. The characteristic Equation 2.41 is then simplified to:

$$s^2 + \omega_n^2 = 0 \tag{2.48}$$

Of course actual systems do not behave in this way. They have to be represented by at least third-order models, as will be shown in later chapters.

GENERAL CONDITIONS FOR STABILITY

When the characteristic equation is a high-order polynomial in s, given by

$$G(s) = 0 \tag{2.49}$$

the general solution of the differential equation will contain many terms. However, a relatively simple expression can be derived for the limit of stability. This is given by the condition that the general solution should contain a term of the form:

$$A \sin(\omega_n t + \beta) \tag{2.50}$$

This is the case when the characteristic equation can be divided by $s^2 + \omega_n^2$, which means that the expression for the limit of stability can be determined by putting the remainder of the division equal to zero:

$$\text{remainder}\left(\frac{G(s)}{s^2 + \omega_n^2}\right) = 0 \tag{2.51}$$

Another way to put this is to say that the characteristic equation should have a pair of imaginary roots $\pm j\omega_n$.

In practice, the value of the damping coefficient should not be too large or too small. A large value means a slow response, and a small value means continued oscillation. A reasonable choice is, for example, a value of ζ where the amplitudes of successive periods in the oscillation have a ratio of 0.20. This means for the difference in phase (see Equation 2.44):

$$\omega_n \sqrt{1 - \zeta^2}\,(t_2 - t_1) = 2\pi \tag{2.52}$$

and for the amplitude ratio:

$$e^{-\omega_n \zeta (t_2 - t_1)} = 0.20 \tag{2.53}$$

Elimination of $\omega_n(t_2 - t_1)$ results in:

$$\zeta \cong 0.25 \tag{2.54}$$

For the level control loop, this means, according to Equation 2.43:

$$\tau_L = \frac{1}{4}\,\tau_m \tag{2.55}$$

from which the period of the damped oscillation:

$$P_0 = \frac{2\pi}{\omega_0} = \frac{2\pi}{\omega_n\sqrt{1-\zeta^2}} \tag{2.56}$$

becomes equal to (see also Equation 2.42):

$$P_0 = 3.24\ \tau_m \tag{2.57}$$

This expression indicates how the speed of control depends on the secondary lag in the control loop, the measurement lag. A small value of P_0 means fast control, a large value means slow control.

For a control loop with the characteristic Equation 2.49, the period P_0 for ζ equal to 0.25 can be determined from the condition:

$$\text{remainder}\left[\frac{G(s)}{\left(\frac{s}{\omega_0}\right)^2 + 0.5\left(\frac{s}{\omega_0}\right) + 1}\right] = 0 \tag{2.58}$$

The roots are of the form: $\omega_0(0.25 \pm j.0.968)$. The ratio between the imaginary and real parts is equal to $15^{1/2} = 3.87$.

APPLICATION OF LEVEL CONTROL

In practice, level control is often installed on buffer tanks with the intention to keep the discharge flow as constant as possible. As long as the tank does not overflow or become empty, the level may have every value. A way to realize this is proportional level control where the control valve is wide open at maximum level and closed at minimum level. The behavior of the buffer tank can then be approximated by ($\tau_L >> \tau_m$):

$$\frac{\delta\phi_{out}}{\delta\phi_{in}} = \frac{1}{1 + \tau_L s} \tag{2.59}$$

with $\tau_L = \Delta V/\Delta\phi_{out}$. ΔV is the volume of the buffer tank corresponding to the span of the level measuring unit and $\Delta\phi_{out}$ is the discharge flow range corresponding to the range of the control valve. However, it may happen in other cases, e.g., in chemical reactors, that the level must be controlled very accurately. Then the controller gain has to be made relatively large.

In this chapter we have assumed that the inlet liquid flow follows the controller output without any time lag nor nonlinearity. However, the controller signal usually changes a valve position, resulting in flow variations. In the next chapter, the valve characteristic and correction lag will be dealt with in more detail.

EXAMPLES

Level Control (Figure 2.7)

A water tank has an area of 1 m^2 and a height of 3 m. The average liquid supply and discharge flow are equal to 0.01 m^3/sec. Under static conditions the liquid height in the tank is equal to 1 m. A level control is installed with a linear control valve in the discharge flow. The flow through the valve is turbulent and can be given by:

$$\phi = u c_v \rho \sqrt{gh} \tag{2.60}$$

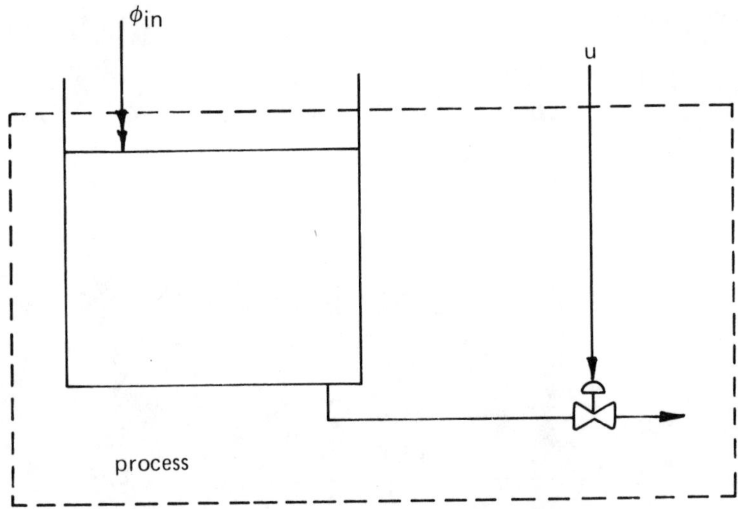

Figure 2.7 Level process for example problem.

where ϕ = the mass flow (kg/sec)
u = the signal from the controller $0 \leqslant u \leqslant 1$
c_v = the valve constant (m^2)
ρ = the density of the liquid (kg/m^3)
g = the acceleration of gravity (m/sec^2)
h = the liquid height (m)

Under the original static condition u is equal to 0.3.

1. With what percentage does the liquid level increase, when the liquid supply is increased 10% and if there is no control?
2. Represent the model with control in an information flow diagram with transfer functions. Measurement and correction lag are both equal to 4 sec.
3. Determine the effect of controller gain on the limit of stability.

Hint: start from a mass balance for the liquid in the tank.

1. In the original static situation we have:

$$\phi_{in} = \phi_{out} = u c_v \rho \sqrt{gh} \tag{2.61}$$

with the following data:

$\phi_{in} = 0.01\ m^3/sec = 10\ kg/sec$
$u = 0.3$
$\rho = 10^3\ kg/m^3$
$g = 9.81\ m/sec^2$
$h = 1\ m$

we can compute c_v:

$$c_v = 0.0106\ m^2$$

In the new static situation when the inlet flow is equal to 11 kg/sec, inlet and discharge flow will again be the same:

$$\phi_{in} = \phi_{out} = 0.3 \times 0.0106 \times 10^3 \sqrt{9.81\ h} = 11\ kg/sec$$

from which,

$$h = 1.21\ m$$

which means that the level increases 21%. This result was to be expected; the quadratic relationship between flow and level height leads to doubling the percentage, or, more precisely, to $(1.1)^2 = 1.21$ for the relative level height change.

2. We start with a dynamic mass balance:

$$\rho A \frac{dh}{dt} = \phi_{in} - \phi_{out} \tag{2.62}$$

where A is the vessel area. Combination of Equations 2.61 and 2.60 results in:

$$\rho A \frac{dh}{dt} = \phi_{in} - uc_v\rho\sqrt{gh} \tag{2.63}$$

Linerarization of Equation 2.63 with d/dt = s gives:

$$\rho As\delta h = \delta\phi_{in} - c_v\rho\sqrt{g\bar{h}}\cdot\delta u - \frac{\bar{u}c_v\rho\sqrt{g}}{2\sqrt{\bar{h}}}\delta h \tag{2.64}$$

or

$$\left[\rho As + \frac{\bar{u}c_v\rho\sqrt{g}}{2\sqrt{\bar{h}}}\right]\delta h = \delta\phi_{in} - c_v\rho\sqrt{g\bar{h}}\cdot\delta u \tag{2.65}$$

Substitution of the given data in Equation 2.65 yields:

$$\delta h = \frac{0.02}{200s + 1}\delta\phi_{in} - \frac{6.67}{200s + 1}\delta u \tag{2.66}$$

When it is assumed that the signal of the measurement device varies between 0 and 1 for level variations between 0 and 3 m, the transfer function for the measurement device is:

$$\frac{0.33}{1 + 4s} \tag{2.67}$$

The information flow diagram is given in Figure 2.8.

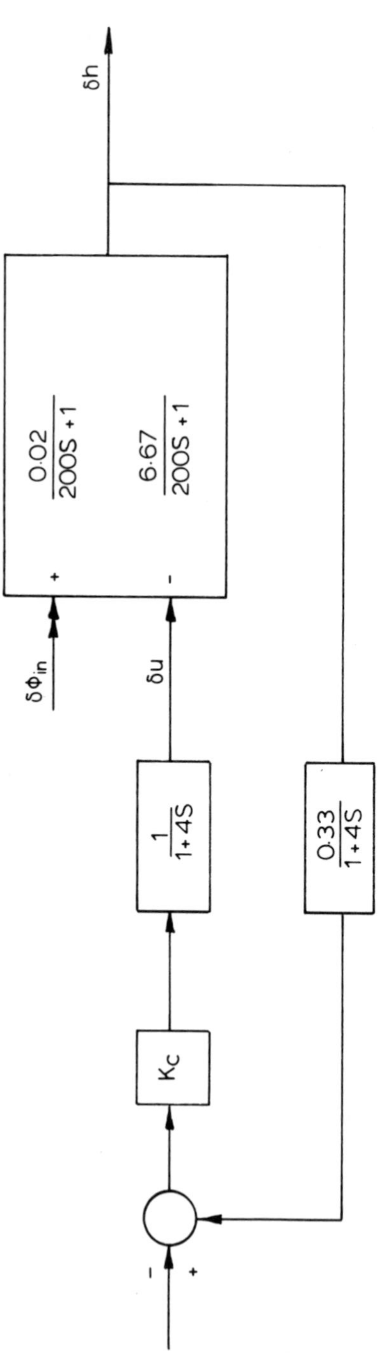

Figure 2.8 Information flow diagram for the controlled process.

3. The characteristic equation is:

$$1 + K_{c,u}\frac{1}{1 + 4s}\,\frac{6.67}{200s + 1}\,\frac{0.33}{4s + 1} = 0 \tag{2.68}$$

When comparing this equation with Figure 2.8, it can be seen that the expression is equal to one plus the product of the transfer functions in the control loop. The plus sign is due to negative feedback; if the product of all signs in the control loop is positive (positive feedback), the expression would be equal to one minus the product of the transfer functions. Equation 2.68 can also be written as:

$$3200s^3 + 1616s^2 + 208s + 1 + 2.2K_{c,u} = 0 \tag{2.69}$$

Dividing by $s^2 + \omega_u^2$ results in two conditions:

$$-3200\omega_u^2 + 208 = 0 \tag{2.70}$$

and

$$-1616\omega_u^2 + 1 + 2.2K_{c,u} = 0 \tag{2.71}$$

From Equation 2.70, the value of ω_u can be calculated: $\omega_u = 0.255$ rad/sec, resulting in an oscillation period of

$$P_0 = \frac{2\pi}{\omega_u} = \frac{2\pi}{0.255} = 24.6 \text{ sec}$$

Substitution of the value of ω_u in Equation 2.71 gives:

$$K_{c,u} = 47.3$$

which is the ultimate proportional gain.

Pressure and Level Control

In a liquid separator, pressure and level are controlled. The vessel diameter is equal to 2 m, the height is 3.3 m, and the average value of the liquid height is 2.75 m. The residence time for the liquid is 1200 sec; for the gas, 30 sec. The gas pressure in the vessel is 0.3 bars gauge. The gas is blown off to the atmosphere through a pressure control valve. The level controller adjusts the valve opening of the control valve in the liquid discharge flow. Both valves have a linear characteristic. Measurement lag and correction lag may be ignored. Flow through the partially opened valves is turbulent. For the gas, $1/\rho \cdot \delta\rho/\delta = 1\ \text{bar}^{-1}$ for $0 \leqslant \rho \leqslant 1$ bar. The opening of the valves is 50% under static condition. The liquid density is 800 kg/m^3.

Determine the damping coefficient of the pressure response when the gain of the level controller is equal to 10 m^{-1} and the gain of the pressure controller equal to 10^{-5} m^2/N.

The situation is shown schematically in Figure 2.9. The liquid height is h, the height of the gas volume, h_2. The model for this problem consists of mass balances for the liquid and gas flows.

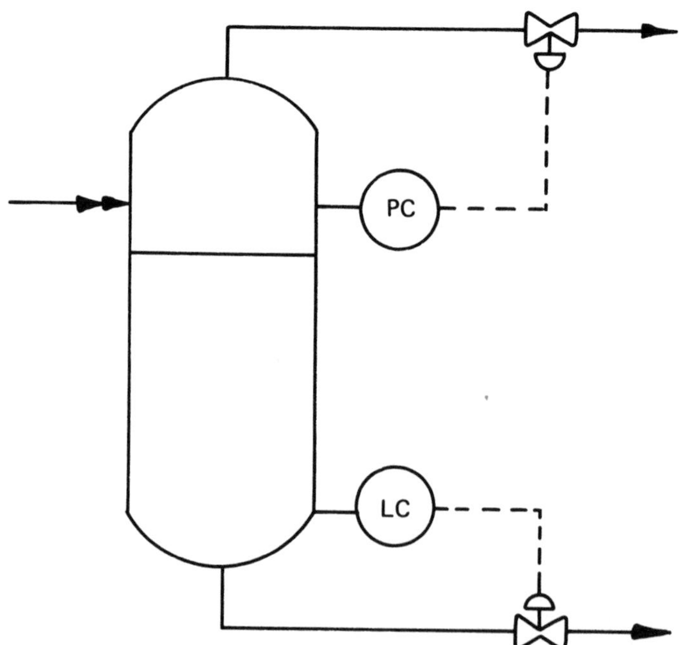

Figure 2.9 Flow and level control.

The mass balance for the liquid flow is:

$$\rho_\ell A \frac{dh_1}{dt} = \phi_{\ell,in} - \phi_{\ell,out} \tag{2.72}$$

where the subscript ℓ stands for liquid. The flow through the valve follows from:

$$\begin{aligned}\phi_{\ell,out} &= c_{v1} \cdot u_1 \sqrt{\Delta p_{valve}} \\ &= c_{v1} u_1 \sqrt{\rho_\ell g h_1 + P}\end{aligned} \tag{2.73}$$

where P = the gas pressure. Combination of Equations 2.72 and 2.73 results in:

$$\rho_\ell A \frac{dh_1}{dt} = \phi_{\ell,in} - c_{v1} u_1 \sqrt{\rho_\ell g h_1 + P} \tag{2.74}$$

Equation 2.74 represents the functional relationship:

$$h_1 = f(\phi_{\ell,in}, u_1, h_1, P) \tag{2.75}$$

For small variations we may write:

$$\delta h_1 = \left(\frac{\partial f}{\partial \phi_{\ell,in}}\right) \delta \phi_{\ell,in} + \left(\frac{\partial f}{\partial u_1}\right) \delta u_1 + \left(\frac{\partial f}{\partial h_1}\right) \delta h_1 + \left(\frac{\partial f}{\partial P}\right) \delta P \tag{2.76}$$

When the inlet flow is constant, Equation 2.74 may be linearized to:

$$\rho_\ell A s \delta h_1 = -c_{v1} \sqrt{\rho_\ell g \bar{h}_1 + \bar{P}} \cdot \delta u_1 - \frac{c_{v1} \bar{u}_1 \rho_\ell g}{2\sqrt{\rho_\ell g \bar{h}_1 + \bar{P}}} \delta h_1 - \frac{c_{v1} \bar{u}_1}{2\sqrt{\rho_\ell g \bar{h}_1 + \bar{P}}} \delta P \tag{2.77}$$

With the aid of Equation 2.73, Equation 2.77 can be written as:

$$\rho_\ell A s \delta h_1 = -\frac{\bar{\phi}_{\ell,out}}{\bar{u}_1} \delta u_1 - \frac{\rho_\ell g \bar{\phi}_{\ell,out}}{2(\rho_\ell g \bar{h}_1 + \bar{P})} \delta h_1 - \frac{\bar{\phi}_{\ell,out}}{2(\rho_\ell g \bar{h}_1 + \bar{P})} \delta P \tag{2.78}$$

Multiplication of this equation by $[2(\rho_\ell g\bar{h}_1 + \bar{P})]/[\rho_\ell g\phi_{\ell,out}]$ gives:

$$\left(\frac{2A(\rho_\ell g\bar{h}_1 + \bar{P})}{g\bar{\phi}_{\ell,out}} s + 1\right)\delta h_1 = -\frac{2(\rho_\ell g\bar{h}_1 + \bar{P})}{\rho_\ell g\bar{u}_1}\delta u_1 - \frac{1}{\rho_\ell g}\delta P \tag{2.79}$$

Now a time constant τ_1 can be defined as:

$$\tau_1 = \frac{2A(\rho_\ell g\bar{h}_1 + \bar{P})}{g\bar{\phi}_{\ell,out}} = \frac{2(\rho_\ell g\bar{h}_1 + \bar{P})}{\rho_\ell g\bar{h}_1} \times \frac{A\bar{h}_1\rho_\ell}{\bar{\phi}_{\ell,out}} = \frac{2(\rho_\ell g\bar{h}_1 + \bar{P})}{\rho_\ell g\bar{h}_1}\tau_{R\ell}$$

$$= \frac{2(800 \times 9.81 \times 2.75 + 0.3 \times 10^5)}{800 \times 9.81 \times 2.75} \times 1200 = 5736 \text{ sec}$$

with $\tau_{R\ell}$, the liquid resistance time, is equal to $(A\bar{h}_1\rho_\ell)/(\bar{\phi}_{\ell,out})$. When the following process gains are defined:

$$K_1 = \frac{2(\rho_\ell g\bar{h}_1 + \bar{P})}{\rho_\ell g\bar{u}_1} = \frac{2 \times (800 \times 9.81 \times 2.75 + 0.3 \times 10^5)}{800 \times 9.81 \times 0.5} = 26.5 \text{ m}$$

$$K_2 = \frac{1}{\rho_\ell g} = \frac{1}{800 \times 9.81} = 1.27 \times 10^{-4} \text{ m}^2\text{sec}^2/\text{kg} = \text{m/Pa}$$

Equation 2.79 can be written as:

$$\delta h_1 = -\frac{K_1}{\tau_1 s + 1}\delta u_1 - \frac{K_2}{\tau_1 s + 1}\delta P \tag{2.80}$$

The other mass balance is the balance for the gas:

$$\frac{dM_g}{dt} = \phi_{g,in} - \phi_{g,out}$$

or

$$\frac{d}{dt}(\rho_g V) = \phi_{g,in} - \phi_{g,out} \tag{2.81}$$

Equation 2.81 can be written as:

$$\bar{\rho}_g \frac{dV}{dt} + \bar{V} \frac{d\rho_g}{dt} = \phi_{g,in} - \phi_{g,out}$$

or

$$A\rho_g \frac{dh_2}{dt} + \bar{V}\left(\frac{\partial \rho_g}{\partial P}\right) \times \frac{dP}{dt} = \phi_{g,in} - \phi_{g,out} \tag{2.82}$$

As $h_2 = h - h_1$ and $\phi_{g,out} = c_{v2}u_2\sqrt{P}$, Equation 2.80 becomes

$$-A\bar{\rho}_g \frac{dh_1}{dt} + A\bar{h}_2\left(\frac{\partial \rho_g}{\partial P}\right) \times \frac{dP}{dt} = \phi_{g,in} - c_{v2}u_2\sqrt{P} \tag{2.83}$$

When ignoring flow variations in the inlet gas flow, Equation 2.83 can be linearized to:

$$A\bar{\rho}_g s\delta h_1 + A\bar{h}_2\left(\frac{\partial \rho_g}{\partial P}\right) s\delta P = -c_{v2}\sqrt{\bar{P}}\delta u_2 - \frac{c_{v2}\bar{u}_2}{2\sqrt{\bar{P}}}\delta P \tag{2.84}$$

This equation can be written as:

$$\left[A\bar{h}_2 \frac{1}{\rho_g}\left(\frac{\partial \rho_g}{\partial P}\right) \frac{2\bar{\rho}_g\sqrt{\bar{P}}}{c_{v2}\bar{u}_2} s + 1\right] \delta P = \frac{2A\bar{\rho}_g\sqrt{\bar{P}}}{c_{v2}\bar{u}_2} s\delta h_1 - \frac{2\bar{P}}{\bar{u}_2}\delta u_2 \tag{2.85}$$

The following parameters are defined:

$$\tau_2 = A\bar{h}_2 \frac{1}{\bar{\rho}_g} \times \left(\frac{\partial \rho_g}{\partial P}\right) \times \frac{2\bar{\rho}_g\sqrt{\bar{P}}}{c_{v2}\bar{u}_2} = \frac{1}{\bar{\rho}_g} \times \left(\frac{\partial \rho_g}{\partial P}\right) \times \tau_{Rg} \times 2\bar{P}$$

$$= 10^{-5} \times 30 \times 0.6 \times 10^5 = 18 \text{ sec}$$

with τ_{Rg} the residence time for the gas.

$$K_3 = \frac{2A\bar{\rho}_g\sqrt{\bar{P}}}{c_{v2}\bar{u}_2} = \frac{2P}{\bar{h}_2} \times \tau_{Rg} = 32.7 \times 10^5 \text{ kg/m}^2\text{-sec}$$

$$K_4 = \frac{2\bar{P}}{\bar{u}_2} = \frac{0.6 \times 10^5}{0.5} = 1.2 \times 10^5 \text{ kg/m-sec}^2$$

Equation 2.85 can finally be rewritten as:

$$\delta P = \frac{K_3 s}{\tau_2 s + 1}\delta h_1 - \frac{K_4}{\tau_2 s + 1}\delta u_2 \tag{2.86}$$

From Equations 2.80 and 2.86, an information flow diagram (Figure 2.10) can be constructed.

Summarizing, we have the model:

$$(\tau_1 s + 1)\delta h_1 = -K_1\delta u_1 - K_2 P \tag{2.87}$$

$$(\tau_2 s + 1)\delta P = K_3 s\delta h_1 - K_4\delta u_2 \tag{2.88}$$

$$\delta u_1 = K_{c1}\delta h_1 \tag{2.89}$$

$$\delta u_2 = K_{c2}\delta P \tag{2.90}$$

Combination of Equations 2.87 and 2.89 results in:

$$\delta h_1 = \frac{-K_2/(1 + K_1 K_{c1})}{\frac{\tau_1}{1 + K_1 K_{c1}}s + 1} = -\frac{K_2'}{\tau_1' s + 1}\delta P \tag{2.91}$$

$$K_2' = \frac{1.27 \times 10^{-4}}{1 + 26.5 \times 10} = 4.77 \times 10^{-7}$$

$$\tau_1' = \frac{5736}{1 + 26.5 \times 10} = 21.6$$

Substitution of Equations 2.90 and 2.91 in Equation 2.88 gives:

$$(\tau_2 s + 1)\delta P = \frac{-K_2' K_3 s}{\tau_1' s + 1}\delta P - K_4 K_{c2}\delta P \tag{2.92}$$

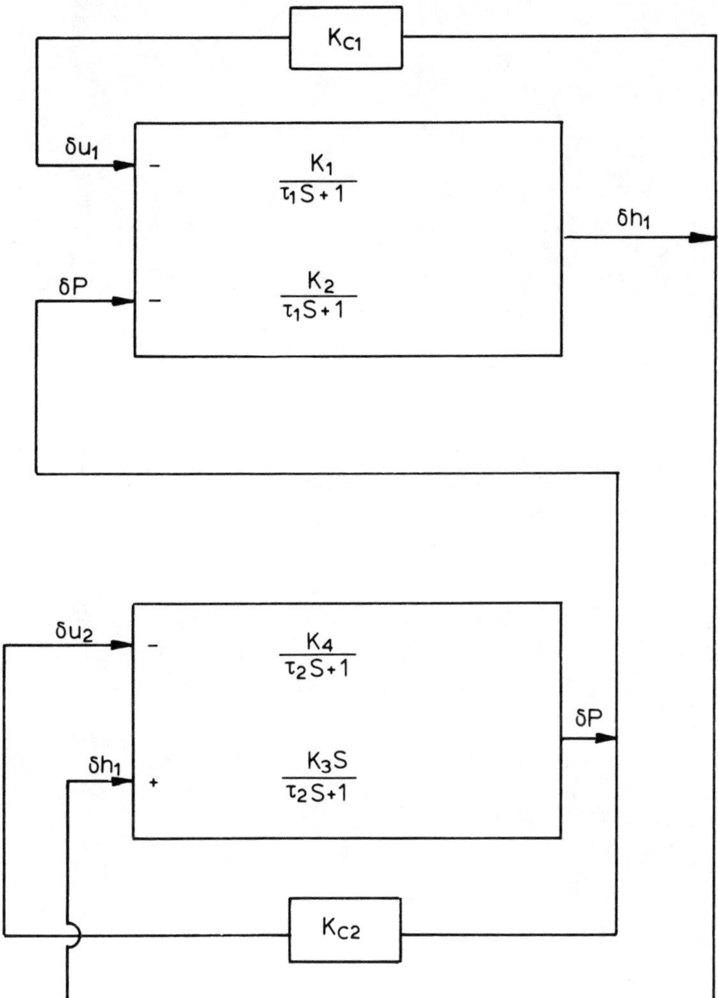

Figure 2.10 Information flow diagram for pressure and level control.

Equation 2.92 may be written as:

$$\tau_1'\tau_2 s^2 + (\tau_1' + \tau_2 + K_2'K_3 + K_4K_{c2}\tau_1')s + K_4K_{c2} + 1 = 0 \qquad (2.93)$$

which is the characteristic equation. Substitution of data gives:

$$176.4s^2 + 30.5s + 1 = 0 \qquad (2.94)$$

which can be written as:

$$\left(\frac{s}{\omega_n}\right)^2 + 2\zeta\left(\frac{s}{\omega_n}\right) + 1 = 0 \qquad (2.95)$$

From a comparison between Equations 2.94 and 2.95, it follows that

$$\omega_n = 0.075 \text{ sec}^{-1} \text{ and } \zeta = 1.15$$

Without level control, the pressure control loop would have been represented by a first-order system. Hence Equation 2.95 shows the effect of interaction between the two loops, which in this case increases the order to two.

CHAPTER 3

VALVE CHARACTERISTIC AND CORRECTION LAG

In the preceding chapter, the behavior of the control valve was idealized to a linear relationship between the output signal of the controller and the flow. In reality, the behavior is complicated by the valve and flow circuit in which the valve is mounted.

In this chapter, no recipe will be given for control valve selection, but the behavior of a valve in a flow circuit will be analyzed. Restriction will be made to two types of valves: linear and exponential (equal percentage, logarithmic). For these valves, the relationship between valve position and flow is linear and exponential, respectively, at constant pressure drop across the valve. In practice, when the control valve is installed in a flow circuit, pressure drop across the valve is usually not constant, and deviations are found of the linear and exponential characteristics.

VALVE CHARACTERISTICS

The flow circuit to be considered is shown in Figure 3.1. A pump moves liquid from a storage tank to a distillation column. The flowrate can be adjusted by a valve. Heat is exchanged in a heat exchanger. For simplicity, it is assumed that the physical properties of the liquid are constant. The pressure drop $P_{in} - P_{out}$ consists of the following terms:

$$P_{in} - P_{out} = -\Delta P_p + \Delta P_v + \Delta P_A + \Delta P_t + \Delta P_g \tag{3.1}$$

where ΔP_p = pressure drop across the pump
ΔP_v = pressure drop across the valve
ΔP_A = pressure drop across the heat exchanger
ΔP_t = pressure drop across the pipe
ΔP_g = pressure drop due to difference in height

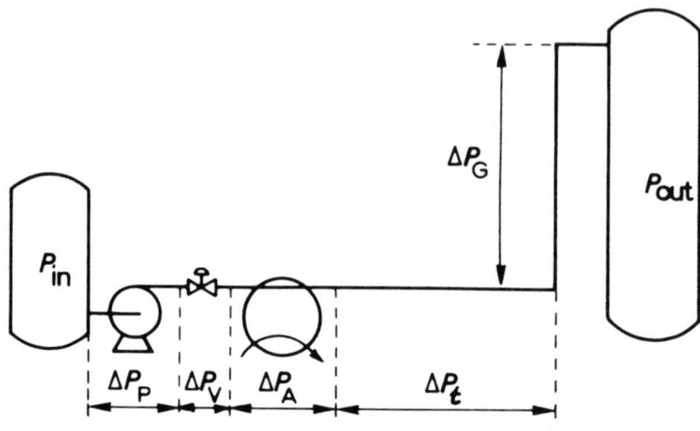

Figure 3.1. Flow circuit.

For the pump the following characteristic is assumed:

$$\Delta P_p = \Delta P_{\phi=0} - \frac{1}{\rho}\left(\frac{\phi}{c_p}\right)^2 \tag{3.2}$$

where $\Delta P_{\phi=0}$ = the pressure increase at zero flow
ρ = the density of the liquid
ϕ = the mass flow
c_p = a constant

In the case of fully developed turbulent flow, we may write for the pipeline:

$$\Delta P_t = \frac{1}{\rho}\left(\frac{\phi}{c_t}\right)^2 \tag{3.3}$$

where c_t is a constant. In the same way for the heat exchanger:

$$\Delta P_A = \frac{1}{\rho}\left(\frac{\phi}{c_A}\right)^2 \tag{3.4}$$

with the constant c_A. For the control valve a similar equation holds:

$$\Delta P_v = \frac{1}{\rho}\left(\frac{\phi}{c_v^*}\right)^2 \tag{3.5}$$

where the parameter c_v^* is adjusted by the position of the valve shaft. Substitution of Equations 3.2 to 3.5 into Equation 3.1 results in:

$$\Delta P_s = P_{in} - P_{out} + \Delta P_{\phi=0} - \Delta P_g = \frac{1}{\rho}\left[\left(\frac{\phi}{c}\right)^2 + \left(\frac{\phi}{c_v^*}\right)^2\right] \tag{3.6}$$

with

$$\frac{1}{c^2} = \frac{1}{c_t^2} + \frac{1}{c_A^2} + \frac{1}{c_p^2} \tag{3.7}$$

ΔP_s is called the static pressure drop. The pressure distribution in the flow circuit is given in Figure 3.2. When there is no control valve in the flow circuit, the liquid flow will be equal to ϕ_0. Equation 3.6 can then be written as:

$$\Delta P_s = \frac{1}{\rho}\left(\frac{\phi_0}{c}\right)^2 \tag{3.8}$$

Combination of Equations 3.6 and 3.8 gives:

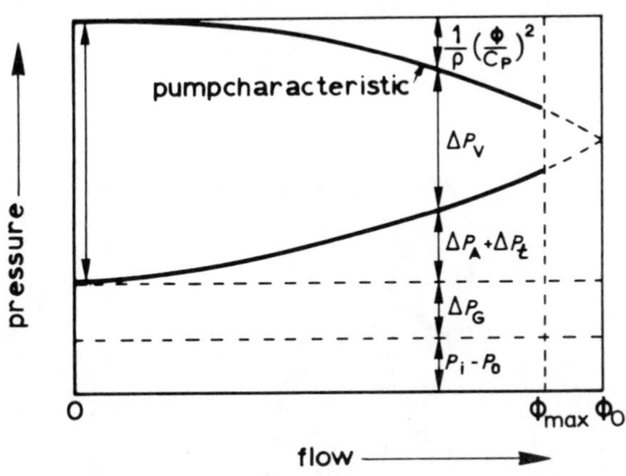

Figure 3.2. Pressure distribution in the flow circuit.

$$\left(\frac{\phi}{\phi_0}\right)^2 \left\{1 + \left[\left(\frac{c_v}{c_v^*}\right)\left(\frac{c}{c_v}\right)\right]^2\right\} = 1 \tag{3.9}$$

where c_v pertains to a fully opened valve, which can be determined from Equation 3.5 at maximum flow through the valve. For a linear valve, ignoring leakage, the parameter c_v is given by:

$$c_v^* = \frac{u}{u_{max}} c_v = u_r c_v \tag{3.10}$$

and for an exponential valve:

$$c_v^* = R_v^{u_r - 1} c_v \tag{3.11}$$

where u_r = relative valve input signal ($0 \leqslant u_r \leqslant 1$). The parameter R_v (rangeability) is defined as the ratio between maximum and minimum adjustable flow for a given pressure drop across the valve. The value of R_v is usually between 20 and 50, depending on the model and size of the valve.

From Equations 3.10 and 3.11, the ratio c_v/c_v^* can be determined. When we introduce:

$$\alpha = \left(\frac{c}{c_v}\right)^2 \tag{3.12}$$

and

$$\psi = \frac{\phi}{\phi_0} \tag{3.13}$$

Equation 3.9 can be rewritten for the linear valve as:

$$\psi^2 \left(1 + \frac{\alpha}{u_r^2}\right) = 1 \tag{3.14}$$

and for the exponential valve as:

$$\psi^2[1 + \alpha \cdot R_v^{2(1-u_r)}] = 1 \tag{3.15}$$

The characteristics calculated with the aid of Equations 3.14 and 3.15 are shown in Figure 3.3 for a R_v value of 50. A rangeability R_v of 50 indicates a leakage in "closed" position of 2% of the maximum flow at constant pressure drop.

VALVE GAIN

Valve gain K_v is defined as the change in flow due to a change in relative valve position u_r:

$$K_v = \frac{\delta\phi}{\delta u_r} \tag{3.16}$$

The expression for K_v can be determined by differentiating Equations 3.14 and 3.15. Differentiation of Equation 3.14 for the linear valve gives:

$$K_v = \frac{\delta\phi}{\delta u_r} = \phi_0 \frac{\delta\psi}{\delta u_r} = c_v \sqrt{\Delta P_s \rho}\,(1 - \psi^2)^{3/2} \tag{3.17}$$

and differentiation of Equation 3.15 for the exponential valve gives:

$$\frac{\delta\psi}{\delta u_r} = \psi(1 - \psi^2)\, ln R_v \tag{3.18}$$

resulting in:

$$K_v = \phi_0 ln R_v \psi(1 - \psi^2) \tag{3.19}$$

In Figure 3.4 the valve gain is given as a function of the reduced flow ϕ/ϕ_0. The curve for the exponential control valve has a maximum near which the gain is relatively constant.

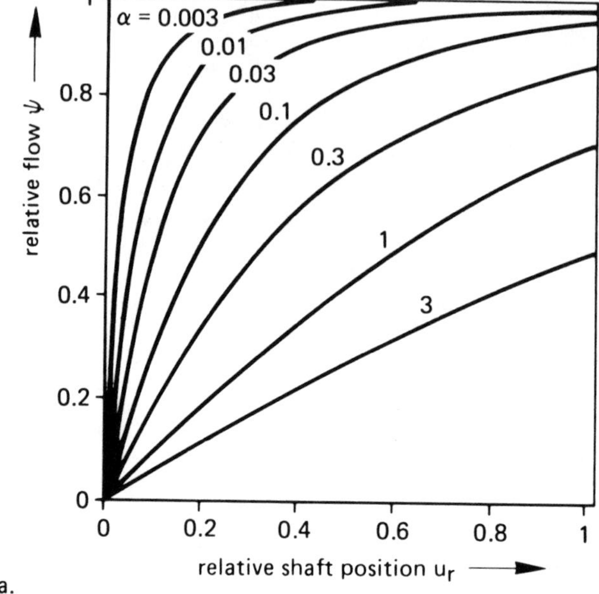

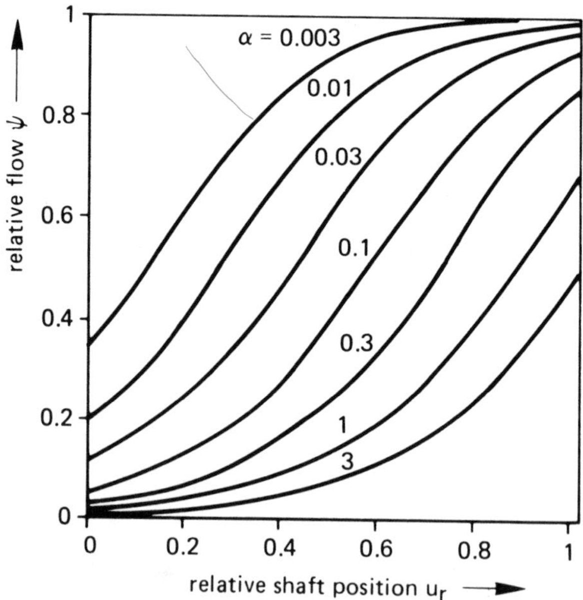

Figure 3.3. Valve characteristics: (a) linear valve; (b) exponential valve, $R_V = 50$.

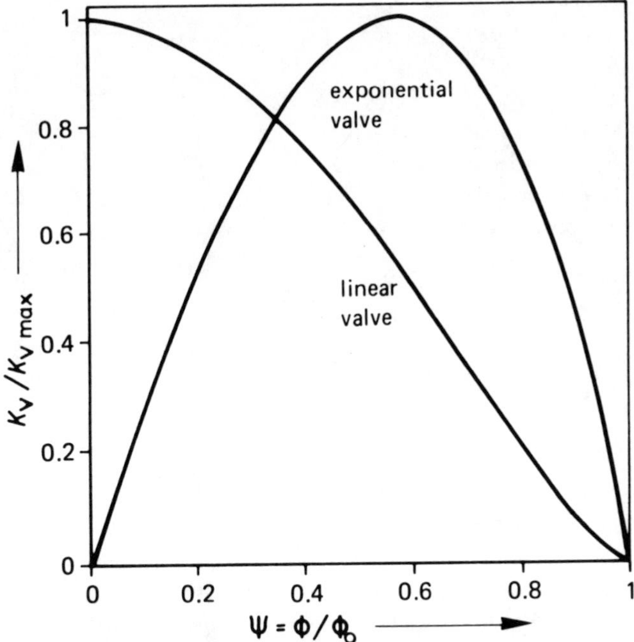

Figure 3.4. Valve gain as a function of flow.

DESIRED VALVE CHARACTERISTIC

We now arrive at the difficult problem: what is the best valve characteristic of the control valve in the flow circuit? A possible criterion is to keep the control loop always at a certain distance from the limit of stability, irrespective of the flowrate. A simple example is given by Eckman [6]. An airflow is electrically heated in an air heater (Figure 3.5). When ideal mixing is assumed, the heat balance is:

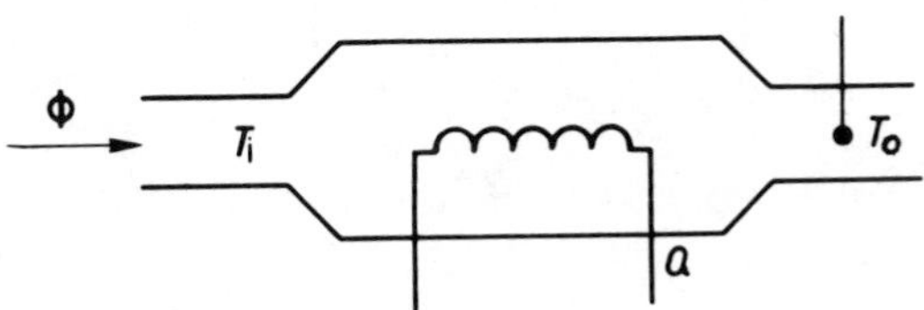

Figure 3.5. Electrical air heater.

$$M\gamma \frac{dT_{out}}{\delta Q} = \phi\gamma(T_{in} - T_{out}) + Q \tag{3.20}$$

where M = mass of air in the heater
γ = specific heat of the air
Q = supplied heat
T = temperature

The transfer function from Q to T_{out} is:

$$\frac{\delta T_{out}}{\delta Q} = \frac{1}{\bar{\phi}\gamma} \times \frac{1}{1 + (M/\bar{\phi})s} \tag{3.21}$$

where $\bar{\phi}$ is the average mass flow of air through the heater. The static process gain is inversely proportional to ϕ, hence, one could think about a controller gain proportional to ϕ. The dynamic behavior, however, changes the picture completely. For instance, with a first-order measurement lag, the open control loop transfer function becomes:

$$K_c \times \frac{1}{\bar{\phi}\gamma} \times \frac{1}{1 + (M/\bar{\phi})s} \times \frac{1}{1 + \tau_m s} \tag{3.22}$$

with τ_m the time constant of the measurement lag. For the closed control loop the transfer function is proportional to:

$$\left(1 + \frac{K_c}{\bar{\phi}\gamma} \times \frac{1}{1 + (M/\bar{\phi})s} \times \frac{1}{1 + \tau_m s}\right)^{-1} \tag{3.23}$$

When this expression is written in the form

$$\left[\left(\frac{s}{\omega_n}\right)^2 + 2\zeta\left(\frac{s}{\omega_n}\right) + 1\right]^{-1} \tag{3.24}$$

and when ζ is chosen equal to 0.25 (see Chapter 2), it follows that:

$$\frac{1}{\omega_n^2} = \frac{\gamma M \tau_m}{K_c + \bar{\phi}\gamma} \tag{3.25}$$

and

$$\frac{0.5}{\omega_n} = \frac{\bar{\phi}\gamma(M/\bar{\phi} + \tau_m)}{K_c + \bar{\phi}\gamma} \tag{3.26}$$

Eliminating ω_n from Equations 3.25 and 3.26 yields the desired relationship between K_c and ϕ:

$$K_c = \frac{4\gamma}{M\tau_m}(M + \tau_m\bar{\phi})^2 - \gamma\bar{\phi} \tag{3.27}$$

This relationship is quite different from the static relationship. In fact, reality is still more complicated because τ_m depends on $\bar{\phi}$, and because the heat transfer from the coil to the air has a flow-dependent lag. If the controller also has an integral action (see Chapter 16), the picture will become even more complicated.

Until now, only the independent variable Q has been varied. Variations in T_{in} also require a relationship between K_c and T_{in}. Finally, K_c has to be a function of all independent variables, which can only be realized in a function generator. Due to all these complications, the problem of the valve characteristic is seldom handled in practice, and controller gains are set for the worst combination of circumstances. The loss of control quality is quite significant. Possibly, adaptive algorithms can be used to keep the controller gain at a suitable value under varying circumstances.

VALVE HYSTERESIS AND VALVE POSITIONER

There must not be any leakage in the gasket along the valve shaft. This requires a relatively large normal force on the valve shaft, resulting in large friction forces in the gasket. The friction force is partially proportional to the speed of movement and partially to the normal force.

If the valve input signal is increased, the shaft movement stays below (Figure 3.6). When subsequently the valve input signal is decreased, the shaft stops first, followed by the movement above the input. The phenomenon that behavior depends on the direction of movement is called hysteresis.

As it causes an additional lag, hysteresis is disadvantageous for control. The effect of hysteresis can be reduced by a valve positioner (Figure 3.7). This instrument compares the position of the valve shaft continuously with

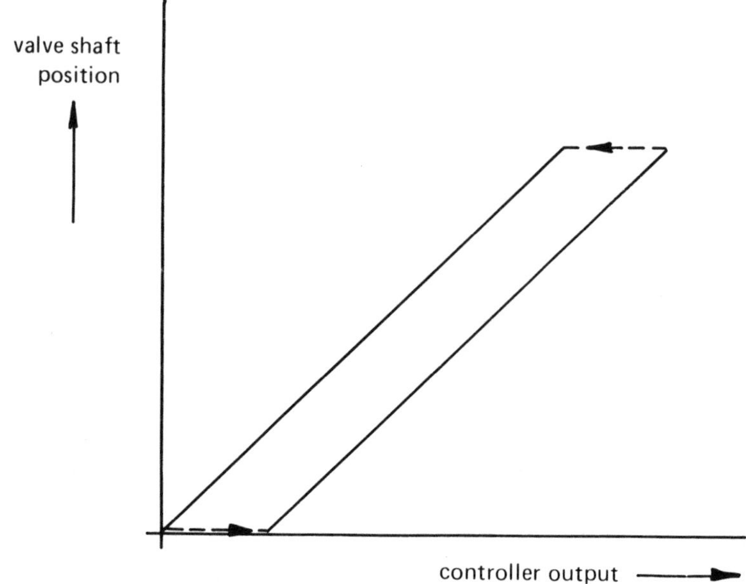

Figure 3.6. Hysteresis in a control valve.

the output signal of the controller. If there is a discrepancy, the input signal to the valve motor is corrected. The factor with which the effect of hysteresis is reduced is equal to the gain of the pneumatic amplifier. In practice, hysteresis can be reduced by a factor of 5–20: from a few percent to a fraction of a percent.

Together with the control valve, the valve positioner forms a type of feedback loop that sometimes can cause instability in the control system. For instance, this can happen when the control loop, of which the control valve is a part, has a fast response, as is the case with flow control loops. In this case the solution is to omit the valve positioner. In fact, control is already fairly fast, hence increasing its speed is not important.

CORRECTION LAG

Control valves are usually driven by pneumatic signals. The valve motor is nothing more than a chamber, enclosed on one side by a flexible diaphragm connected to the valve shaft (Figure 3.8). The air pressure in the chamber pushes against the reacting force of a spring, which opposes diaphragm motion. In this way, the movement of the diaphragm (hence, that of the valve shaft) is proportional to the air pressure.

Figure 3.7. Foxboro control valve with valve positioner.

The volume of the chamber plays a role in the correction lag. This lag can be appreciable when pneumatic tubing is used between the controller (in the central control room) and the control valve (without valve positioner).

The dynamic behavior of pneumatic transmissions is rather complicated, as it depends on the geometric coordinate along the tubing. Strictly speaking, it should be described by partial differential equations.

Figure 3.8. Foxboro F8 Stabiflo valve with section of the P50 valve motor.

A reasonable approximation can be obtained when the tubing is "discretized," e.g., as done in Figure 3.9. Here the volume of the tubing is divided into two equal parts, and the resistance for flow in the tubing is concentrated in between these two parts. In the case of laminar flow, the pressure drop in the tubing is:

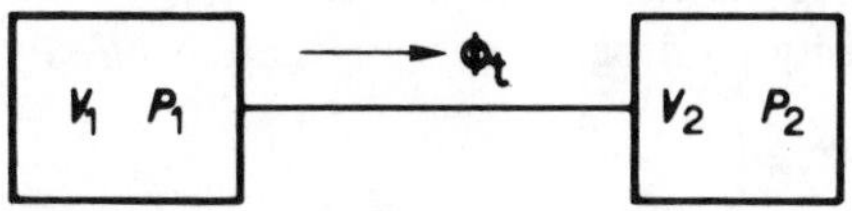

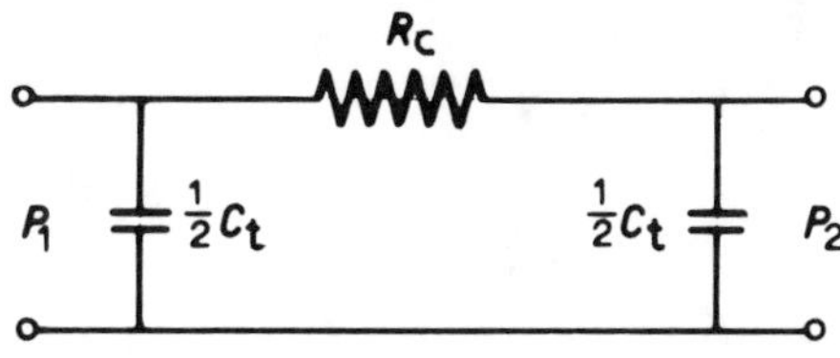

Figure 3.9. Pneumatic tubing with electrical analog.

$$P_1 - P_2 = R_t \phi_t \tag{3.28}$$

where

$$R_t = \frac{128\mu\ell}{\pi\rho d^4} \tag{3.29}$$

For air at an average valve pressure of 1.6×10^5 N/m^2, a temperature of 293°K and a tube diameter of 5 mm, R_t has a value of

$$R_t \approx 0.62 \times 10^6 \, \ell \text{ m}^{-1}\text{-sec}^{-1} \tag{3.30}$$

In practice, R_t is larger, due to bends, joints, etc., so that we assume:

$$R_t = 10^6 \, \ell \text{ m}^{-1}\text{-sec}^{-1} \tag{3.31}$$

The mass balance for the volume V_2 is:

$$\frac{dM_2}{dt} = \phi_t \tag{3.32}$$

As the elasticity of the tubing can be ignored, V_2 is constant. If, by approximation, the density ρ_2 depends only on the pressure P_2, Equation 3.32 can be written as:

$$C_2 \frac{dP_2}{dt} = \phi_t \tag{3.33}$$

where

$$C_2 = V_2 \frac{d\rho_2}{dP_2} \tag{3.34}$$

Volume V_2 is related to the capacity C_2. Similarly, the volume V_1 is related to the capacity C_1. As the tubing has been divided into two equal parts $C_1 = C_2 = \frac{1}{2}C_t$, where:

$$C_t = V_t \frac{d\rho}{dP} = \frac{1}{4} \pi d^2 \ell \frac{d\rho}{dP} \tag{3.35}$$

For air at 293°K, which behaves polytropically (between isothermal and adiabatic), $d\rho/dP$ is equal to:

$$\frac{d\rho}{dP} \cong 10^{-5} \text{ sec}^2/\text{m}^2 \tag{3.36}$$

For a tube diameter of 5×10^{-3} m, Equation 3.35 can now be rewritten as:

$$C_t \cong 2 \times 10^{-10} \ \ell\text{-m-sec}^2 \tag{3.37}$$

The electrical analog for the tubing is given in Figure 3.9c. Combination of Equations 3.28 and 3.33, linearization and replacing d/dt by s give:

$$\frac{\delta P_2}{\delta P_1} = \frac{1}{\frac{1}{2} R_t C_t s + 1} \tag{3.38}$$

Usually, the inlet of the pneumatic tubing is connected to a pneumatic amplifier. Its output pressure decreases at increasing output air flow. This effect can be characterized by an output resistance R_c. Actual values of R_c are $5-25 \times 10^6$ Pa/kg-sec (m^{-1}-sec^{-1}).

The outlet of the tubing can be connected directly to the pneumatic motor of the control valve. The input resistance of the valve motor is usually very small. The capacitative action, however, should be taken into account. Its value follows from the mass balance for the valve motor volume:

$$\frac{dM_v}{dt} = \phi_v \tag{3.39}$$

where M_v is the mass of the air in the valve motor and ϕ_v the air flow to the valve. This equation can also be written as:

$$\frac{dM_v}{dt} = \frac{d\rho_v V_v}{dt} = \left(V_v \frac{d\rho_v}{dP_v} + \rho_v \frac{dV_v}{dP_v}\right) \frac{dP_v}{dt} \tag{3.40}$$

Equations 3.39 and 3.40 are now modified to:

$$C_v \frac{dP_v}{dt} = \phi_v \tag{3.41}$$

where

$$C_v = V_v \frac{d\rho_v}{dP_v} + \rho_v \frac{dV_v}{dP_v} \tag{3.42}$$

An approximate expression for C_v will be given. At a pressure on the control valve between the minimum (1.2×10^5 N/m^2) and the maximum (2×10^5 N/m^2) we assume:

$$V_v \approx \frac{V_{min} + V_{max}}{2} \tag{3.43}$$

where V_{min} and V_{max} are the minimum and maximum volume of the valve motor. Furthermore dV_v/dP_v is approximated by:

$$\frac{dV_v}{dP_v} \cong \frac{V_{max} - V_{min}}{P_{max} - P_{min}} \cong \frac{V_{max} - V_{min}}{0.8 \times 10^5} \tag{3.44}$$

substitution into Equation 3.42 yields:

$$C_v = 10^{-5}(2.9V_{max} - 1.9V_{min}) \tag{3.45}$$

e.g., when $V_{min} = 10^{-3}$ m^3 and $V_{max} = 2.5 \times 10^{-3}$ m^3, the value of C_v is equal to 5.35×10^{-8} m-sec^2.

As the output resistance of the valve and output volume of the controller can be ignored, the capacitance C_v is in parallel with the capacitance $\frac{1}{2}C_L$ and the resistance R_c is in series with the resistance R_t. The total electrical analog is shown in Figure 3.10. From this scheme, the transfer function for the pneumatic system can be derived. The result is:

$$\frac{\delta P_v}{\delta P_c} = \frac{1}{\tau_1\tau_2 s^2 + (\tau_1 + \tau_2 + \tau_3)s + 1} \tag{3.46}$$

where

$$\tau_1 = \frac{1}{2} R_c C_t \tag{3.47}$$

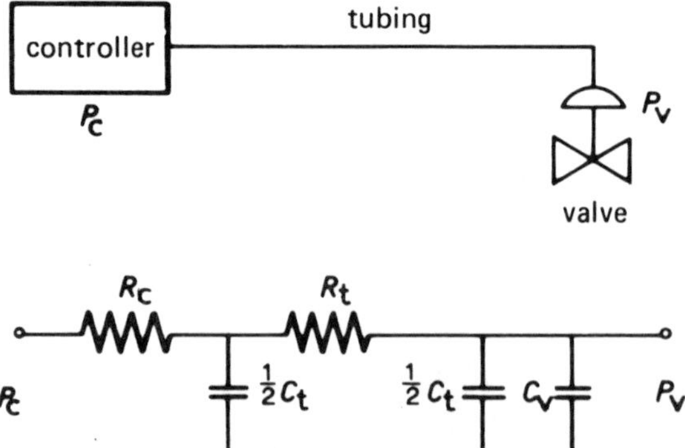

Figure 3.10. Total analog for pneumatics.

$$\tau_2 = R_t\left(\frac{1}{2}C_t + C_v\right) \tag{3.48}$$

$$\tau_3 = R_c\left(\frac{1}{2}C_t + C_v\right) \tag{3.49}$$

Starting from an average value of R_c and C_v, for example $R_c = 15 \times 10^6$ m^{-1}-sec^{-1} and $C_v = 5 \times 10^{-8}$ m-sec^2, the values of τ_1, τ_2 and τ_3 can be calculated as functions of the length of the pneumatic tubing, with the aid of Equations 3.32 and 3.39. The result is:

$$\tau_1 = 1.5 \times 10^{-3}\,\ell \tag{3.50}$$

$$\tau_2 = 10^{-4}\,\ell^2 + 5 \times 10^{-2}\,\ell \tag{3.51}$$

$$\tau_3 = 1.5 \times 10^{-3}\,\ell + 0.75\,\ell \tag{3.52}$$

For short tubing, e.g., 10 m, the capacitance of the tubing can be ignored; hence $\tau_1 \approx 0$, $\tau_2 \approx R_tC_v$ and $\tau_3 \approx R_cC_v$, and Equation 3.46 becomes:

$$\frac{\delta P_v}{\delta P_c} = \frac{1}{(R_c + R_t)C_v s + 1} \tag{3.53}$$

For long tubing, e.g., 500 m, τ_1 and τ_3 are much smaller than τ_2, resulting in:

$$\frac{\delta P_v}{\delta P_c} = \frac{1}{R_t\left(\frac{1}{2}C_t + C_v\right)s + 1} \tag{3.54}$$

Here approximation by a first-order lag is inadequate, if the other lags in the control loop are small. The pneumatic tube should then be represented by an electrical network with two or more sections.

A valve positioner separates the tubing from the valve motor. Terms with C_v should then be omitted. Of course, the valve positioner loop introduces an additional lag.

EXAMPLE: FLOW CONTROL

Process

A reflux pump pumps liquid with a density of 900 kg/m^3 through a pipeline with an internal diameter of 0.15 m to the top of a distillation column. The flow is turbulent (the roughness factor for the pipe wall may be chosen freely). The maximum pressure yield of the pump (1 MPa at closed discharge valve) is utilized in the case of an entirely open valve as follows:

- 200 kPa in the pump
- 400 kPa for difference in height
- 100 kPa in valves, bends, losses due to inflow and outflow
- 200 kPa in the pipe line
- 100 kPa in the control valve

Then the flow velocity is 5 m/sec.

The characteristic of the control valve is linear for constant pressure drop across the valve. In fact, the pressure drop depends on the flow velocity. The momentum balance, which describes the dynamic behavior of the process, has a derivative corresponding to the inertia of the liquid mass in the pipeline.

Control

The flow is measured by a differential pressure transmitter measuring the pressure drop across an orifice, which has a range of 0–4.5 m/sec. Both the measurement device and controller have an output resistance corresponding to 20 m of pneumatic tubing. The measurement device is connected to a controller by pneumatic tubing with an internal diameter of 5 mm and a length of 100 m.

The same connection exists between controller and control valve. The compressibility of the air in these tubings should not be ignored. For simplicity, the compressibility may be equally divided over the ends of the tubing with the viscous friction in between.

1. Make a dynamic model of the process and show the transfer functions in an information flow diagram.
2. Determine the limit of stability for proportional control.
3. Approximate the closed-loop transfer function by a dead time (delay) and one or more first-order time lags.

Development of the Model

First, the model parameters in static condition will be calculated. For maximum flow under turbulent conditions we have for the pipeline:

$$\Delta P_t = c_t v^2$$
$$2 \times 10^5 = c_t 5^2 \qquad (3.55)$$

hence $c_t = 8 \times 10^3$ kg/m^3

The length of the pipeline follows from the pressure drop relation:

$$\Delta P_t = f \frac{1}{2} \rho v^2 \frac{\ell}{d} \qquad (3.56)$$

When f = 0.04 at 5 m/sec and ΔP_t = 200 kPa, the length of the pipeline equals 66.7 m.

For the pump, it is assumed that the characteristic can be apprximated by a quadratic relationship:

$$\Delta P_p = \Delta P_0 - c_p v^2$$
$$2 \times 10^5 = 10 \times 10^5 - c_p \times 25 \qquad (3.57)$$

hence $c_p = 32 \times 10^3$ kg/m^3. For difference in height:

$$\Delta P_h = \rho g h = \text{constant} = 4 \times 10^5 \text{ N/m}^2 \qquad (3.58)$$

for valves, bends, etc.

$$\Delta P_a = c_a v^2$$
$$10^5 = c_a \times 25 \qquad (3.59)$$

hence $c_a = 4 \times 10^3$ kg/m^3. For a linear control valve the following equation holds:

$$\phi = u c_v' \sqrt{\Delta P_v} \qquad (3.60)$$

It can be seen that at constant pressure drop ΔP_v, the flow ϕ is proportional to the signal from the controller u. From the previous equation it can be derived that:

$$\Delta P_v = c_v \frac{v^2}{u^2} \tag{3.61}$$

For an open control valve, u = 1 and the constant c_v is calculated to be 4×10^3 kg/m^3.

Model Equation for the Flow Circuit

The total available pressure drop is used to accelerate the liquid mass in the pipeline (Newton' s law) and to overcome pressure drop due to friction. When it is assumed that the pressure drop across the orifice is small compared to other pressure drops and can be ignored we may write:

$$P_{in} - P_{out} = \rho \ell \frac{dv}{dt} + \Delta P_h - \Delta P_p + \Delta P_a + \Delta P_t + \Delta P_v \tag{3.62}$$

Substitution of Equations 3.55, 3.61 and 3.57 to 3.59 into Equation 3.62 results in:

$$P_{in} - P_{out} = \rho \ell \frac{dv}{dt} - \rho gh - \Delta P_0 + \left(c_p + c_a + c_t + \frac{c_v}{u^2}\right) v^2 \tag{3.63}$$

If the pressure drop across the orifice is not small compared to other pressure drops, a term $c_0 v^2$ can easily be added to the right side of Equation 3.63.

Linearization of Equation 3.63 gives:

$$\delta P_{in} - \delta P_{out} = \rho \ell s \delta v + 2\bar{v} \left(c_p + c_a + c_t + \frac{c_v}{\bar{u}^2}\right) \delta v - \frac{2 c_v \bar{v}^2}{\bar{u}^2} \delta u \tag{3.64}$$

where a bar over a variable refers to the static condition. When variations in in- and outlet pressure can be ignored, Equation 3.64 can be simplified to:

$$\frac{\delta v}{\delta u} = \frac{K_p}{\tau_p s + 1} \tag{3.65}$$

where

$$K_p = \frac{c_v \bar{v}}{\bar{u}^3 \left(c_p + c_a + c_t + \dfrac{c_v}{\bar{u}^2}\right)} \tag{3.66}$$

$$\tau_p = \frac{\rho \ell}{2\left(c_p + c_a + c_t + \dfrac{c_v}{\bar{u}^2}\right)\bar{v}} \tag{3.67}$$

Assuming an average flow velocity of 2.5 m/sec, the pipeline pressure drop becomes 50 kPa, the pump pressure drop 50 kPa, the pressure drop in bends 25 kPa, resulting in a pressure drop across the valve equal to 475 kPa. According to Equation 3.61, the signal from the controller is then $\bar{u} = 0.23$.

Calculation of process gain and time constant result in:

$$K_p = 6.85 \quad ; \quad \tau_p = 0.10 \text{ sec} \tag{3.68}$$

When the output signal of the orifice varies between 0 and 1 (corresponding to the span of 0–4.5 m/sec), the measurement equation is:

$$\Delta m = \alpha v^2$$

$$1 = \alpha 4.5^2$$

hence $\alpha = 0.049$. Linearization of Equation 3.69 gives:

$$\delta(\Delta m) = 2\alpha\bar{v}\delta v = 2 \times 0.049 \times 2.5\delta v = 0.25\delta v$$

hence,

$$\frac{\delta(\Delta m)}{\delta v} = 0.25 \tag{3.70}$$

Instrumentation

The electrical analog shown in Figure 3.11 can also be used here. It has already been shown (Equation 3.46) that:

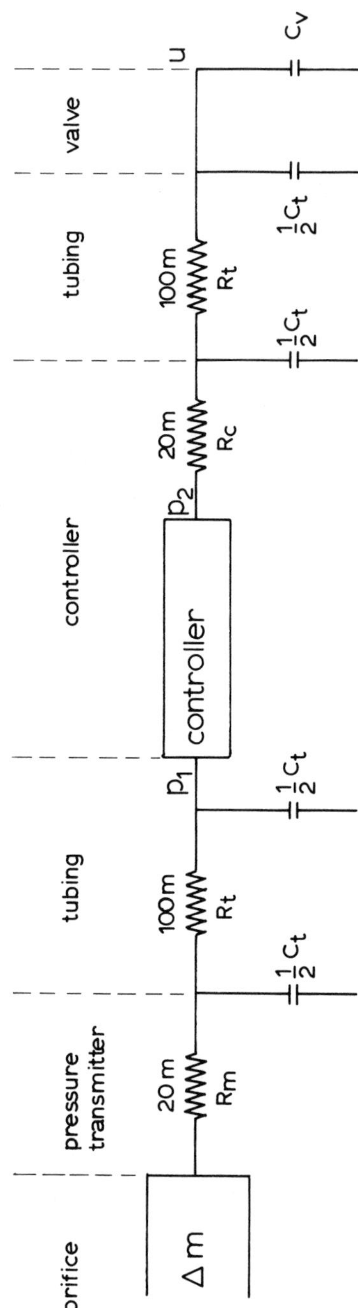

Figure 3.11. Electrical analog for the pneumatic part of the flow process.

$$\frac{\delta p_1}{\delta(\Delta m)} = \frac{1}{\tau_1\tau_2 s^2 + (\tau_1 + \tau_2 + \tau_3)s + 1} \qquad (3.71)$$

with

$$\tau_1 = \frac{1}{2} R_m C_t \ ; \quad \tau_2 = \frac{1}{2} R_t C_t \ ; \quad \tau_3 = \frac{1}{2} R_m C_t \qquad (3.72)$$

and

$$R_m = 0.62 \times 10^6 \quad \ell = 0.62 \times 10^6 \quad \times \quad 20 = 12.4 \times 10^6 \ \text{m}^{-1}\text{-sec}^{-1}$$
$$R_t = 0.62 \times 10^6 \quad \ell = 0.62 \times 10^6 \quad \times 100 = 62 \quad \times 10^6 \ \text{m}^{-1}\text{-sec}^{-1}$$
$$C_t = 2 \quad \times 10^{-10}\, \ell = 2 \quad \times 10^{-10} \times 100 = 2 \quad \times 10^{-8} \ \text{m-sec}^2$$

The values of τ_1 to τ_3 are now:

$$\tau_1 = \tau_3 = 0.124 \text{ sec}$$
$$\tau_2 = 0.62 \text{ sec}$$

from which:

$$\frac{\delta p_1}{\delta(\Delta m)} = \frac{1}{0.077s^2 + 0.87s + 1}$$
$$= \frac{1}{(1 + 0.77s)(1 + 0.10s)} \qquad (3.73)$$

For the section between controller and control valve:

$$R_c = 12.4 \times 10^6 \ \text{m}^{-1}\text{-sec}^{-1}$$
$$R_t = 62 \quad \times 10^6 \ \text{m}^{-1}\text{-sec}^{-1}$$
$$C_t = \ 2 \quad \times 10^{-8} \ \text{m-sec}^2$$

If a value of C_v is assumed equal to 5×10^{-8} m-sec^2, we have:

$$\tau_1 = \frac{1}{2} R_c C_t \qquad = 0.124 \text{ sec}$$

$$\tau_2 = \left(\frac{1}{2} C_t + C_v\right) R_t = 3.72 \text{ sec}$$

$$\tau_3 = \left(\frac{1}{2} C_t + C_v\right) R_c = 0.744 \text{ sec}$$

resulting in:

$$\frac{\delta u}{\delta p_2} \simeq \frac{1}{(1 + 0.10s)(1 + 4.5s)} \tag{3.74}$$

The model can now be represented in an information flow diagram as shown in Figure 3.12.

Determination of the Limit of Stability

The limit of stability can be determined from the fifth-order characteristic equation, which is equal to one plus the product of the closed-loop transfer functions.

$$1 + K_c \times \frac{1}{(1 + 0.10s)(1 + 4.5s)} \times \frac{6.85}{(1 + 0.10s)} \times \frac{0.25}{(1 + 0.10s)(1 + 0.775s)} = 0 \tag{3.75}$$

which can be written as:

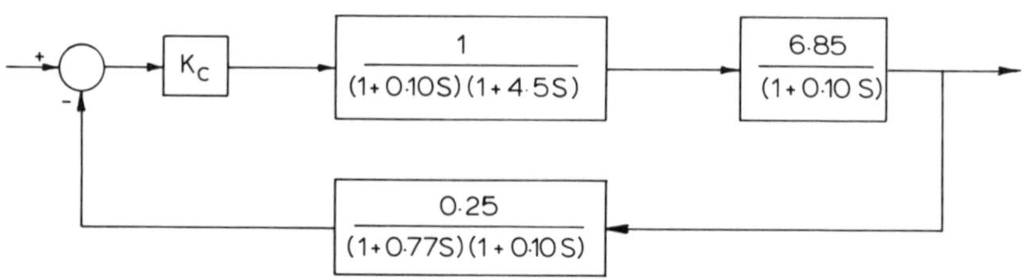

Figure 3.12. Information flow diagram for the flow process.

$$0.00347s^5 + 0.10937s^4 + 1.2001s^3 + 5.081s^2 + 5.570s + 1 + 1.70K_c = 0 \tag{3.76}$$

Dividing by $s^2 + \omega_n^2$ and setting the remainder equal to zero results in an oscillation period and ultimate (at the limit of stability) gain of:

$$P_0 = \frac{2\pi}{\omega_u} = 2.9 \text{ sec}$$

$$K_{c,u} = 12.1$$

Model Approximation

Many small time constants can be approximated by a delay. When we approximate:

$$\frac{1}{(1 + 0.10s)^3} \approx e^{-0.30s} \tag{3.77}$$

the characteristic equation becomes:

$$1 + \frac{6.85 \times 0.25\ K_c e^{-0.30s}}{(1 + 4.5s)(1 + 0.77s)} = 0 \tag{3.78}$$

When approximating the exponential by a Taylor series:

$$e^{-0.30s} \approx 1 - 0.30s + 0.045s^2 \tag{3.79}$$

the characteristic equation becomes:

$$(3.47 + 0.0765K_c)s^2 + (5.27 - 0.51K_c)s + 1 + 1.70K_c = 0 \tag{3.80}$$

From Equation 3.80, the oscillation period and ultimate gain become:

$$P_0 = 3 \text{ sec}$$

$$K_{c,u} = 10.3$$

It can be seen that this result deviates somewhat from the one of the fifth-order equation: the oscillation period deviates 3% and the gain about 20%. In those cases, however, where we have to deal with many small time constants, the calculation via a time delay is faster than solution of the high-order characteristic equation. Moreover, the approximation becomes better when the number of small time constants increases.

CHAPTER 4

DYNAMICS OF LONG PIPELINES AND CONTROL OF FLOW

In the preceding chapter attention was paid to the dynamics of flow processes in short pipes. There are, however, cases where pipelines are so long that the delay in the propagation of flow variations becomes significant. This can also be important for shorter pipelines, when one wants to know the pressure fluctuations due to closing of valves, etc. (water hammer).

DYNAMICS OF FLOW IN PIPES

Consider a pipe through which a liquid-gas mixture flows at constant temperature (Figure 4.1). Let the gas fraction equal ϵ. The pressure and velocity of the fluid in the pipe are a function of distance along the pipe and time; the dynamics can therefore be described by partial differential equations. However, here we shall follow another approach: the total length

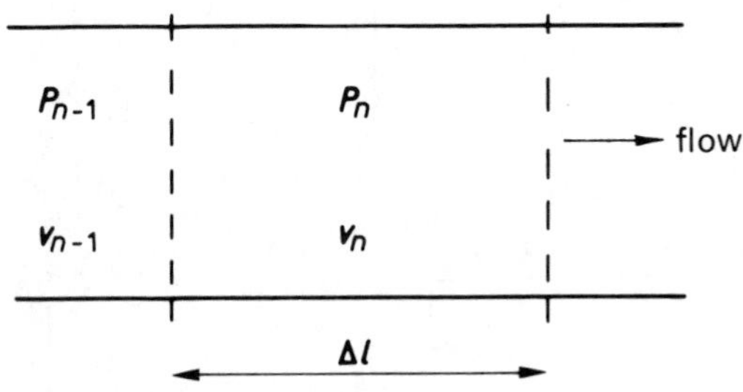

Figure 4.1 Process flow diagram.

ℓ of the pipe is divided into N equal segments with length $\Delta\ell$. Let the nth segment be characterized by a pressure P_n and a velocity v_n. When the pressure at the entrance of the pipe changes, the pressure and velocity at the end will also change. However, when the velocity of the fluid at the end is changed, the pressure and velocity at the pipe entrance will change as well. Obviously there is a flow of information inside the pipe in both directions. Therefore the arrows in the information diagram of Figure 4.2 have opposite directions.

To study the dynamics, momentum and mass balances must be set up. For reasons of simplicity it is assumed that the mass of the gas can be ignored compared to the mass of the liquid. The compressibility of the liquid and the elasticity of the tube wall will also be ignored. The pressure difference $P_n - P_{n+1}$ is used to accelerate the mass (the inertia effect) and to overcome the friction due to (turbulent) flow. Hence the momentum balance can be written as:

driving force = mass times acceleration + friction forces

or

$$A_c(P_n - P_{n+1}) = (1 - \epsilon_n)A_c\Delta\ell\rho_\ell \frac{dv_n}{dt} + A_c \frac{\Delta\ell}{d} f \cdot \frac{1}{2} \rho_\ell(1 - \epsilon_n)v_n^2 \tag{4.1}$$

where A_c = the cross-sectional area of the pipe (m^2)
ρ_ℓ = the liquid density (kg/m^3)
v_n = the average velocity of the liquid-gas mixture (m/sec)
f = the friction factor for the liquid-gas mixture

If conditions are assumed to be isothermal, the gas fraction is related to the pressure by:

$$\frac{\epsilon_n}{\epsilon_0} = \frac{P_0}{P_n} \tag{4.2}$$

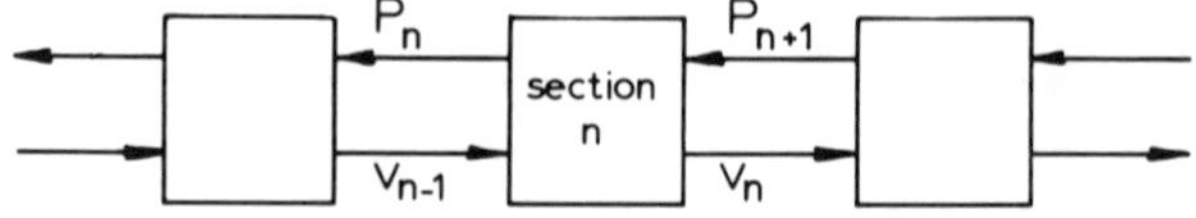

Figure 4.2 Simple information flow diagram.

where ϵ_0 and P_0 refer to entrance conditions. The mass balance for an element of the pipe is:

$$\frac{d}{dt}(M_g + M_\ell)_n = \phi_{n-1} - \phi_n \tag{4.3}$$

where M = mass

ϕ = mass flow

As the mass of the gas is ignored compared to the mass of the liquid and as it is assumed that the liquid is incompressible, Equation 4.3 may be written in the form:

$$\frac{d}{dt}[(1 - \epsilon_n)A_c\rho_\ell \Delta\ell] = v_{n-1}(1 - \epsilon_{n-1})\rho_\ell A_c - v_n(1 - \epsilon_n)\rho_\ell A_c$$

or

$$-\Delta\ell \frac{d\epsilon_n}{dt} = v_{n-1}(1 - \epsilon_{n-1}) - v_n(1 - \epsilon_n) \tag{4.4}$$

The mathematical model now consists of Equations 4.1, 4.2 and 4.4.

LINEARIZATION

Combination of Equations 4.1 and 4.2, linearization and replacing d/dt by s results in:

$$-\delta P_{n+1} + \left(1 - \frac{\Delta\ell}{d}\cdot f\cdot\frac{1}{2}\rho_\ell \bar{v}_n^2 \frac{\epsilon_0 P_0}{\bar{P}_n^2}\right)\delta P_n$$
$$= \left[(1 - \bar{\epsilon}_n)\Delta\ell\rho_\ell s + \frac{\Delta\ell}{d}\cdot f\cdot\rho_\ell(1 - \bar{\epsilon}_n)\bar{v}_n\right]\delta v_n \tag{4.5}$$

This equation can be written as:

$$\delta v_n = -\frac{K_1}{\tau_1 s + 1}\delta P_{n+1} + \frac{K_2}{\tau_1 s + 1}\delta P_n \tag{4.6}$$

where the time constant

$$\tau_1 = \frac{d}{f\bar{v}_n} \tag{4.7}$$

The process gains K_1 and K_2 can be determined from Equations 4.5 and 4.6. They are:

$$K_1 = \frac{1}{\frac{\Delta \ell}{d} \cdot f \cdot \rho_\ell (1 - \bar{\epsilon}_n)\bar{v}_n} \tag{4.8}$$

$$K_2 = \frac{1 - \frac{\Delta \ell}{d} \cdot f \cdot \frac{1}{2} \rho_\ell v_n^2 \frac{\epsilon_0 P_0}{\bar{P}_n^2}}{\frac{\Delta \ell}{d} \cdot f \cdot \rho_\ell (1 - \bar{\epsilon}_n)\bar{v}_n} \tag{4.9}$$

In a similar way Equations 4.2 and 4.4 give:

$$\left(\Delta \ell \frac{\epsilon_0 P_0}{\bar{P}_n^2} s + \bar{v}_n \frac{\epsilon_0 P_0}{\bar{P}_n^2}\right) \delta P_n$$

$$= (1 - \bar{\epsilon}_{n-1})\delta v_{n-1} - (1 - \bar{\epsilon}_n)\delta v_n + \bar{v}_{n-1} \frac{\epsilon_0 P_0}{\bar{P}_{n-1}^2} \delta P_{n-1} \tag{4.10}$$

or

$$\delta P_n = \frac{K_3}{\tau_2 s + 1} \delta v_{n-1} - \frac{K_4}{\tau_2 s + 1} \delta v_n + \frac{K_5}{\tau_2 s + 1} \delta P_{n-1} \tag{4.11}$$

with the time constant

$$\tau_2 = \frac{\Delta \ell}{\bar{v}_n} \tag{4.12}$$

Gains K_3 to K_5 can now be given by

$$K_3 = \frac{(1 - \bar{\epsilon}_{n-1})\bar{P}_n^2}{\bar{v}_n \epsilon_0 P_0} \tag{4.13}$$

$$K_4 = \frac{(1 - \bar{\epsilon}_n)\bar{P}_n^2}{\bar{v}_n \epsilon_0 P_0} \tag{4.14}$$

$$K_5 = \frac{\bar{v}_{n-1}\bar{P}_n^2}{\bar{v}_n \bar{P}_{n-1}^2} \tag{4.15}$$

The information flow diagram for Equations 4.6 and 4.11 is given in Figure 4.3.

The variations δv_n and δP_n can be obtained by combining Equations 4.6 and 4.11. The result for δv_n is:

$$\begin{aligned} \delta v_n = & \frac{K_2K_3}{\tau_1\tau_2 s^2 + (\tau_1 + \tau_2)s + 1 + K_2K_4}\,\delta v_{n-1} \\ & + \frac{K_2K_5}{\tau_1\tau_2 s^2 + (\tau_1 + \tau_2)s + 1 + K_2K_4}\,\delta P_{n-1} \\ & - \frac{K_1(\tau_2 s + 1)}{\tau_1\tau_2 s^2 + (\tau_1 + \tau_2)s + 1 + K_2K_4}\,\delta P_{n+1} \end{aligned} \tag{4.16}$$

The result for δP_n is:

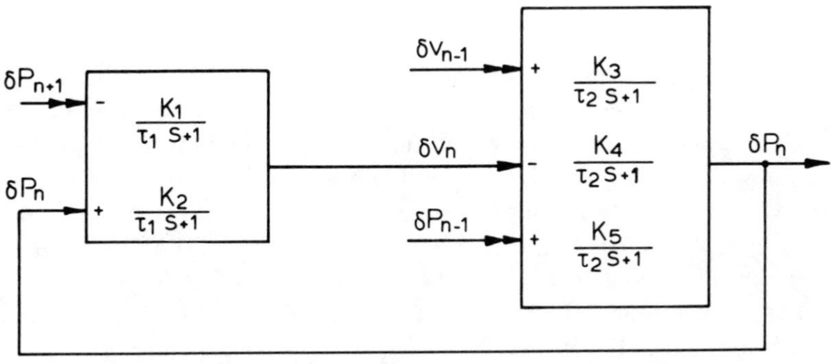

Figure 4.3 Detailed information flow diagram for section n of the pipe.

$$\delta P_n = \frac{K_3(\tau_1 s + 1)}{\tau_1\tau_2 s^2 + (\tau_1 + \tau_2)s + 1 + K_2K_4}\,\delta v_{n-1}$$

$$+ \frac{K_1K_4}{\tau_1\tau_2 s^2 + (\tau_1 + \tau_2)s + 1 + K_2K_4}\,\delta P_{n+1}$$

$$+ \frac{K_5(\tau_1 s + 1)}{\tau_1\tau_2 s^2 + (\tau_1 + \tau_2)s + 1 + K_2K_4}\,\delta P_{n-1} \qquad (4.17)$$

From Equations 4.16 and 4.17 it can be seen that the information flow diagram of Figure 4.2 has to be extended.

SPECIAL CASES

Two special cases can be distinguished. The first case is the one where only liquid flows through the pipe, thus ϵ is equal to zero. Equation 4.10 then reduces to:

$$\delta v_{n-1} = \delta v_n = \delta v \qquad (4.18)$$

In this case, the flow propagation seems to have infinite speed. From Equation 4.5, the following transfer function is determined:

$$\frac{\delta v}{\delta(\Delta P)} = \frac{d/\Delta\ell\, f\rho_\ell \bar{v}}{\frac{d}{f\bar{v}}\, s + 1} \qquad (4.19)$$

where ΔP is the pressure drop over the total length of the pipe. For instance, when d = 0.1 m, $\bar{v}$ = 1 m/sec and f = 0.04, the time constant is equal to 2.5 sec. It should be noted, however, that the speed of propagation of flow (and pressure) variations is limited by the compressibility of the liquid and elasticity of the tube wall. In practice, a flow circuit will consist of different lengths of pipeline, heat exchangers, a pump, a valve, etc. For those elements, inertia can be ignored, resulting in a large contribution to the friction term in the momentum balance. This results in extra terms in the denominator of the time constant in Equation 4.19, thus reducing this time constant.

The second special case is gasflow. The gas fraction ϵ is now equal to one, and the gas holdup may not be ignored. Equation 4.1 becomes:

$$P_n - P_{n+1} = \Delta\ell\rho_{g,n}\frac{dv_n}{dt} + \frac{\Delta\ell}{d}\cdot f\cdot\frac{1}{2}\rho_{g,n}v_n^2 \tag{4.20}$$

where ρ_g refers to the gas density. The mass balance can now be written as:

$$\frac{d}{dt}(\rho_{g,n}V_g) = v_{n-1}A_c\rho_{g,n-1} - v_nA_c\rho_{g,n} \tag{4.21}$$

or

$$\Delta\ell\left(\frac{\partial\rho_g}{\partial P}\right)_n\frac{dP_n}{dt} = v_{n-1}\rho_{g,n-1} - v_n\rho_{g,n} \tag{4.22}$$

Note that the gas density also has a subscript which refers to the section number, because pressure and therefore the gas density vary over the length of the pipeline.

It is possible to linearize Equations 4.20 and 4.22 again, resulting in a detailed information flow diagram for this situation. In the literature, however, often another (less accurate) approach is given. Equations 4.20 and 4.22 are then written in terms of the volumetric gasflow ϕ_v. After linearization and replacing d/dt by s, Equation 4.20 is written in the form:

$$\delta P_n - \delta P_{n+1} = Ls\delta\phi_{v,n} + R\delta\phi_{v,n} \tag{4.23}$$

where L is called the coefficient of self-induction and R the resistance to flow.

Usually, L and R are taken constant, although they both depend on the section number. In the case of turbulent flow, as was assumed here, R even depends on the average value of the flow. Equation 4.22 is written in the form:

$$Cs\delta P_n = \delta\phi_{v,n-1} - \delta\phi_{v,n} \tag{4.24}$$

where C corresponds to a certain capacitive effect. Equations 4.23 and 4.24 can be represented by an electrical analog as shown in Figure 4.4.

In the case of laminar flow through a pipe, the description is more complicated. The velocity profile is parabolic for that case and Equations 4.20

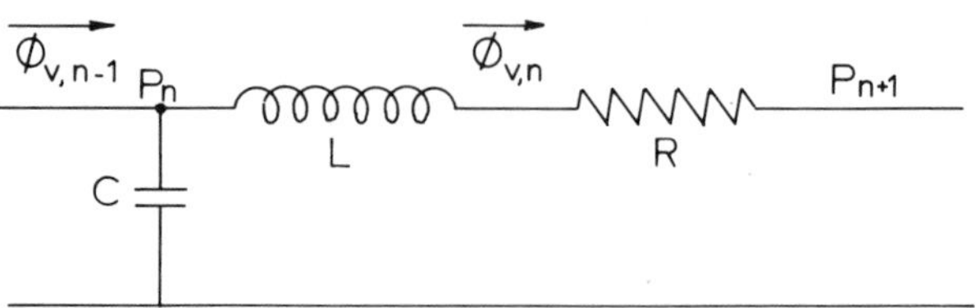

Figure 4.4 Analog for gasflow in a pipe section.

and 4.22 are not valid anymore. Then the Navier–Stokes equations have to be solved [7].

A somewhat more accurate approach in the case of turbulent flow is obtained, when the capacity C is divided equally over both ends of the circuit. For that case the analog of Figure 4.5 is obtained for n sections.

COMPLETE FLOW CIRCUIT

The dynamics of flow in pipe sections were described above. Usually, the pipe is connected at both ends to some kind of apparatus or instrument. Assume that the inlet is connected to a pump and the outlet to a control valve. When the pipe is divided into four sections, the information flow diagram of Figure 4.5 results. The model equations are, for the pump:

$$v_0 = f(p_1, n) \tag{4.25}$$

where n is the speed of rotation, for the pipeline:

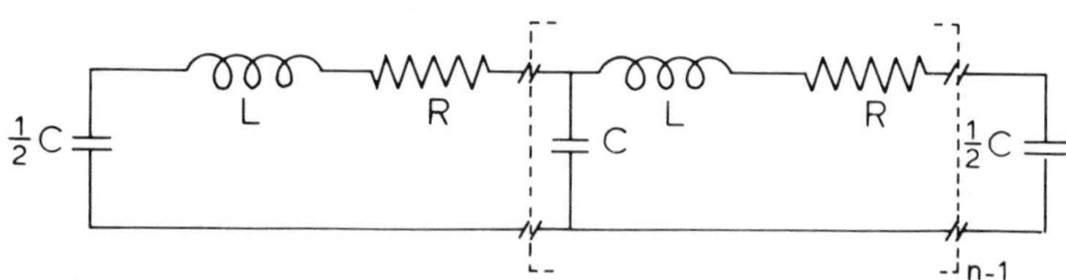

Figure 4.5 Analog for a pipe of n sections.

$$\left.\begin{aligned} p_1 &= f(v_0, p_2); & v_1 &= f(v_0, p_2) \\ p_2 &= f(v_1, p_3, p_1); & v_2 &= f(v_1, p_1, p_3) \\ p_3 &= f(v_2, p_4, p_2); & v_3 &= f(v_2, p_2, p_4) \\ p_4 &= f(v_3, p_5, p_3); & v_4 &= f(v_3, p_3, p_5) \end{aligned}\right\} \quad (4.26)$$

and for the control valve:

$$p_5 = f(v_4, p_6, u) \quad (4.27)$$

where u = position of the valve shaft
f = a static or dynamic dependence (for every equation this dependence may be different)

Equations 4.25 to 4.27 represent 10 equations with 13 unknown variables; hence, 3 variables may be selected independently. In Figure 4.6, the speed of rotation of the pump, the valve shaft position and the valve discharge pressure are obvious choices.

When the flow in the circuit is, for example, liquid flow, the information flow diagram of Figure 4.6 can be simplified, because the arrows pointing to the right and denoted with p_1, p_2 and p_3 can be omitted in that case.

COMPRESSORS

In the previous chapter, flow control by a control valve was discussed extensively. However, by varying the speed of rotation of compressors and pumps, flow can also be varied. Compressors and pumps usually have a small gas or liquid content compared to the connected pipeline. Therefore, we are only interested in the static behavior of the rotating equipment.

Compressors must be protected against a variety of undesired conditions: speed too high or too low, flowrates too low, etc. The last one corresponds

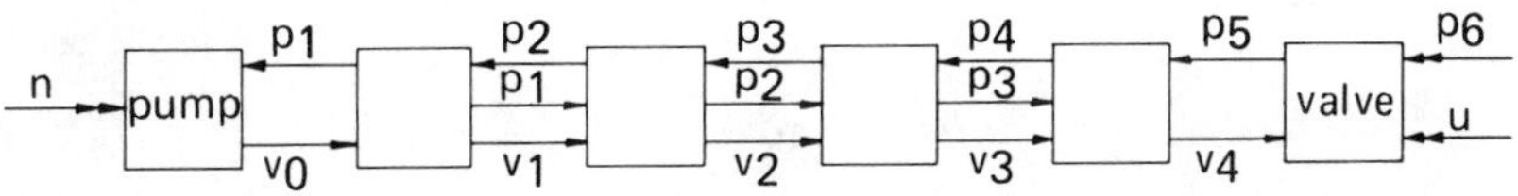

Figure 4.6 Information flow diagram of a flow circuit with pump and control valve.

to the surging or pumping limit, where the flow through the compressor starts to be unstable. Pumping can destroy a compressor within seconds.

The static behavior of a compressor can be expressed in three dimensionless numbers:

the inlet pressure to discharge pressure ratio (p_i/p_0) (4.28)

the ratio of the rotor circumferential velocity to sound velocity (ND/c_s) (4.29)

where N = the speed of rotation
D = the rotor diameter
c_s = the sound velocity

the ratio between suction velocity to sound velocity (v_i/c_s) (4.30)

With the aid of the gas law:

$$\frac{p_i}{\rho_i} = \frac{z_i R_g T_i}{m} \tag{4.31}$$

where z_i = compressibility
R_g = universal gas constant
T_i = temperature
m = molecular weight

With the expression for the sound velocity

$$c_s = \sqrt{\kappa_i p_i/\rho_i} \tag{4.32}$$

with κ_i = the ratio of specific heats, Equations 4.29 and 4.30 can be transformed to a modified speed of rotation:

$$N^* = N\sqrt{m/\kappa_i z_i T_i} \tag{4.33}$$

and a modified volumetric flow:

$$\phi^* = \frac{\phi}{p_i} \sqrt{z_i T_i / \kappa_i m} \tag{4.34}$$

where ϕ is the mass flow.

Figure 4.7 shows the diagram in which the static behavior of the compressor can be represented. The area of operation is limited by:

1. The maximum speed of rotation. N^* is dependent on the process conditions (particularly on the molecular weight and inlet temperature); this limit is a shifting curve in Figure 4.7.
2. The maximum power. This limit is also a shifting curve in Figure 4.7, and can hardly be drawn because the location is strongly dependent on the pressure level.
3. The minimum speed of rotation. This limit is related to the highest critical speed of rotation of the rotor.
4. The stalling limit. In this case the sound velocity in the compressor is reached, and increase of flow is impossible.
5. The pumping limit.

The primary reason for this behavior is the shape of the pressure flow curve which shows a maximum at 50% of the maximum flow and decreases for lower and higher flows. Operation close to the maximum point can lead to a rapid fluctuation of flow between the points for which the pressure has the same value. Because low values of the flow can occur during normal operation, a safety device which puts the compressor out of operation is not attractive. It is more convenient to install an automatic control which adjusts

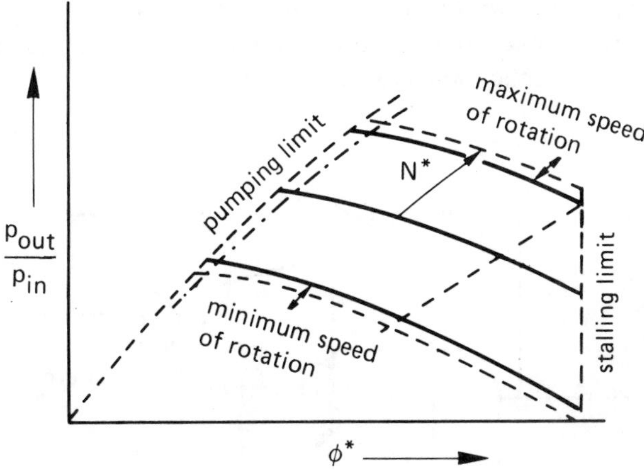

Figure 4.7 Compressor diagram.

a bypass valve and keeps the flow above the pumping limit (Figure 4.8). The controlled condition can be calculated from flow and pressure measurements, to approach the pumping limit with a constant margin.

Because the compressor usually operates above the pumping limit, integral control action will lead to saturation. In that case the controller "lies on one ear," and it will take a long time before the controller becomes active in case of a passing of the pumping limit. This can be avoided by not using integral action at all. But then it must be possible to make the proportional gain large enough (see Figure 4.7). One can also use a controller with anti-reset windup. Then the controller output is kept artificially at the saturation limit during normal operation of the compressor. Staroselsky and Ladin [8] discuss surge control of centrifugal compressors in detail.

SPEED CONTROL OF PUMPS

The flow through a compressor is usually controlled by variation of the speed of rotation. For pumps, this type of flow control is not yet used very often. Usually, the flow is still adjusted by a control valve. However, from a point of view of energy saving, pump speed adjustment can be an attractive alternative to adjustment by a control valve.

Assume that a flow is adjusted by a control valve. When the valve is closed somewhat more, the pump pressure will increase, and the pressure drop in apparatus and pipes will decrease, while the pressure drop across the valve increases. This situation is shown in Figure 4.9, where the pump pressure and the energy consumption of the pump are plotted as a function of the flow through the pump. Suppose the flow is decreased by closing the control valve somewhat more (from A_2 to C_2). The pump pressure will increase and the energy consumption will decrease from A_1 to C_1.

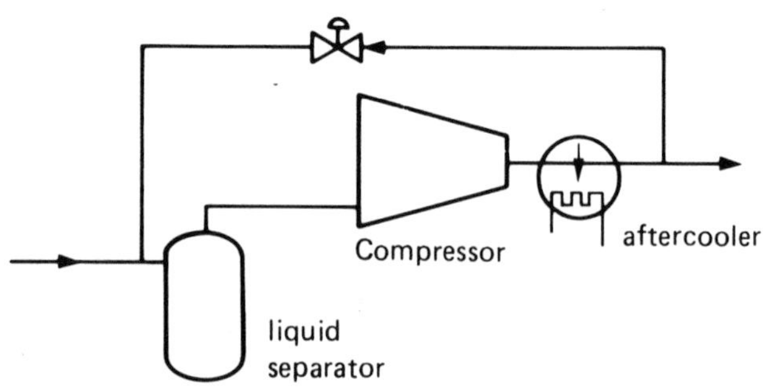

Figure 4.8 Compressor with bypass.

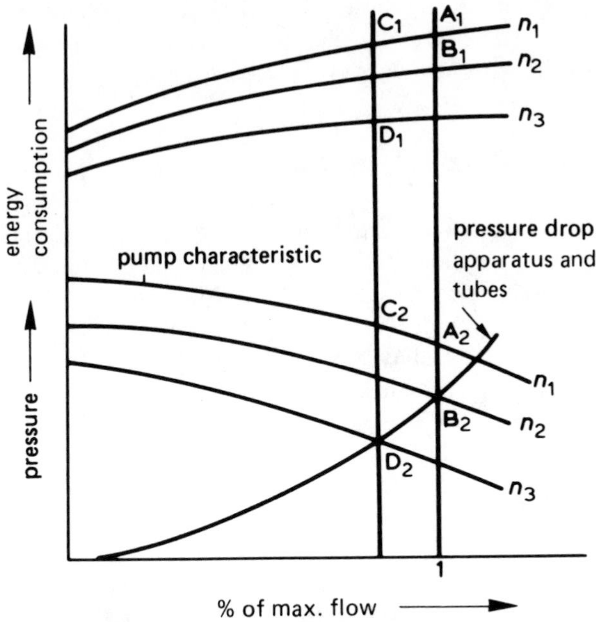

Figure 4.9 Pressure and energy consumption as a function of flow.

If the control valve is now removed (from A_2 to B_2), the speed of rotation may be decreased to get the same flow because the valve pressure drop (A_2B_2) need not be overcome anymore. If the speed of rotation is decreased further (from n_2 to n_3), resulting in a decrease in pump pressure from B_2 to D_2, then the energy consumption will decrease further from B_1 to D_1.

A decrease in flow results in an energy saving from A_1 to C_1 in the case of a valve; in the case of a reduction of the speed of rotation a similar flow reduction results in an energy saving from A_1 to D_1. It is evident that a not-too-expensive speed control of a pump is an attractive alternative for a control valve. There are several possibilities to control the pump speed, depending on the pump drive. For direct current motors the field voltage may be varied with the aid of thyristor circuits.

Asynchronous motors, however, have a more or less constant speed of rotation, which is proportional to the frequency of the electrical supply. In this case, a frequency converter is required. The voltage of the alternating current supply is converted into direct current, after which a conversion is made to a three-phase alternating current with adjusted amplitude and frequency. The frequency determines the speed of rotation of the motor; the voltage is adjusted to maintain the maximum torque over the whole range.

Asynchronous motors are more robust than direct-current motors; hence, they need less maintenance. Furthermore, an asynchronous motor has no collector, which results in a reduced explosion danger.

EXAMPLE: PRESSURE CONTROL

A gas-fired furnace is supplied with combustion air by a ventilator. The distance between ventilator and burner is 40 m, the rectangular pipe has dimensions equal to 0.65 × 0.65 m. The airflow in the pipe is 2.21 m^3/sec under normal conditions; the air pressure at the burner is 2.2 kPa gauge. This pressure is controlled by adjustment of the speed of rotation of the ventilator. The measurement lag is equal to 1 sec, the correction lag equal to 4 sec.

The relationship between pressure increase, flow and speed of rotation of the ventilator is given by:

$$p = (3.00 - 0.082\,\phi^2)\,(n/n_0)^{1.7} \tag{4.35}$$

where n_0 = the speed of rotation under static conditions
p = the discharge pressure in kPa
ϕ = the flow (m^3/sec)

The friction factor for the pipe has a value of 0.02. In- and outflow losses and friction in bends have to be taken into consideration, and are approximately equal to four times the pipe friction losses. Calculate the limit of stability for proportional control.

Static Analysis

We shall start with a static analysis of the system. The area of the pipe is 0.65 × 0.65 = 0.423 m^2. As the airflow is equal to 2.21 m^3/sec, the air velocity is 5.22 m/sec. The pressure drop in the pipe can be calculated according to:

$$\Delta p = 4f \cdot \frac{\ell}{d} \cdot \frac{1}{2} \rho v^2 \tag{4.36}$$

When $\rho = 1.2$ kg/m^3, this pressure drop is:

$$\Delta p = 4 \cdot 0.02 \cdot \frac{40}{0.65} \cdot 0.5 \cdot 1.2 \cdot (5.22)^2 = 80.5 \text{ Pa} \tag{4.37}$$

It has been assumed that the losses due to in- and outflow and in bends are about four times the pipe friction losses. This results in a pressure drop of 322 Pa. The total pressure drop now becomes 402.5 Pa. The air pressure at the burner is 2.2 kPa, which means that the discharge pressure of the ventilator must be equal to 2.603 kPa. With Equation 4.35 we can compute the speed of rotation:

$$\frac{n}{n_0} = \left(\frac{p}{3.00 - 0.082\phi^2}\right)^{0.59}$$
$$= \left(\frac{2.603}{3.00 - 0.082\,(2.21)^2}\right)^{0.59} = 1.0 \qquad (4.38)$$

which means that n is equal to n_0, the speed of rotation under static conditions.

Dynamic Analysis

A dynamic analysis of the situation can be made, starting from a mass balance and a momentum balance. The mass balance for the pipe is:

$$\frac{dM}{dt} = \phi_{in} - \phi_{out} \qquad (4.39)$$

where M = mass
ϕ = mass flow

In terms of the volumetric flow, Equation 4.39 can be written as:

$$\frac{V}{\bar{\rho}}\left(\frac{\partial \rho}{\partial p}\right)_0 \frac{dp_0}{dt} = \phi_i - \phi_0 \qquad (4.40)$$

or

$$C\,\frac{dp_0}{dt} = \phi_i - \phi_0 \qquad (4.41)$$

where p_0 = pressure at the end of the pipe (just upstream of the burner)
V = volume of the pipe
$\bar{\rho}$ = average density

The volume of the air in the pipe is 0.65 X 0.65 X 40 = 16.90 m^3. When $\partial\rho/\partial p = 10^{-5}$ sec^2-m^{-2} and $\bar{\rho} = 1.2$ kg/m^3, capacity C is equal to:

$$C = \frac{16.90}{1.2} \cdot 10^{-5} = 14.08 \times 10^{-5}\ m^4\text{-}sec^2/kg \qquad (4.42)$$

Equation 4.41 can be written in terms of variations as:

$$Cs\delta p_0 = \delta\phi_i - \delta\phi_0 \qquad (4.43)$$

A momentum balance related to the inlet conditions can be written as:

$$p_i - p_0 = \frac{\bar{\rho}\ell}{A_c}\frac{d\phi_i}{dt} + 4f \cdot \frac{\ell}{d}\frac{1}{A_c^2} \cdot \frac{1}{2}\bar{\rho}\phi_i^2 + c_L\phi_i^2 \qquad (4.44)$$

in which c_L is a coefficient for losses due to in- and outflow and losses in bends. From static data it is known that these losses are equal to 322 Pa at a flow of 2.21 m^3/sec, from which a value of c_L results:

$$c_L = \frac{322}{(2.21)^2} = 65.93 \qquad (4.45)$$

linearization of Equation 4.44 gives:

$$\delta p_i - \delta p_0 = \frac{\rho\ell}{A_c} s\delta\phi_i + 4f \cdot \frac{\ell}{d} \cdot \frac{1}{A_c^2}\overline{\rho\phi_i}\delta\phi_i + 2c_L\bar{\phi}_i\delta\phi_i \qquad (4.46)$$

or

$$\delta p_i - \delta p_0 = (Ls + R)\delta\phi_i \qquad (4.47)$$

with

$$L = \frac{\rho\ell}{A_c} = \frac{1.2 \times 40}{0.423} = 113.5$$

$$R = 4f \cdot \frac{\ell}{d} \cdot \frac{1}{A_c^2} \overline{\rho \phi_i} + 2c_L \bar{\phi}_i$$

$$= 4 \cdot 0.02 \cdot \frac{40}{0.65} \frac{1}{(0.423)^2} \cdot 1.2 \cdot 2.21 + 2 \cdot 65.93 \cdot 2.21$$

$$= 72.97 + 291.41 = 364.38$$

The model can be completed by linearizing the equation for the ventilator. In N/m^2 (Pa), this equation becomes:

$$\delta p_i = -164\, \bar{\phi}_i \delta\phi_i + 1.7 \times 10^3\ (3.00 - 0.082\, \bar{\phi}_i^2) \frac{\delta n}{n_0}$$

$$= -a_0 \delta\phi_i + a_1 \frac{\delta n}{n_0} \qquad (4.48)$$

with

$$a_0 = 164 \bar{\phi}_i = 164 \cdot 2.21 = 362.4$$

$$a_1 = 1.7 \times 10^3\ [3.00 - 0.082\ (2.21)^2] = 4420$$

The mathematical model for the process now consists of Equations 4.43, 4.47 and 4.48. Combination of these equations results in:

$$\delta p_0 (LCs^2 + (RC + a_0 C)s + 1) = a_1 \frac{\delta n}{n_0} - (Ls + a_0 + R)\delta\phi_0 \qquad (4.49)$$

from which:

$$\frac{\delta p_0}{(\delta n / n_0)} = \frac{a_1}{LCs^2 + (R + a_0)Cs + 1} \qquad (4.50)$$

with

$$LC = 113.5 \times 14.08 \times 10^{-5} = 0.016$$

$$(a_0 + R)C = (362.4 + 364.4)\ 14.08 \times 10^{-5} = 0.102$$

From these data it can be seen that the process dynamics are very fast. The response is undercritically damped.

The following equation holds for the adjustment of the speed of rotation:

$$\frac{n}{n_0} = 1 - K_c(p_0 - p_{set}) \tag{4.51}$$

where K_c is the gain of the proportional controller and p_{set} the set value for the pressure upstream the burner. For variations, Equation 4.51 can be written as:

$$\frac{\delta n}{n_0} = -K_c \delta p_0 \tag{4.52}$$

The information flow diagram can now be constructed and is given in Figure 4.10. The characteristic equation can easily be derived from this figure. It is equal to one plus the product of the closed loop transfer functions:

$$1 + \frac{a_1 K_c}{(1 + \tau_c s)(1 + \tau_m s)(LCs^2 + a_0 Cs + RCs + 1)} = 0 \tag{4.53}$$

After some rearrangements and substitution of parameters, Equation 4.53 becomes:

$$0.064s^4 + 0.488s^3 + 4.526s^2 + 5.102s + 1 + 4420K_c = 0 \tag{4.54}$$

Dividing by $s^2 + \omega_u^2$ results in:

$$\omega_u^2 = \frac{5.102}{0.488} = 10.45$$

from which the oscillation period becomes:

$$P_u = \frac{2\pi}{\omega_u} = 1.95 \text{ sec}$$

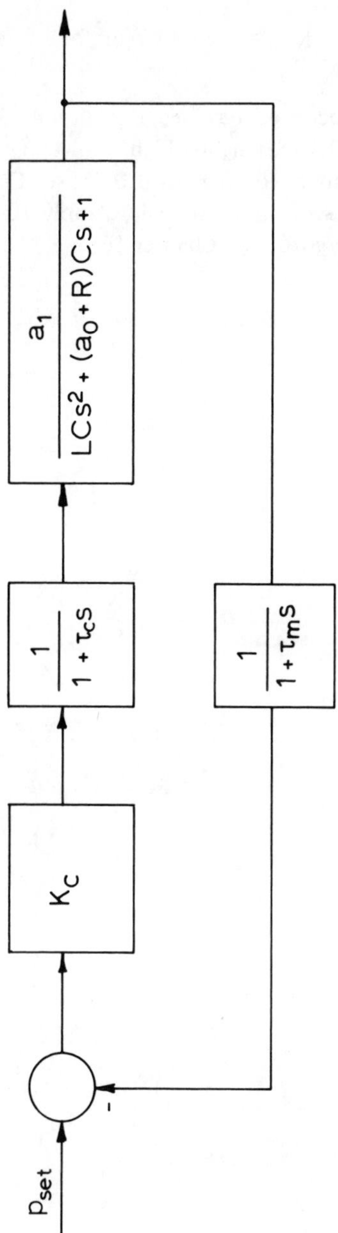

Figure 4.10 Information flow diagram for burner pressure control.

The value of the proportional gain becomes:

$$K_{c,u} = 8.9 \times 10^{-3}\ m^2/N$$

In combination with process gain a_1 (see Equation 4.53), the ultimate loop gain is 39.3, which looks reasonably high. However, reasonable damping (corresponding to a damping coefficient of 0.25; see Chapter 2) leads to an oscillation period of 3.8 sec and a loop gain of only 10.4. The latter value is too low for satisfactory control (see Chapter 16).

CHAPTER 5

DYNAMICS AND CONTROL OF HEAT TRANSFER PROCESSES

In this chapter attention will be focused on the dynamics and control of evaporators, condensers and furnaces.

STEAM-HEATED EVAPORATOR

Evaporators are usually parts of larger process units, e.g., distillation columns. The temperature in such a process unit often should be controlled from the heating side of the evaporator. As this temperature reacts approximately as a first-order lag with large time constant to variations in the evaporation rate, the dynamics of an evaporator have to be combined with a transfer function of the type

$$\frac{K}{\tau s + 1} \tag{5.1}$$

Dynamic Behavior

Figure 5.1 shows a steam-heated evaporator. The liquid flows (often by thermosiphon action) to the top of the pipes and evaporates partially by condensing steam at the outside of the pipes. If there is a condensate level in the evaporator, there are two mass balances for the shell side; one for the steam and one for the condensate:

$$\frac{d}{dt}(\rho_s V_s) = \phi_s - \phi_c \tag{5.2}$$

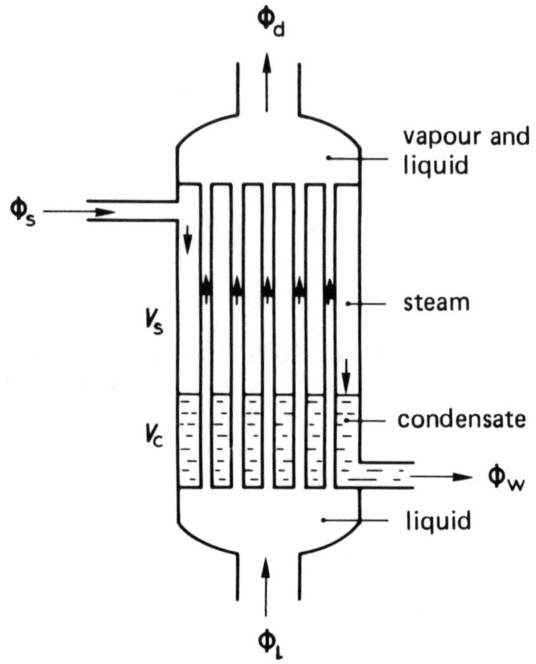

Figure 5.1 Steam-heated evaporator.

$$\frac{d}{dt}(\rho_c V_c) = \phi_c - \phi_w \tag{5.3}$$

where ρ = the density (kg/m^3)
V = the volume (m^3)
ϕ = the flow (kg/sec)

The subscript s refers to steam, c to condensate and w to water.

As the total shell volume V_m is constant:

$$V_s + V_c = V_m \tag{5.4}$$

The steam density ρ_s in the evaporator will be a function of the steam pressure P_s:

$$\rho_s = \rho_s(P_s) \tag{5.5}$$

Elimination of V_s and ρ_s in Equation 5.2 with the aid of Equations 5.4 and 5.5 results in:

$$(V_m - V_c)\frac{\partial \rho_s}{\partial P_s} \cdot \frac{dP_s}{dt} - \rho_s \frac{dV_c}{dt} = \phi_s - \phi_c \qquad (5.6)$$

Equations 5.6 and 5.3 are shown in the information flow diagram of Figure 5.2. The internal condensate flow ϕ_c is determined by the heat transfer to the tube side, and is therefore represented as an independent variable in Figure 5.2. The steam pressure is also an independent variable, determined by the steam supply system. Equation 5.6 determines the steam quantity, and Equation 5.3 the condensate volume and consequently the condensate level.

Due to the difference in heat transfer coefficients, heat transfer is mainly determined by condensation of steam and only for a small part by cooling of condensate, resulting in an energy balance:

change of heat content = heat supply − heat discharge

$$\frac{d}{dt}(\rho_s V_s c_s T_s) = \phi_c H_c - h_s(1 - V_c/V_m)A_w(T_s - T_w) \qquad (5.7)$$

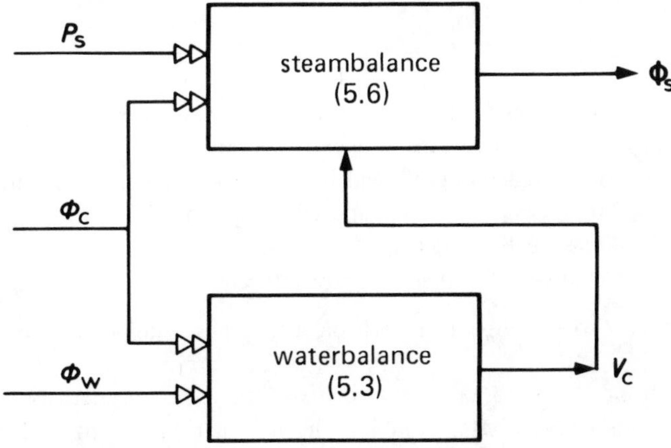

Figure 5.2 Information flow diagram of the shell side.

where c_s = the specific heat of the steam (J/kg-°K)
T_s = the steam temperature (°K)
T_w = the wall temperature (°K)
H_c = the heat of condensation (J/kg)
A_w = the outerside heat transfer area of the pipes (m^2)
h_s = the heat transfer coefficient of the condensing steam (W/m^2-°K)

For simplicity, the temperature in the pipes has been assumed to be the same at all heights.

In many cases the left side of Equation 5.7 is so small that this differential equation can be approximated by an algebraic equation:

$$\phi_c = (T_s - T_w)\left(1 - \frac{V_c}{V_m}\right) A_w \frac{h_s}{H_c} \tag{5.8}$$

The heat conductivity of the pipe material is usually so large that temperature differences between in- and outside can be ignored. Then the energy balance for the pipe wall is:

$$M_w c_w \frac{d}{dt}\left[\left(1 - \frac{V_c}{V_m}\right) T_w\right] = \phi_c H_c - Q_p \tag{5.9}$$

where

$$Q_p = (T_w - T_p)\left(1 - \frac{V_c}{V_m}\right) A_p h_p \tag{5.10}$$

where Q_p = heat flow to the evaporating liquid inside the tubes (W)
A_p = internal heat transfer area of the pipes (m^2)
h_p = heat transfer coefficient to the evaporating liquid (W/m^2-°K)
T_p = temperature of the evaporating liquid (°K)
M_w = mass of the wall (kg)
c_w = heat capacity of the wall (J/kg-°K)

For simplicity the temperature within the pipes will be assumed to be constant.

The shell of the evaporator also contributes to the dynamics, but this contribution is ignored here. Combination of Equations 5.8 to 5.10 together with Figure 5.2 results in the information flow diagram of Figure 5.3. The

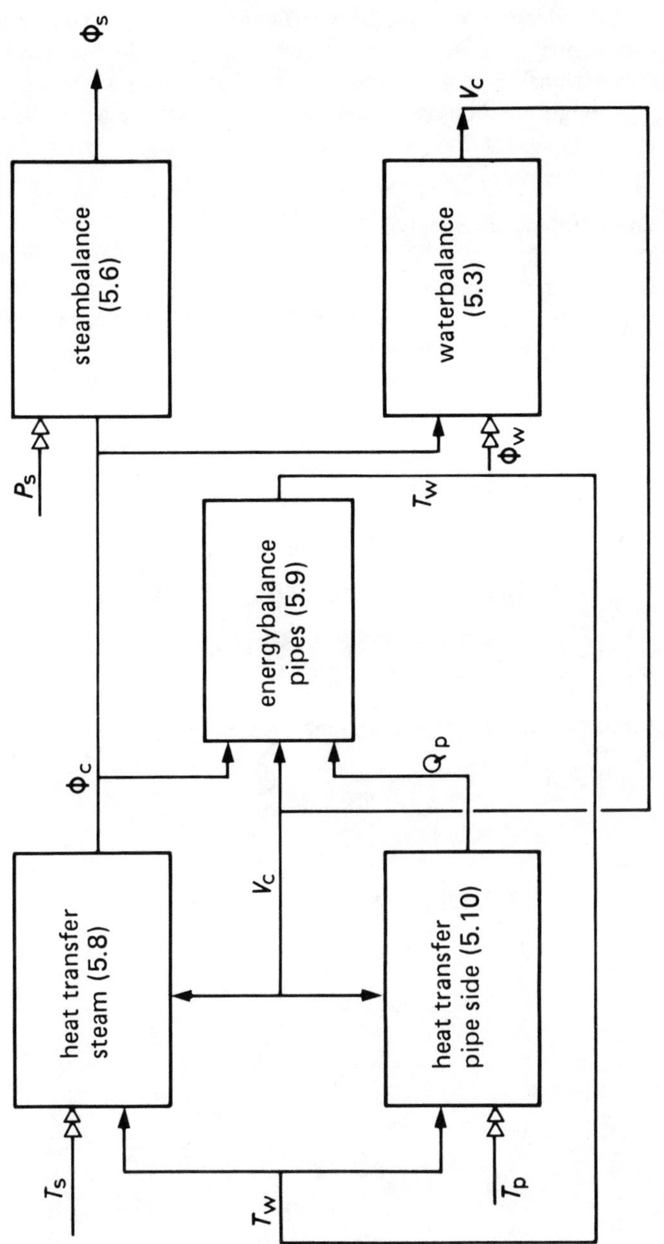

Figure 5.3 Information flow diagram of the evaporator.

temperature of the saturated steam (T_s) depends directly on P_s. Equations 5.8 and 5.10 can be substituted into Equations 5.6, 5.3 and 5.9, thus reducing the number of blocks to three (which is equal to the number of differential equations). Two models will be developed, corresponding to the possible correcting conditions: the steam supply ϕ_s and the water discharge flow ϕ_w.

Control Valve in the Steam Supply

In the first case, a control valve is used in the steam supply for adjustment of the heat transfer, and a steam trap takes care of water discharge. Due to the use of a steam trap, the water content V_c can be ignored and the water balance can be omitted. Substitution of $V_c = 0$ in Equations 5.6, 5.8, 5.9 and 5.10 results in:

$$V_m(\partial \rho_s/\partial P_s)dP_s/dt = \phi_x - \phi_c \qquad (5.11)$$

$$\phi_c = (T_s - T_w)A_w(h_s/H_c) \qquad (5.12)$$

$$M_w c_w(dT_w/dt) = \phi_c H_c - Q_p \qquad (5.13)$$

$$Q_p = (T_w - T_p)A_p h_p \qquad (5.14)$$

The behavior of the control valve can be represented by:

$$\phi_s = c_v f(u)\sqrt{(P_0 - P_s)\rho_s} \qquad (5.15)$$

where P_0 represents the steam pressure in the steam supply line. For a linear valve the expression for f(u) is:

$$f(u) = 1/R_v + (1 - 1/R_v)u_r \qquad (5.16)$$

and for an exponential valve:

$$f(u) = R_v^{u_r - 1} \qquad (5.17)$$

where u_r = relative valve stem position, $0 \leqslant u_r \leqslant 1$
R_v = rangeability, the ratio between maximum and minimum nominal flow through the valve (see Chapter 3).

We shall now determine the transfer function from f(u) to Q_p by linearization of Equations 5.11 to 5.15:

$$sV_m(\overline{\partial\rho_s/\partial P_s})\cdot\delta P_s = \delta\phi_s - \delta\phi_c \tag{5.18}$$

$$\delta\phi_c = (\delta T_s - \delta T_w)(A_w\bar{h}_s/\bar{H}_c) + (\bar{T}_s - \bar{T}_w)A_w(\delta h_s/\bar{H}_c) + (\bar{T}_s - \bar{T}_w)A_w\bar{h}_s(\delta H_c/\bar{H}_c^2) \tag{5.19}$$

$$sM_wc_w\delta T_w = \bar{H}_c\delta\phi_c + \bar{\phi}_c\delta H_c - \delta Q_p \tag{5.20}$$

$$\delta Q_p = (\delta T_w - \delta T_p)A_p\bar{h}_p + (\bar{T}_w - \bar{T}_p)A_p\delta h_p \tag{5.21}$$

$$\delta\phi_s = \delta f c_v\sqrt{(P_0 - \bar{P}_s)\bar{\rho}_s} + \frac{1}{2}c_v\overline{f(u)}[\sqrt{(P_0 - \bar{P}_s)\bar{\rho}_s}]^{-1} \times [(P_0 - \bar{P}_s)(\overline{\partial\rho_s/\partial T_s}) - \bar{\rho}_s(\overline{dP_s/dT_s})]\delta T_s \tag{5.22}$$

The terms with δh_s, δH_c and δh_p can be taken into account by using empirical relationships between material properties and flow velocities. In order not to complicate the equations too much, all of these terms and the term with δT_p will be ignored. (This is based on the large time constant in the response of this temperature; see Equation 5.1.) With the aid of the static relationships:

$$\bar{\phi}_s = \bar{\phi}_c = (\bar{T}_s - \bar{T}_w)A_w(\bar{h}_s/\bar{H}_c) \tag{5.23}$$

$$\bar{\phi}_c\bar{H}_c = \bar{Q}_p = (\bar{T}_w - \bar{T}_p)A_ph_p \tag{5.24}$$

we can eliminate $\bar{h}_s$ and $\bar{h}_p$ from Equations 5.19 and 5.21. The result is:

$$\frac{\delta\phi_c}{\bar{\phi}_c} = \frac{\bar{H}_c\delta\phi_c}{\bar{Q}_p} = \frac{\delta T_s - \delta T_w}{\bar{T}_s - \bar{T}_w} \tag{5.25}$$

$$\frac{\delta Q_p}{\bar{Q}_p} = \frac{\delta T_w}{\bar{T}_w - \bar{T}_p} \tag{5.26}$$

The equation for the valve can be transformed to:

$$\delta\phi_s/\bar{\phi}_s = \bar{H}_c\delta\phi_s/\bar{Q}_p = \delta f/\overline{f(u)} - G_v\delta T_s \tag{5.27}$$

where

$$G_v = \frac{1}{2}\{[1/(P_0 - \bar{P}_s)](\overline{dP_s/dT_s}) - (1/\bar{\rho}_s)(\overline{\partial\rho_s/\partial T_s})\} \tag{5.28}$$

Substitution of Equations 5.25 and 5.27 into Equation 5.18 and Equations 5.25 and 5.26 into Equation 5.20 results in:

$$sC_s\delta T_s = \frac{\delta f}{\overline{f(u)}} - G_v\delta T_s - \frac{\delta T_s - \delta T_w}{\bar{T}_s - \bar{T}_w} \tag{5.29}$$

$$sC_w\delta T_w = \frac{\delta T_s - \delta T_w}{\bar{T}_s - \bar{T}_w} - \frac{\delta T_w}{\bar{T}_w - \bar{T}_p} \tag{5.30}$$

where

$$C_s = V_m\left(\overline{\frac{\partial\rho_s}{\partial T_s}}\right)\bar{\phi}_s^{-1} \tag{5.31}$$

$$C_w = M_w c_w \bar{Q}_p^{-1} \tag{5.32}$$

C_s is related to the residence time of the steam τ_s:

$$C_s = \tau_s \cdot \frac{1}{\bar{\rho}_s}\left(\overline{\frac{\partial\rho_s}{\partial T_s}}\right) \tag{5.33}$$

This residence time usually is small (a few seconds), thus it can be expected that the dynamics are fast.

Example

The following data are available:

P_0 = 3.5 bar (abs),
$\bar{T}_s$ = 394°K,
τ_s = 7.3 sec,
M_w = 680 kg,
$\bar{Q}_p$ = 2.4×10^6 W,
$\bar{T}_p$ = 370°K,
$\bar{T}_w$ = 390°K,
$\bar{P}_s$ = 2.1 bar (abs) (from steam tables),
$\overline{dP_s/dT_s}$ = 0.066 bar/°K, and
$(1/\bar{\rho}_s)(\overline{\partial\rho_s/\partial T_s})$ = 0.030°K^{-1}.

Substitution of these data in the expressions for G_v, C_s and C_w results in:

$G_V = 0.0086$,
$C_S = 0.22$ sec/°K, and
$C_W = 0.14$ sec/°K.

Equations 5.29 and 5.30 now become:

$$0.22s\,\delta T_s = \frac{\delta f}{f(u)} - 0.0086\delta T_s - 0.25(\delta T_s - \delta T_w) \tag{5.34}$$

$$0.14s\,\delta T_w = 0.25(\delta T_s - \delta T_w) - 0.05\delta T_w \tag{5.35}$$

Elimination of δT_s results in:

$$\frac{\delta T_W}{\delta f} = \frac{16.5}{f(u)} \cdot \frac{1}{(6.4s + 1)(0.3s + 1)} \tag{5.36}$$

With the aid of Equation 5.26 we now find:

$$\frac{\delta Q_P}{\delta f} = \frac{\overline{Q}_P}{\overline{T}_W - \overline{T}_P} \cdot \frac{\delta T_W}{\delta f} = \frac{16.5\,\overline{Q}_P}{\overline{f(u)}(\overline{T}_W - \overline{T}_P)} \; \frac{1}{(6.4s + 1)(0.3s + 1)} \tag{5.37}$$

When using an exponential valve with a rangeability $R_v = 50$ and a relative valve stem position $u_r = 0.75$, Equation 5.17 gives:

$$\delta f = R_v^{u_r - 1} \ln R_v \delta u_r = 1.47\delta u_r \tag{5.38}$$

For the evaporating vapor:

$$\delta Q_p = H_v \delta \phi_d \tag{5.39}$$

where H_v is the heat of vaporization of the liquid (J/kg) and ϕ_d is the vapor flow (kg/sec) leaving the top of the condenser. Combination of Equations 5.37, 5.38 and 5.39, with $H_v = 5 \times 10^5$ J/kg, results in:

$$\frac{\delta \phi_d}{\delta u_r} = \frac{15.5}{(6.4s + 1)(0.3s + 1)} \tag{5.40}$$

The dynamics are rather fast. For the description of the dynamic behavior of the control loop, the large time constant of Equation 5.1 and the instrumentation dynamics (particularly the control valve) also have to be taken into account.

Control Valve in the Condensate

Another possibility is to install the control valve in the water discharge flow. Then there will be a level in the evaporator, above which the steam supply pressure P_0 is present. In this case there are some practical advantages:

1. The steam temperature is higher ($T_s = T_0$), resulting in a higher temperature difference for the heat transfer
2. There is no need for a steam trap
3. The control valve is smaller and cheaper.

Equation 5.6 is not relevant for the dynamic behavior because ϕ_s does not appear in the other equations. The resulting set of equations is:

$$\rho_c(dV_c/dt) = \phi_c - \phi_w \tag{5.3}$$

$$\phi_c = (T_0 - T_w)(1 - V_c/V_m)A_w(h_s/H_c) \tag{5.8}$$

$$Q_p = (T_w - T_p)(1 - V_c/V_m)A_p h_p \tag{5.10}$$

$$M_w c_w \frac{d(1 - V_c/V_m)T_w}{dt} = \phi_c H_c - Q_p \tag{5.9}$$

The pressure drop across the control valve is relatively constant, hence:

$$\phi_w = f(u)\,\phi_{wmax} \tag{5.45}$$

with ϕ_{wmax} the maximum water flow. To simplify the analysis, the heat capacity of the wall $M_w c_w$ will be ignored. Linearization of the five preceding equations results in:

$$s\bar{\rho}_c \delta V_c = \delta\phi_c - \phi_{wmax}\cdot\delta f \tag{5.46}$$

$$\frac{\delta\phi_c}{\bar{\phi}_c} = -\frac{\delta T_w}{(\bar{T}_0 - \bar{T}_w)} - \frac{\delta V_c}{(V_m - \bar{V}_c)} \tag{5.47}$$

$$\frac{\delta Q_p}{\bar{Q}_p} = \frac{\delta T_w}{(\bar{T}_w - \bar{T}_p)} - \frac{\delta V_c}{(V_m - \bar{V}_c)} \tag{5.48}$$

$$\overline{H}_c \delta\phi_c = \delta Q_p \tag{5.49}$$

$$\delta T_w = 0 \tag{5.50}$$

Substitution of Equation 5.50 into Equation 5.47 results in:

$$\delta V_c = -\frac{V_m - \overline{V}_c}{\overline{\phi}_c} \delta\phi_c \tag{5.51}$$

Substitution of Equation 5.51 into Equation 5.46 gives:

$$\{s[\rho_c(V_m - \overline{V}_c)/\overline{\phi}_c] + 1\} \; \delta\phi_c = \phi_{wmax} \cdot \delta f \tag{5.52}$$

from which, after combination with Equation 5.49:

$$\frac{\delta Q_p}{\delta f} = \frac{\overline{H}_c \phi_{wmax}}{[\rho_c(V_m - \overline{V}_c)/\overline{\phi}_c]s + 1} \tag{5.53}$$

In the same way, as was done in Equations 5.38 and 5.39, the transfer function from u_r to ϕ_d can be determined. According to Equation 5.53, the transfer function has a first-order character. The time constant is equal to the residence time that the condensate would have if it would occupy the steam volume. Due to the large density of water with respect to steam, the value of the time constant is rather large. The heat capacity of the pipes still adds a small time constant to the large one. As the process unit also introduces another large time constant (see Equation 5.1), the speed of control becomes low.

Evaporator Shell

The evaporator shell will also contribute to the dynamics, when the evaporator is small. The dynamics of the shell are very complicated, because different parts of the shell may behave dynamically in a different way. When the steam pressure increases, steam condenses at the inner side of the shell. Hence the time constant will be small, as h_m has a reasonable value:

$$\tau_m = \frac{M_m c_m}{h_m A_m} \tag{5.54}$$

where M_m = the mass of the shell (kg)
c_m = the specific heat of the shell (J/kg-°K)
h_m = the heat transfer coefficient to the shell (W/m^2-°K)
A_m = the shell area (m^2)

When the steam pressure decreases the condensate film will evaporate. Consequently, the heat transfer coefficient to the shell wall will become much smaller, resulting in a much larger value of τ_m.

Below the condensate level in the evaporator, the value of h_m is extremely low due to the low velocity of the condensate flow. Therefore the time constant will be very large for the part below the condensate level; above the condensate level the time constant is small. Fortunately, large fluctuations in the shell response only have a small effect on the dynamics of the evaporator, as the capacitive action of the shell is parallel to the capacitive action of the liquid and the pipes. This is shown in the electrical analog of Figure 5.4 where T_a is the atmospheric temperature and T_m the shell temperature. With the aid of this analog one can easily determine the influence of the lag of the shell on the pipe temperature of the evaporator.

WATER-COOLED CONDENSER

The dynamics of water-filled condensers and tubular heaters are more complicated than that of the steam-heated evaporator, since the temperature of the cooling water, or of the fluid flowing through the pipes, depends on the location. As a result, the energy balances contain not only a differential quotient with respect to time, but also a differential with respect to location, so that partial differential equations result. We shall not go further into this here, but will divide the pipe circuit into a number of segments, assuming ideal mixing in each segment.

Figure 5.5 gives a schematic diagram of a water-cooled condenser. Water flows through the pipes, on the outer surfaces of which vapor condenses.

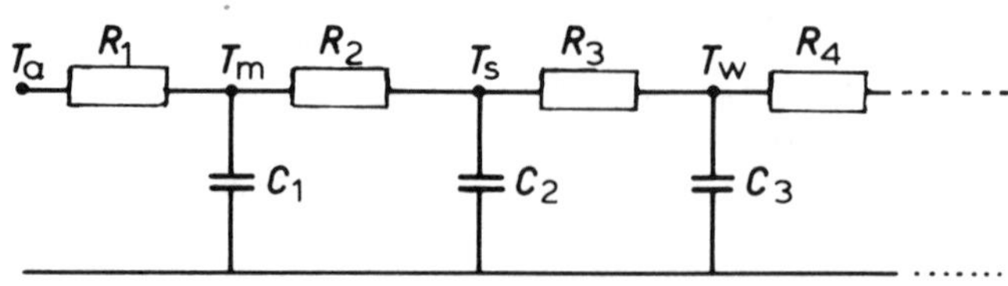

Figure 5.4 Simplified electrical analog for the evaporator without condensate level.

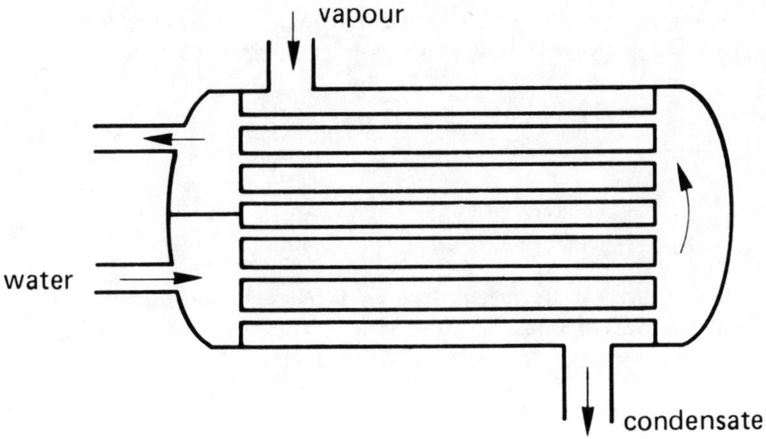

Figure 5.5 Water-cooled condenser.

Instead of two partial differential equations (for the water and for the pipe wall temperature) we have, using the described approximation, a set of simultaneous differential equations, which take the form:

$$\frac{M_p c_p}{N} \frac{dT_{p,n}}{dt} = \phi_p c_p (T_{p,n-1} - T_{p,n}) + \frac{h_p A_p}{N} (T_{w,n} - T_{p,n}) \qquad (5.55)$$

$$\frac{M_w c_w}{N} \frac{dT_{w,n}}{dt} = \frac{h_p A_p}{N} (T_{w,n} - T_{p,n}) + \frac{h_m A_m}{N} (T_m - T_{w,n}) \qquad (5.56)$$

where the symbols have the same meaning as in the previous section, and N is the number of segments.

The first term on the right side of Equation 5.55 represents the heat transfer by flow, the second term the heat transfer from the wall to the water. The second term on the right side of Equation 5.56 represents the heat transfer from the condensing vapor to the wall. There is a separate differential equation for the head, where the water flow reverses direction.

The heat transfer coefficient on the pipe side depends on the boundary layer coefficient $h_{p,A}$ of the cooling water flow, and also on the coefficient of the deposit $h_{p,g}$ which settles on the pipe wall:

$$h_p^{-1} = h_{p,A}^{-1} + h_{p,g}^{-1} \qquad (5.57)$$

The boundary layer coefficient can be determined from an empirical correlation between the Nusselt, Reynolds and Prandtl numbers:

$$Nu = 0.0265\,Re^{0.8}Pr^{1/3}(\mu/\mu_w)^{1/7} \tag{5.58}$$

where μ = dynamic viscosity
μ_w = dynamic viscosity at the pipe wall

The dependence on Re means that an increase in the quantity of water directly improves the heat transfer. If it is assumed that the coefficient of the deposit is constant, linearization of Equations 5.57 and 5.58 gives:

$$\frac{1}{\bar{h}_p^2}\,\delta h_p = \frac{1}{\bar{h}_{pA}^2}\,\delta h_{pA} \tag{5.59}$$

$$\frac{\delta h_{pA}}{\bar{h}_{pA}} = 0.8\,\frac{\delta\phi_p}{\bar{\phi}_p} \tag{5.60}$$

Combination of Equations 5.57, 5.59 and 5.60 results in:

$$\frac{\delta h_p}{\bar{h}_p} = 0.8\,\frac{\bar{h}_{pg}}{\bar{h}_{pA} + \bar{h}_{pg}}\,\frac{\delta\phi_p}{\bar{\phi}_p} \tag{5.61}$$

Linearization of Equation 5.55, substitution of s for d/dt, and combination with Equation 5.61 gives:

$$\delta T_{p,n} = -\frac{K_1}{\tau_1 s + 1}\delta\phi_p + \frac{K_2}{\tau_1 s + 1}\delta T_{p,n-1} + \frac{K_3}{\tau_1 s + 1}\delta T_{w,n} \tag{5.62}$$

The gains and time constant of this equation are given in Table 5.1. The expression for K_1 can be derived with the aid of the static version of Equation 5.55.

Linearization of Equation 5.56, replacement of d/dt by s, and combination with Equation 5.61 gives, on the condition that $\delta h_m = 0$:

$$\delta T_{w,n} = \frac{K_4}{\tau_2 s + 1}\delta T_{p,n} + \frac{K_5}{\tau_2 s + 1}\delta T_m - \frac{K_6}{\tau_2 s + 1}\delta\phi_p \tag{5.63}$$

Table 5.1 Parameters in Equations 5.62 and 5.63

$$\tau_1 = \frac{M_p c_p/N}{\bar{h}_p A_p/N + \bar{\phi}_p c_p}$$

$$K_1 = -c_p(\bar{T}_{p,n-1} - \bar{T}_{p,n}) + 0.8\,\frac{\bar{h}_p A_p \bar{h}_{p,s}(\bar{T}_{w,n} - \bar{T}_{p,n})}{N\bar{\phi}_p(\bar{h}_{p,f} + \bar{h}_{p,s})}\,[(\bar{h}_p A_p/N + \bar{\phi}_p c_p)]^{-1}$$

$$= c_p(\bar{T}_{p,n} - \bar{T}_{p,n-1})(1 - 0.8(\bar{h}_{p,s}/\bar{h}_{p,f} + h_{p,s})$$

$$\cdot (\bar{h}_p A_p/N + \bar{\phi}_p c_p)^{-1}$$

$$K_2 = \bar{\phi}_p c_p/(\bar{h}_p A_p/N + \bar{\phi}_p c_p)$$

$$K_3 = (\bar{h}_p A_p/N)(\bar{h}_p A_p/N + \bar{\phi}_p c_p)$$

$$\tau_2 = M_w c_w/(\bar{h}_p A_p + \bar{h}_m A_m)$$

$$K_4 = \bar{h}_p A_p/(\bar{h}_p A_p + \bar{h}_m A_m)$$

$$K_5 = \bar{h}_m A_m/(\bar{h}_p A_p + \bar{h}_m A_m)$$

$$K_6 = 0.8(\bar{h}_p A_p/N)(\bar{T}_{w,n} - \bar{T}_{p,n})[\bar{h}_{p,s}/(\bar{h}_{p,f} + h_{p,s})](1/\bar{\phi}_p)$$

$$= 0.8 c_p(\bar{T}_{p,n} - \bar{T}_{p,n-1})[\bar{h}_{p,s}/(\bar{h}_{p,f} + h_{p,s})]$$

The information flow diagram for Equations 5.62 and 5.63 is given in Figure 5.6.

A problem with water-cooled condensers is the restriction which must be put on the water temperature to avoid heavy fouling. This means that in many cases the water flow may not become too small. The permissible variation in the heat flow by a control valve in the cooling water circuit is therefore limited, so that the control range is small. The following numerical example gives an idea.

Suppose $T_m = 375°K$, $T_{in} = 290°K$, $T_{p,max} < 330°K$. The maximum heat transfer is proportional to $T_m - T_{p,in} = 85°K$. The minimum heat transfer ($T_{p,out} = T_{p,max}$) is proportional to:

$$\frac{(T_m - T_{p,in}) - (T_m - T_{p,out})}{ln\,\dfrac{T_m - T_{p,in}}{T_m - T_{p,out}}} = 62.5°K \tag{5.64}$$

The power of control is only a factor of about 1.4, which is too small to take up large disturbances and heavy contamination. This calculation gives a better result if account is taken of the dependence of the boundary layer coefficients on the water flow.

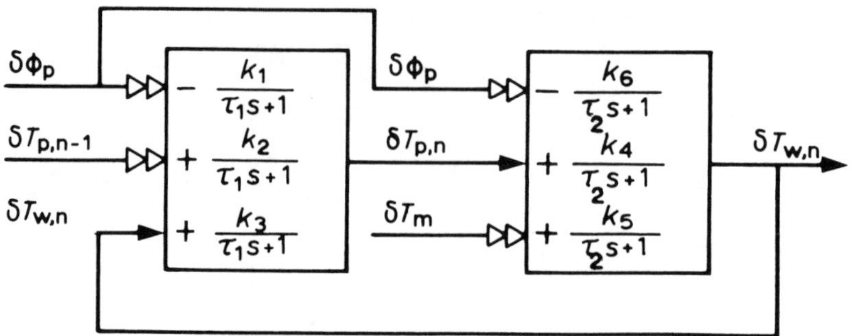

Figure 5.6 Information flow diagram for one section of the water-cooled condenser.

LIQUID-LIQUID HEAT EXCHANGERS

The tube side of liquid-liquid heat exchangers can be treated in the same manner as the tube side of the water-cooled condenser. The shell side should now also be split into the same number of segments as the tube side.

The dynamic model can be simplified by distributing the heat capacity of the pipes over the pipe and shell sides. Also, one must not forget the heat capacity of the outer wall. This can have an appreciable effect in small heat exchangers [9].

DYNAMICS OF FURNACES

Figure 5.7 shows the radiating section of a tubular heater. The feed material to be heated flows through pipes which are placed around the burner(s). The heat transfer is largely effected by radiation, partly directly to the tubes and partly by adsorption and reradiation from the furnace walls. Apart from the radiation section, there are one or more convection sections in which the latent heat of the flue gases is used for heating the feed and/or for steam generation. In the latter case, the dynamics are complicated, since the radiation section must be studied in conjunction with the convection sections.

It is not simple to give an accurate description of the dynamic behavior. There is a fast path from fuel flow to tube temperature. A change in radiation Q is almost directly followed by a considerable radiation which instantaneously increases the heat flow to the tubes. For a small part of the boiler tube (of length $\Delta 1$) the following energy balance applies:

$$\left(\frac{\Delta \ell}{\ell}\right) M_w c_w \frac{dT_{w,n}}{dt} = \left(\frac{\Delta \ell}{\ell}\right) Q + \left(\frac{\Delta \ell}{\ell}\right) A_p h_p (T_{w,n} - T_{p,n}) \qquad (5.65)$$

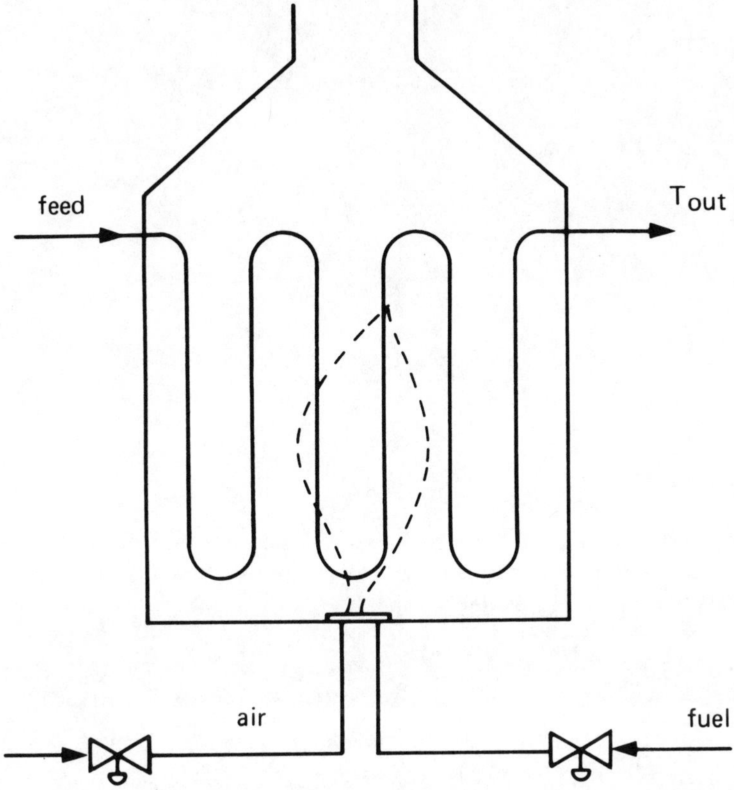

Figure 5.7 Radiation section of a furnace.

where $T_{p,n}$ is the temperature of the segment of the pipe. For the substance flowing through the tube, the energy balance is:

$$\left(\frac{\Delta \ell}{\ell}\right) M_p c_p \frac{dT_{p,n}}{dt} = \phi_p c_p (T_{p,n-1} - T_{p,n}) + \left(\frac{\Delta \ell}{\ell}\right) A_p h_p (T_{w,n} - T_{p,n}) \quad (5.66)$$

As an approximation, all the internal segment temperatures will initially change in the same way. Therefore, the variation of the first term of the right side of Equation 5.66 can be neglected. After linearization and introduction of the operational notation, formulas 5.65 and 5.66 become:

$$(sM_w c_w + A_p h_p)\delta T_{w,n} = \delta Q + A_p h_p \delta T_{p,n} \quad (5.67)$$

$$(sM_pc_p + A_ph_p)\delta T_{p,n} = A_ph_p\,\delta T_{w,n} \tag{5.68}$$

From this follows the transfer function from δQ to $\delta T_{p,n}$:

$$\frac{\delta T_{p,n}}{\delta Q} = \frac{1}{s(M_pc_p + M_wc_w)}\,\frac{1}{1 + s\tau_{wp}} \tag{5.69}$$

where

$$\tau_{wp} = \frac{(M_pc_p)(M_wc_w)}{(M_pc_p) + (M_wc_w)}\,\frac{1}{(A_ph_p)} \tag{5.70}$$

Numerical Example

Stainless steel furnace tubes with external diameter 0.15 m and wall thickness 0.01 m; through the tube flows oil with specific heat 2×10^3 J/kg-°K and density 800 kg/m^3. Per meter of tube:

$$M_wc_w = \left(\frac{\pi}{4} \times 0.15^2 - \frac{\pi}{4} \times 0.13^2\right) \times 8 \times 10^3 \times 500 = 1.76 \times 10^4 \text{ J/°K-m}$$

$$M_pc_p = \frac{\pi}{4} \times 0.13^2 \times 8 \times 10^2 \times 2 \times 10^3 = 2.13 \times 10^4 \text{ J/°K-m}$$

$$A_p = 0.13\pi = 0.41 \text{ m}^2\text{/m}$$

$$\tau_{wp} = \frac{23.5 \times 10^3}{h_p} \text{ J/K-m}^2$$

With $h_p = 10^3$ W/m^2-°K we have a time constant of 24 sec. Formula 5.69 also contains an integration. This is a result of the assumption that all internal temperatures change to the same extent. In reality there is a braking effect, which is propagated from the inlet of the furnace to the outlet, and which leads to a final static gain.

Instead of the integration we can roughly expect a first-order with a time constant of the same order as, or greater than, the residence time of the feed substance in the radiant section [10]. This view must be supplemented by the slow influence of the furnace wall, and perhaps of the convection section. Summarizing, the transfer function of fuel feed to outlet temperature can be approximated by a second-order with a large time constant (at least equal to the residence time of the feed material in the radiation section) and a small time constant (of the order of tens of seconds, see Equation 5.70).

DYNAMICS OF TEMPERATURE MEASUREMENT

In industrial processes, the temperature sensor is usually fitted in a thermowell, to enable replacement during normal operation (Figure 5.8). The dynamics of the temperature measurement consists of two parts:

1. the resistance to heat transfer in the liquid boundary layer (which must possibly be increased by the thermal resistance of a ceramic or glass coating), combined with the thermal capacity of the thermowell itself, and
2. the thermal resistance between thermowell and sensor, combined with the thermal capacity of the sensor.

The differential equation for the first part is:

$$M_w c_w \frac{dT_w}{dt} = A_w h_w (T_p - T_w) - A_0 h_0 (T_w - T_0) \quad (5.71)$$

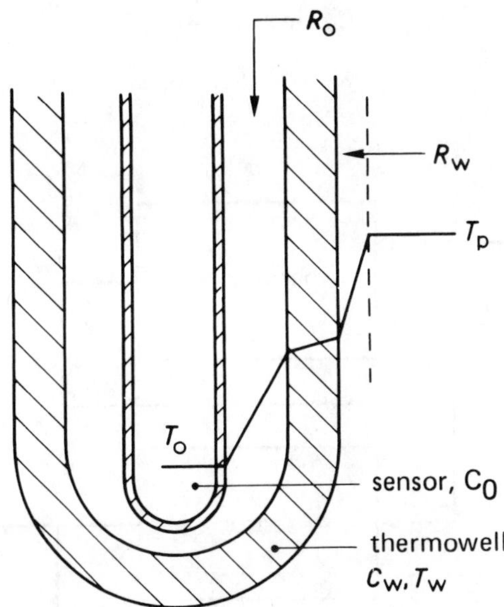

Figure 5.8 Thermocouple with thermowell.

where the suffix w now refers to the thermowell. For the second part the differential equation is:

$$M_0 c_0 \frac{dT_0}{dt} = A_0 h_0 (T_w - T_0) \tag{5.72}$$

We can write Equations 5.71 and 5.72 as:

$$C_w \frac{dT_w}{dt} = \frac{T_p - T_w}{R_w} - \frac{T_w - T_0}{R_0} \tag{5.73}$$

$$C_0 \frac{dT_0}{dt} = \frac{T_w - T_0}{R_0} \tag{5.74}$$

where C is a thermal capacity and R a thermal resistance.

The information flow diagram for Equations 5.73 and 5.74 is shown in Figure 5.9. It is an example of interaction between first-order systems. If we now put:

$$\tau_w = R_w C_w \text{ and } \tau_0 = R_0 C_0 \tag{5.75}$$

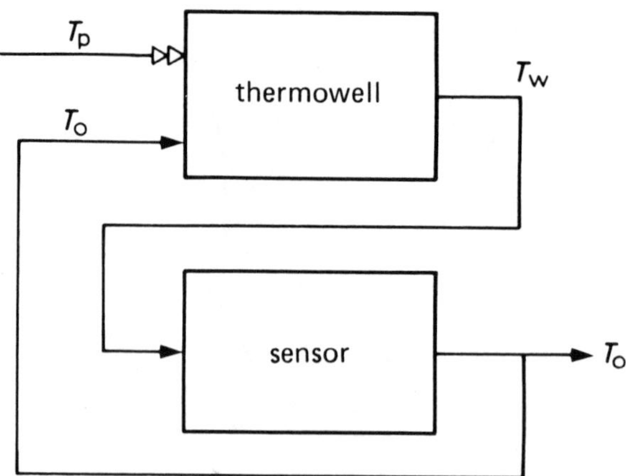

Figure 5.9 Interaction in the dynamics of temperature measurement.

then

$$\frac{\delta T_0}{\delta T_p} = \frac{1}{\tau_0 \tau_w s^2 + (\tau_0 + \tau_w + R_w C_0)s + 1} \tag{5.76}$$

As this corresponds to an overcritically damped second-order system, Equation 5.76 can be written as:

$$\frac{\delta T_0}{\delta T_p} = \frac{1}{(\tau_a s + 1)(\tau_b s + 1)} \tag{5.77}$$

In practice, thermocouples are often used that have a small thermal capacity, and so show negligible interaction. Then Equation 5.71 reduces to:

$$M_w C_w \frac{dT_0}{dt} = A_w h_w (T_p - T_0) \tag{5.78}$$

Numerical Example

Consider a stainless steel thermowell of length 0.2 m, external diameter 25.4 mm and internal diameter 20.3 mm. Then $M_w = 0.3$ kg, $A_w = 0.016$ m^2, $c_w = 500$ J/kg-°K so that:

$$\tau_w \simeq \frac{10^5}{\bar{h}_w} \tag{5.79}$$

With reasonably fast-flowing liquids, we find values of the order of seconds, with gases of the order of minutes.

If a resistance thermometer is used, M_0 is not negligible, and Equation 5.76 should be used.

Example

If $R_w = 1$, $R_0 = 2$, $C_w = 2$ and $C_0 = 1$, $\delta T_0/\delta T_p$ is, according to Equation 5.76:

$$\frac{\delta T_0}{\delta T_p} = \frac{1}{4s^2 + 5s + 1} = \frac{1}{(s+1)(4s+1)} \tag{5.80}$$

If we neglect the term R_wC_0, Equation 5.76 becomes:

$$\frac{\delta T_0}{\delta T_p} = \frac{1}{(2s+1)^2} \tag{5.81}$$

As a result of the interaction the ratio of the effective time constants τ_b/τ_a is always greater than the ratio τ_0/τ_w.

EXAMPLE: CONTROL OF A HEAT EXCHANGER

A gas-oil flow (10 kg/sec) has to be cooled from 220 to 160°C by an airflow. The inlet air temperature is 18°C, the outlet air temperature is equal to 48°C. The tubes of the heat exchanger have radial fins at the air side. The heat exchanger is dimensioned in such a way that the product of area and heat transfer coefficient at both sides is the same under static conditions. The air velocity may be taken as constant over the cross-sectional area. The outlet temperature of the gas oil is controlled by adjustment of the speed of rotation of the air ventilator motor.

The following data are given: the product of the mass and specific heat of the heat exchanger is equal to 15×10^6 J/°K, the specific heat of the oil is $c_o = 2 \times 10^3$ J/kg-°K, the specific heat of air $c_a = 10^3$ J/kg-°K, the product of the mass and specific heat of the gas oil is equal to 2×10^6 J/°K. These data may be taken as constants.

For the description of the dynamic behavior, the heat exchanger is divided into three sections. For simplicity, the dependence of the heat transfer coefficients on the velocity is ignored. The disturbance is an increase in the inlet air temperature with 5°C. A detailed dynamic model should be made, and the transfer functions should be presented in a detailed information flow diagram. The measurement lag is equal to 10 sec; the correction lag is 3 sec.

Static Analysis

The heat exchanger is given schematically in Figure 5.10. The static overall heat balance is:

$$\phi_o c_o (T_{o0} - T_{o3}) = \phi_a c_a (T_{a3} - T_{a0}) \tag{5.82}$$

where o refers to oil and a to air.

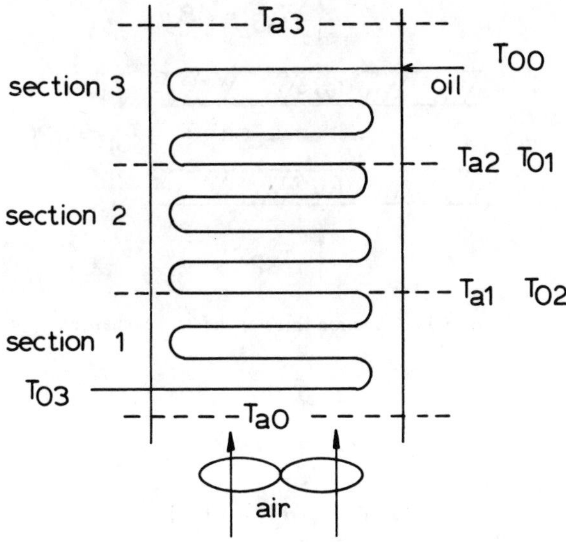

Figure 5.10 Schematic diagram of the heat exchanger.

Substitution of data gives $\phi_a c_a = 4 \times 10^4$ W/°K. The product of overall heat transfer coefficient and area, UA, can be calculated from:

$$Q = \phi_o c_o (T_{o0} - T_{o3}) = UA\Delta T_{ln} \tag{5.83}$$

where Q is the heat flow and ΔT_{ln} the logarithmic mean temperature difference, defined as:

$$\Delta T_{ln} = \frac{(T_{o0} - T_{o3})(T_{o3} - T_{a0})}{ln \dfrac{T_{o0} - T_{a3}}{T_{o3} - T_{a0}}} = \frac{(220 - 48) - (160 - 18)}{ln \dfrac{220 - 48}{160 - 18}} = 156.5\text{°C} \tag{5.84}$$

The calculated value is UA = 7.67 × 10^3 W/°K. Now the intermediate temperatures will be calculated. Section three is shown in Figure 5.11. For this section one may write for the heat balance:

$$Q_3 = \frac{UA}{3}\Delta T_{ln3} = \phi_o c_o (T_{o0} - T_{o1}) = \phi_a c_a (T_{a3} - T_{a2}) \tag{5.85}$$

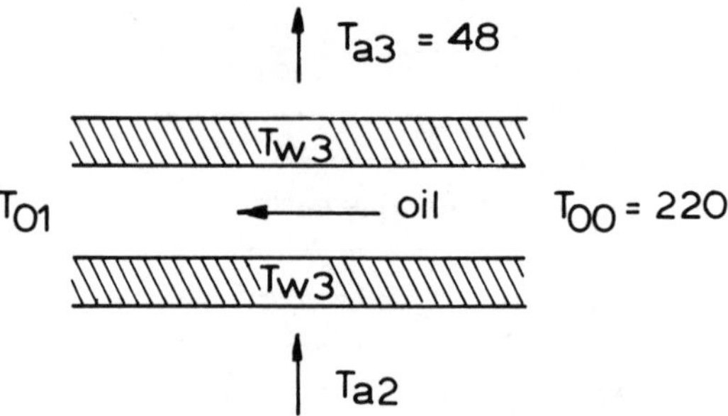

Figure 5.11 Section three of the heat exchanger.

with

$$\Delta T_{ln3} = \frac{(T_{o0} - T_{a3})(T_{o1} - T_{a2})}{ln \dfrac{T_{o0} - T_{a3}}{T_{o1} - T_{a2}}} \tag{5.86}$$

Substitution of data in Equations 5.90 and 5.91 results in the following values: T_{o1} = 198.7°C and T_{a2} = 37.4°C. In a similar way this calculation can be set up for section one. The results are T_{o2} = 178.8°C and T_{a1} = 27.4°C.

Model Consistency

We like to approximate the heat exchanger by three sections, each with uniform temperature. Hence, instead of logarithmic average temperature differences, there are now arithmetic averages. This would lead to errors in the static behavior, unless some parameters are adjusted. For the latter, the products of area and heat transfer coefficients are chosen. The heat balance for section three now becomes:

$$Q_3 = \phi_o c_o(T_{o0} - T_{o1}) = (UA)_3(T_{o1} - T_{a3}) \tag{5.87}$$

where T_{o1} and T_{a3} are the "ideally mixed" temperatures. Substitution of data gives a value for $(UA)_3 = 2.83 \times 10^3$ W/°K, where $(UA)_3$ is the product of heat transfer coefficient and area for section three. This product is now somewhat higher than when using logarithmic mean temperature differences.

Since the product of area and heat transfer coefficient for both sides is the same, one may write:

$$\frac{1}{(UA)_3} = \frac{1}{(\alpha_a A_a)_3} + \frac{1}{(\alpha_o A_o)_3} = \frac{2}{(\alpha_a A_a)_3} \tag{5.88}$$

or

$$(\alpha_a A_a)_3 = (\alpha_o A_o)_3 = 2(UA)_3 = 5.65 \times 10^3 \text{ W/°K} \tag{5.89}$$

with α the heat transfer coefficient at one side. The wall temperature for section three follows from:

$$Q_3 = (\alpha_a A_a)_3(T_{w3} - T_{a3}) = 5.65 \times 10^3 (T_{w3} - 48) \tag{5.90}$$

from which, with the value for Q_3, T_{w3} = 123.4°C. In a similar way, calculations may be made for sections two and one. The results are $(\alpha A)_2$ = 5.65×10^3 W/°K and T_{w2} = 108.1°C. For section one, the calculations give $(\alpha A)_1 = 5.65 \times 10^3$ W/°K and T_{w1} = 93.7°C.

Dynamic Analysis

Since the heat capacity of the air $M_a c_a$ is very small compared to the heat capacities of the metal $M_w c_w$ and the oil $M_o c_o$, the dynamic variations in the heat content of the air are ignored. The energy balances become, for the first section:

$$\text{oil:} \quad \frac{1}{3} M_o C_o \frac{dT_{o3}}{dt} = \phi_o c_o (T_{o2} - T_{o3}) - (\alpha_o A_o)_1 (T_{o3} - T_{w1}) \tag{5.91}$$

where $(\alpha_o A_o)_1$ refers to the adjusted product of heat transfer coefficient and area at the oil side in section one.

$$\text{wall:} \quad \frac{1}{3} M_w c_w \frac{dT_{w1}}{dt} = (\alpha_o A_o)_1 (T_{o3} - T_{w1}) - (\alpha_a A_a)_1 (T_{w1} - T_{a1}) \tag{5.92}$$

$$\text{air:} \quad \phi_a c_a (T_{a1} - T_{a0}) = (\alpha_a A_a)_1 (T_{w1} - T_{a1}) \tag{5.93}$$

Similar equations hold for the two other sections. Linearization of Equation 5.91 gives:

$$\frac{1}{3} M_o c_o s \delta T_{o3} = \bar{\phi}_o c_o(\delta T_{o2} - \delta T_{o3}) - (\alpha_o A_o)_1(\delta T_{o3} - \delta T_{w1}) \tag{5.94}$$

or

$$\delta T_{o3}\left[\frac{1}{3} M_o c_o s + \bar{\phi}_o c_o + (\alpha_o A_o)_1\right] = \bar{\phi}_o c_o \delta T_{o2} + (\alpha_o A_o)_1 \delta T_{w1} \tag{5.95}$$

which can be written as:

$$\delta T_{o3} = \frac{K_1}{\tau_1 s + 1} \delta T_{o2} + \frac{K_2}{\tau_1 s + 1} \delta T_{w1} \tag{5.96}$$

where

$$\tau_1 = \frac{\frac{1}{3} M_o c_o}{\bar{\phi}_o c_o + (\alpha_o A_o)_1} = \frac{\frac{2}{3} \times 10^6}{2 \times 10^4 + 5.65 \times 10^3} = 26 \text{ sec}$$

$$K_1 = \frac{\bar{\phi}_o c_o}{\bar{\phi}_o c_o + (\alpha_o A_o)_1} = \frac{2 \times 10^4}{2 \times 10^4 + 5.65 \times 10^3} = 0.78$$

$$K_2 = \frac{(\alpha_o A_o)_1}{\bar{\phi}_o c_o + (\alpha_o A_o)_1} = \frac{5.65 \times 10^3}{2 \times 10^4 + 5.65 \times 10^3} = 0.22$$

Linearization of Equation 5.92 gives:

$$\frac{1}{3} M_w c_w s \delta T_{w1} = (\alpha_o A_o)_1(\delta T_{o3} - \delta T_{w1}) - (\alpha_a A_a)_1(\delta T_{w1} - \delta T_{a1}) \tag{5.97}$$

resulting in:

$$\delta T_{w1} = \frac{K_3}{\tau_2 s + 1} \delta T_{o3} + \frac{K_4}{\tau_2 s + 1} \delta T_{a1} \tag{5.98}$$

where

$$\tau_2 = \frac{\frac{1}{3} M_w c_w}{(\alpha_o A_o)_1 + (\alpha_a A_a)_1} = \frac{\frac{15}{3} \times 10^6}{(5.65 + 5.65) \times 10^3} = 442 \text{ sec}$$

$$K_3 = \frac{(\alpha_o A_o)_1}{(\alpha_o A_o)_1 + (\alpha_a A_a)_1} = 0.50$$

$$K_4 = \frac{(\alpha_a A_a)_1}{(\alpha_o A_o)_1 + (\alpha_a A_a)_1} = 0.50$$

Linearization of Equation 5.93 gives:

$$\delta\phi_a c_a(\bar{T}_{a1} - \bar{T}_{a0}) + \bar{\phi}_a c_a(\delta T_{a1} - \delta T_{a0}) = (\alpha_a A_a)_1(\delta T_{w1} - \delta T_{a1}) \tag{5.99}$$

or

$$\delta T_{a1} = K_5 \delta T_{w1} + K_6 \delta T_{a0} - K_7 \delta\phi_a \tag{5.100}$$

where

$$K_5 = \frac{(\alpha_a A_a)_1}{\bar{\phi}_a c_a + (\alpha_a A_a)_1} = \frac{5.65 \times 10^3}{4 \times 10^4 + 5.65 \times 10^3} = 0.12$$

$$K_6 = \frac{\bar{\phi}_a c_a}{\bar{\phi}_a c_a + (\alpha_a A_a)_1} = \frac{4 \times 10^4}{4.565 \times 10^4} = 0.88$$

$$K_7 = \frac{c_a(\bar{T}_{a1} - \bar{T}_{a0})}{\bar{\phi}_a c_a + (\alpha_a A_a)_1} = \frac{10^3(27.4 - 18)}{4.565 \times 10^4} = 0.21$$

Combination of Equations 5.96 and 5.98 results in:

$$\delta T_{o3} = \frac{K_1}{\tau_1 s + 1} \delta T_{o2} + \frac{K_2}{\tau_1 s + 1}\left(\frac{K_3}{\tau_2 s + 1} \delta T_{o3} + \frac{K_4}{\tau_2 s + 1} \delta T_{a1}\right) \tag{5.101}$$

or

$$\delta T_{o3}[\tau_1\tau_2 s^2 + (\tau_1 + \tau_2)s + 1 - K_2K_3] = K_1(\tau_2 s + 1)\delta T_{o2} + K_2K_4\delta T_{a1} \tag{5.102}$$

The terms on the left side of this equation can be written as:

$$11{,}492s^2 + 468s + 1 - 0.11 \simeq 0.89(25.8s + 1)(500s + 1)$$

Substitution of the other data in Equation 5.102 results in:

$$\delta T_{o3} = \frac{0.78(442s + 1)}{0.89(25.8s + 1)(500s + 1)}\delta T_{o2} + \frac{0.11}{0.89(25.8s + 1)(500s + 1)}\delta T_{a1} \qquad (5.103)$$

Now the ratio (442s + 1)/(500s + 1) is approximated by a constant equal to $(442/500)^{1/2}$ which introduces a small inconsistency in the static behavior (about 6%).

Accurate and approximated step responses are shown in Figure 5.12. When the approximation is introduced in Equation 5.103, the result is:

$$\delta T_{o3} \simeq \frac{0.82}{(25.8s + 1)}\delta T_{o2} + \frac{0.12}{(25.8s + 1)(500s + 1)}\delta T_{a1} \qquad (5.104)$$

When combining Equations 5.98 and 5.100, the result is:

$$\delta T_{a1} = K_5\left(\frac{K_3}{\tau_2 s + 1}\delta T_{o3} + \frac{K_4}{\tau_2 s + 1}\delta T_{a1}\right) + K_6\delta T_{a0} - K_7\delta\phi_a \qquad (5.105)$$

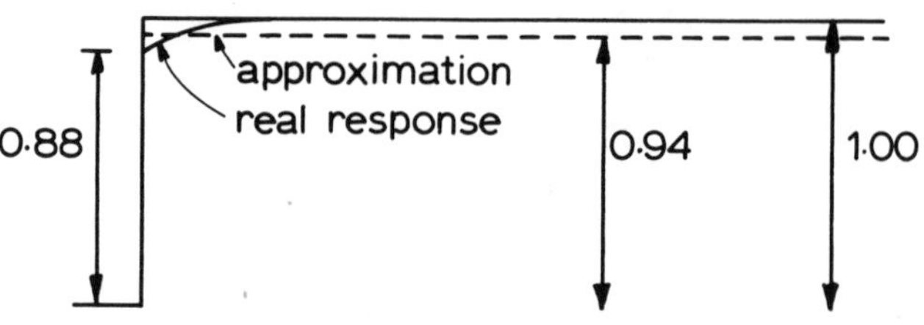

Figure 5.12 Step response and approximation of (442s + 1)/(500s + 1).

or

$$\delta T_{a1}(\tau_2 s + 1 - K_4 K_5) = K_3 K_5 \delta T_{o3} + K_6(\tau_2 s + 1)\delta T_{a0} - K_7(\tau_2 s + 1)\delta\phi_a \tag{5.106}$$

The term on the left side becomes:

$$442s + 1 - 0.06 = 0.94(470s + 1)$$

Substitution of this result together with other data into Equation 5.106 results in:

$$\delta T_{a1} = \frac{0.06}{0.94(470s + 1)}\delta T_{o3} + \frac{0.88(442s + 1)}{0.94(470s + 1)}\delta T_{a0} - \frac{0.21(442s + 1)}{0.94(470s + 1)}\delta\phi_a \tag{5.107}$$

which can be approximated by:

$$\delta T_{a1} \simeq \frac{0.06}{(470s + 1)}\delta T_{o3} + 0.91\delta T_{a0} - 0.22\delta\phi_a \tag{5.108}$$

when $(442s + 1)/(470s + 1)$ is approximated by $(442/470)^{1/2}$.

Figure 5.13 shows the information flow diagram for Equations 5.104 and 5.108. From this figure it can be seen that the influence of the feedback loop is very small: the closed-loop gain is only equal to 0.0072. Ignoring this loop only causes an error of 0.7%. The simplified information flow diagram is shown in Figure 5.14. For sections two and three, similar equations and flow diagrams can be derived. The parameter values for all three sections are the same, except the value of K_7. For the second section, K_7 has the value 0.22; for the third section K_7 is equal to 0.23. As this difference is so small we shall assume that Equations 5.104 and 5.108 also hold for the other two sections. Then the total information flow diagram can be constructed shown in Figure 5.15, where all the feedback loops are ignored. The transfer function from the airflow $\delta\phi_a$ to the outlet oil temperature δT_{o3} can be easily determined from Figure 5.15. It is given by:

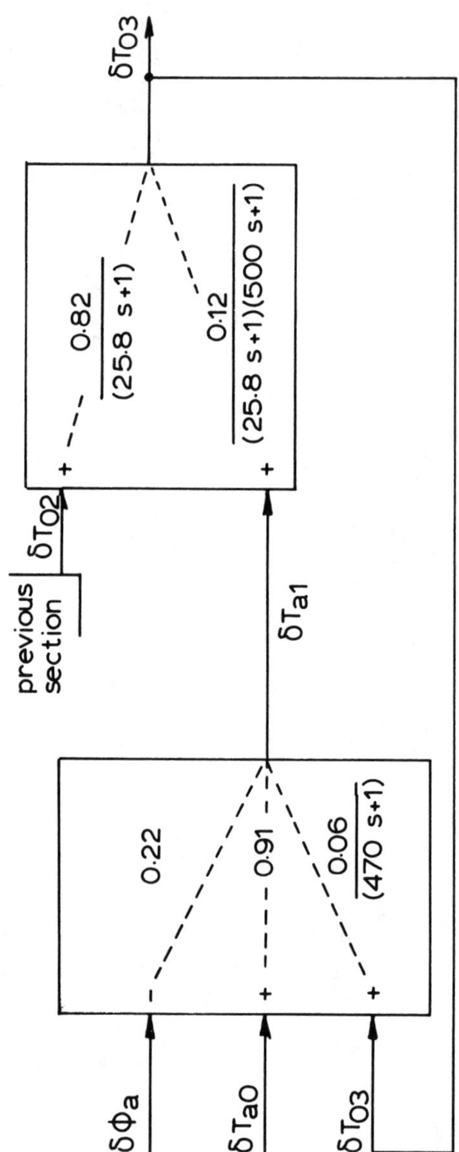

Figure 5.13 Information flow diagram for section one.

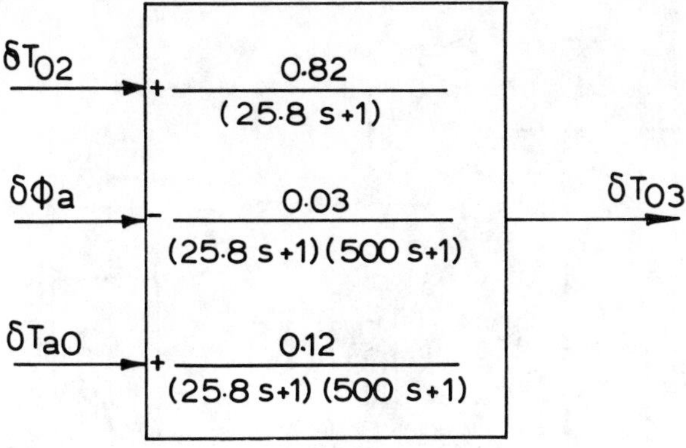

Figure 5.14 Simplified information flow diagram for section one.

$$\frac{T_{o3}}{\delta\phi_a} = \frac{-0.22 \times 0.12}{(25.8s + 1)(500s + 1)} - \frac{0.22 \times 0.12 \times 0.82}{(25.8s + 1)^2(500s + 1)}$$

$$- \frac{0.22 \times 0.12 \times 0.82 \times 0.82}{(25.8s + 1)^3(500s + 1)} - \frac{0.22 \times 0.91 \times 0.12 \times 0.82}{(25.8s + 1)^2(500s + 1)}$$

$$- \frac{0.22 \times 0.91 \times 0.91 \times 0.12 \times 0.82 \times 0.82}{(25.8s + 1)^3(500s + 1)}$$

$$- \frac{0.22 \times 0.91 \times 0.12 \times 0.82 \times 0.82}{(25.8s + 1)^3(500s + 1)} \tag{5.109}$$

After some rearrangement, Equation 5.109 can be reduced to:

$$\frac{\delta T_{o3}}{\delta\phi_a} = - \frac{0.12(151s^2 + 20.9s + 1)}{(28.5s + 1)^3(500s + 1)} \tag{5.110}$$

A second-order approximation for this expression is:

$$\frac{\delta T_{o3}}{\delta\phi_a} \simeq \frac{0.12}{(58s + 1)(498s + 1)} \tag{5.111}$$

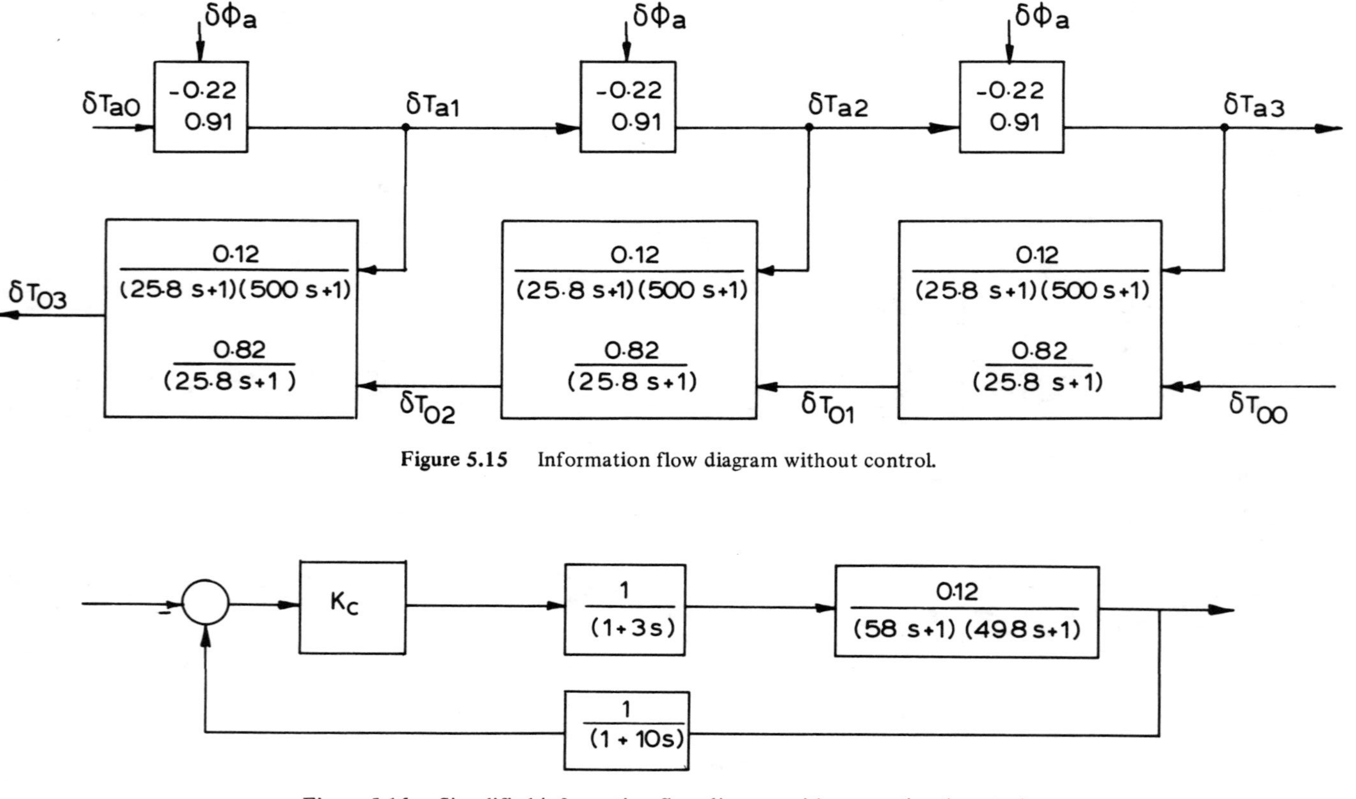

Figure 5.15 Information flow diagram without control.

Figure 5.16 Simplified information flow diagram with proportional control.

The information flow diagram together with correction lag and measurement lag are shown in Figure 5.16. The limit of stability is at $K_{c,u}$ = 400 and ω_u = 0.038 rad/sec (or period at limit of stability equals 165 sec). However, a more accurate calculation, based on Equation 5.110, yields $K_{c,u}$ = 220 and ω_u = 0.024 rad/rev (period = 260 sec). Evidently, one has to be careful with the introduction of approximations.

CHAPTER 6

DYNAMICS AND CONTROL OF CONTINUOUS FLOW REACTORS

This chapter discusses the dynamic behavior and control of chemical reactors. The reaction mechanism is first-order, with or without equilibrium. We will start with a study of the dynamics of an isothermal ideally mixed tank reactor with constant volume. Next, a cascade of ideally mixed reactors will be considered, which approaches the tubular flow reactor in the limit case.

The dynamics and control of an isothermal ideally mixed tank reactor with variable volume will also be studied, after which an analysis will be made of the dynamics and control of reactors with a fixed catalyst bed. Finally, the dynamics of reactors with cooling will be dealt with, and conditions will be derived for the stability of the controlled and uncontrolled process.

IDEALLY MIXED TANK REACTOR

Figure 6.1 shows a continuous flow reactor. If the reactor volume V is constant, inflow and outflow are equal:

$$\phi_{in} = \phi_{out} = \phi \tag{6.1}$$

We start with the assumption that ϕ is constant. If the mixing is ideal, concentrations in the reactor are the same at any location, also at the outlet:

$$C_A = C_{A,out} \ldots \text{etc.} \tag{6.2}$$

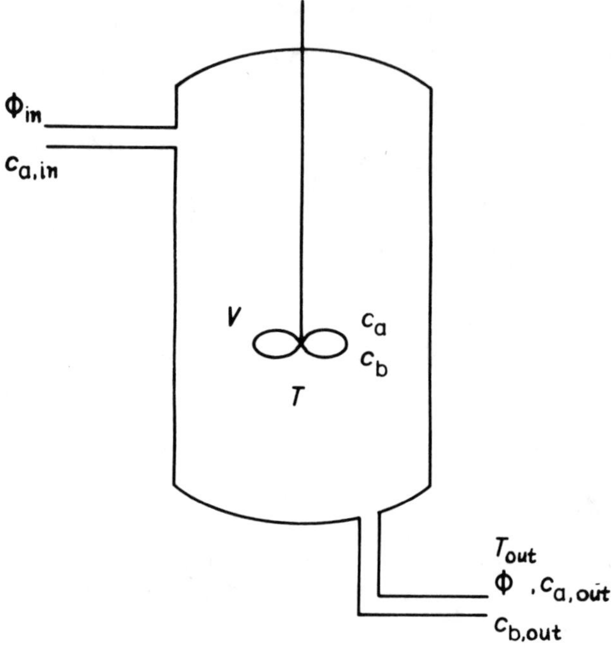

Figure 6.1 Continuous flow reactor.

The same holds for the temperatures, which are independent of time for isothermal behavior:

$$T = T_{out} = \text{constant} \tag{6.3}$$

The dynamic behavior of the reactor is then determined by partial mass balances:

Change in quantity of component per unit time

= Quantity that enters the reactor per unit time

− Quantity that leaves the reactor per unit time

− Quantity that disappears per unit time by reaction

For reactant A this balance is:

$$V\frac{dC_A}{dt} = \phi(C_{A,in} - C_A) - r_A \cdot V \tag{6.4}$$

where r_A = the rate of reaction per unit volume
V = the reactor volume
ϕ = the volumetric flow.

First-Order Reaction

Let us assume that the reaction is first-order:

$$A \xrightarrow[k_1]{} B \tag{6.5}$$

Under isothermal conditions, k_1 is constant, and the rate of reaction can be given by:

$$r_A = k_1 C_A \tag{6.6}$$

Substitution of Equation 6.6 into Equation 6.4, linearization and introduction of the operational notation results in:

$$\frac{\delta C_A}{\delta C_{A,in}} = \frac{K_p}{\tau_p s + 1} \tag{6.7}$$

where process gain K_p and process time constant τ_p are given by:

$$K_p = \frac{1}{1 + k_1 \tau_R} \tag{6.8}$$

$$\tau_p = \frac{\tau_R}{1 + k_1 \tau_R} \tag{6.9}$$

$$\tau_R = \frac{V}{\phi} \text{ (residence time)} \tag{6.10}$$

Because $k_1\tau_R$ is greater than zero, the chemical reaction leads to a reduction of the time constant and an attenuation of concentration variations.

To determine the transfer function from $C_{A,in}$ to C_B, the partial mass balance for component B has to be introduced:

$$V \frac{dC_B}{dt} = -\phi C_B + r_A \cdot V \tag{6.11}$$

Substitution of Equation 6.6 into Equation 6.11, linearization and introduction of s results in:

$$\frac{\delta C_B}{\delta C_A} = \frac{k_1 \tau_R}{\tau_R s + 1} \tag{6.12}$$

Combination of Equations 6.7 and 6.12 gives:

$$\frac{\delta C_B}{\delta C_{A,in}} = \frac{\delta C_B}{\delta C_A} \cdot \frac{\delta C_A}{\delta C_{A,in}} = \frac{\dfrac{k_1 \tau_R}{1 + k_1 \tau_R}}{\left(\dfrac{\tau_R}{1 + k_1 \tau_R} s + 1\right)(\tau_R s + 1)} \tag{6.13}$$

A comparison of Equations 6.7 and 6.13 shows that control of reactant B is more difficult than that of reactant A.

Equilibrium Reaction

For a first-order equilibrium reaction

$$A \underset{k_{-1}}{\overset{k_1}{\rightleftharpoons}} B \tag{6.14}$$

the rate of reaction is given by:

$$r_A = k_1 C_A - k_{-1} C_B \tag{6.15}$$

The mathematical model now consists of Equations 6.4, 6.11 and 6.15. Linearization results in:

$$\frac{\delta C_A}{\delta C_{A,in}} = \frac{\dfrac{\tau_R s + k_{-1} \tau_R + 1}{1 + (k_{-1} + k_1)\tau_R}}{\left(\dfrac{\tau_R}{1 + (k_{-1} + k_1)\tau_R} s + 1\right)(\tau_R s + 1)} \tag{6.16}$$

and

$$\frac{\delta C_B}{\delta C_{A,in}} = \frac{\dfrac{k_1\tau_R}{1 + (k_{-1} + k_1)\tau_R}}{\left(\dfrac{\tau_R}{1 + (k_{-1} + k_1)\tau_R} s + 1\right)(\tau_R s + 1)} \tag{6.17}$$

The transfer functions have the same denominator, which is second-order in s. The numerators, however, are different, as the numerator of the transfer function from $\delta C_{A,in}$ to δC_A is also a function of s, resulting in a generalized first-order character of this transfer function.

CASCADE OF IDEALLY MIXED REACTORS

We shall only consider a first-order reaction as given in Equation 6.5. For the transfer function from $\delta C_{A,in}$ to δC_A Equation 6.7 was found:

$$\frac{\delta C_A}{\delta C_{A,in}} = \frac{K_p}{\tau_p s + 1} \tag{6.7}$$

For the transfer function from $C_{A,in}$ to the outlet concentration of the nth reactor, we have then:

$$\frac{\delta C_A}{\delta C_{A,in}} = \frac{K_p^n}{(\tau_p s + 1)^n} \tag{6.18}$$

The response to a step disturbance in $\delta C_{A,in}$ is shown in Figure 6.2 for different values of n. As n increases, the response shows an increasing apparent "dead time." In the limit the response approaches a pure dead time (also called transportation lag or pure time delay) hence:

$$\lim_{n\to\infty} \frac{\delta C_{A,n}}{\delta C_{A,in}} = \lim_{n\to\infty} \left(\frac{1}{1 + k_1\tau_R}\right)^n \cdot \lim_{n\to\infty} \left(\frac{1}{\tau_p s + 1}\right)^n$$

$$= e^{-nk_1\tau_R} \cdot e^{-n\tau_p s} \tag{6.19}$$

where $n\tau_R$ = total residence time

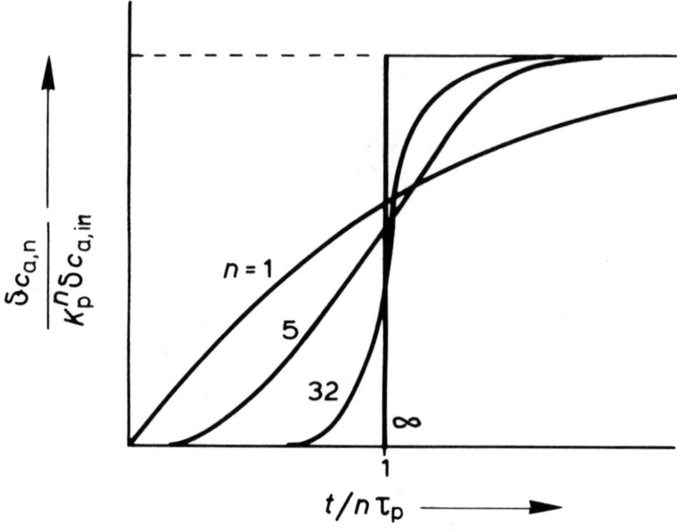

Figure 6.2 Step response of a cascade of n ideally mixed reactors.

The correctness of this result can also be understood in the following manner. For a pure time delay with a value of τ the relationship between the input signal $\delta C_{A,in}$ and the output signal $\delta C_{A,n}$ is:

$$\delta C_{A,n}(t) = K \cdot \delta C_{A,in}(t - \tau) \tag{6.20}$$

Development of $\delta C_{A,in}$ in a Taylor series and introduction of operational notation results in:

$$\begin{aligned} \delta C_{A,n}(t) &= K\left[\delta C_{A,in}(t) - \frac{\tau}{1!}\,\delta C'_{A,in}(t) + \frac{1}{2!}\,\tau^2 \delta C''_{A,in}(t) - \ldots\right] \\ &= K\left[1 - \frac{\tau s}{1!} + \frac{(\tau s)^2}{2!} - \ldots\right]\delta C_{A,in}(t) \\ &\approx K e^{-\tau s} \delta C_{A,in} \end{aligned} \tag{6.21}$$

which corresponds to Equation 6.19. In this equation, the first term of the right side represents the static attenuation of concentration variations; the second term represents the dynamics in the form of a dead time. Hence a cascade of an infinite number infinitely small, ideally mixed reactors is equivalent to a tubular flow reactor. It is interesting to analyze proportional control of the tubular flow reactor. Figure 6.3 shows the information flow diagram for the case that measurement and correction lag can be ignored. Suppose the variation in $C_{A,in}$ to be of sinusoidal form:

$$\delta C_{A,in} = A \sin \omega_n t \tag{6.22}$$

where A = amplitude
ω_n = the frequency.

We can write for the variation in the outlet $\delta C_{A,out}$, bij replacing t by $t - n\tau_p$ and, taking into account the attenuation factor from Equation 6.19:

$$\begin{aligned}\delta C_{A,out} &= Ae^{-nk_1\tau_R} \sin \omega_n(t - n\tau_p) \\ &= A\, e^{-nk_1\tau_R}(\sin \omega_n t \cos \omega_n n\tau_p + \cos \omega_n t \sin \omega_n n\tau_p)\end{aligned} \tag{6.23}$$

Feedback corresponds to multiplication by $-K_c$. The system is on the limit of stability when the input sine is equal to the feedback sine, or:

$$\delta C_{A,in} = -K_c \delta C_{A,out} \tag{6.24}$$

or

$$A \sin \omega_n t = -AK_c e^{-nk_1\tau_R}(\sin \omega_n t \cos \omega_n n\tau_p - \cos \omega_n t \sin \omega_n n\tau_p) \tag{6.25}$$

This equation is valid when

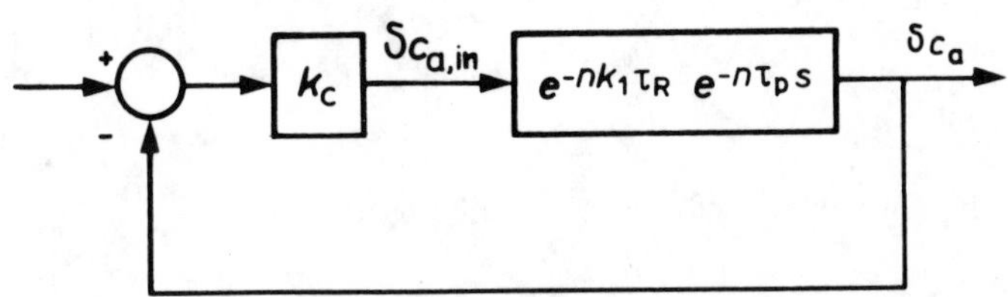

Figure 6.3 Information flow diagram of a tubular flow reactor with control.

$$\omega_n n\tau_p = k\pi (k = 1, 3, 5 \ldots) \tag{6.26}$$

and

$$K_c = e^{nk_1\tau_R} \tag{6.27}$$

The speed of control is determined by the lowest frequency (k = 1), which results in an oscillation period:

$$P_n = \frac{2\pi}{\omega_n} = 2n\tau_p = \frac{2n\tau_R}{1 + k_1\tau_R} \tag{6.28}$$

Note that the oscillation period is twice the process dead time. Evidently, the loop gain is very low (about 0.5 for a reasonable damping); hence, proportional control is unsatisfactory. For a solution, see Chapter 16.

DYNAMICS OF THE IDEALLY MIXED, ISOTHERMAL TANK REACTOR WITH VARIABLE CONTENT

Until now we have assumed that the reactor volume is constant and we have ignored flow variations. In practice, however, reactors are operated with varying liquid levels. Then the dynamic behavior is not only determined by the partial mass balance(s), but an overall mass balance has also to be taken into account. This mass balance is:

$$\frac{d(\rho V)}{dt} = \rho_{in}\phi_{in} - \rho\phi_{out} \tag{6.29}$$

If the density is constant ($\rho = \rho_{in}$) then:

$$\frac{dV}{dt} = \phi_{in} - \phi_{out} \tag{6.30}$$

The partial mass balance is different from Equation 6.4:

$$\frac{d}{dt}(VC_A) = \phi_{in}C_{A,in} - \phi_{out}C_A - r_AV \tag{6.31}$$

When we write the left side of Equation 6.31 as the sum of two terms and subtract Equation 6.30 multiplied by C_A, the result is:

$$V \frac{dC_A}{dt} = \phi_{in}(C_{A,in} - C_A) - Vr_A \tag{6.32}$$

For a first-order reaction, the mathematical model consists of Equations 6.6, 6.30 and 6.32. Linearization and introduction of the operational notation gives:

$$s\delta V = \delta\phi_{in} - \delta\phi_{out} \tag{6.33}$$

$$Vs\delta C_A = \delta\phi_{in}(\bar{C}_{A,in} - \bar{C}_A) + \bar{\phi}(\delta C_{A,in} - \delta C_A) - \bar{V}k_1\delta C_A - k_1\bar{C}_A\delta V \tag{6.34}$$

where $\bar{\phi}_{in} = \bar{\phi}_{out} = \bar{\phi}$

From Equation 6.34 k_1 can be eliminated by combining Equations 6.6 and 6.32 for the static condition:

$$\bar{\phi}(\bar{C}_{A,in} - \bar{C}_A) = \bar{V}k_1\bar{C}_A \tag{6.35}$$

from which:

$$k_1 = \frac{\bar{C}_{A,in} - \bar{C}_A}{\bar{C}_A} \cdot \frac{1}{\tau_R} = \frac{\bar{C}}{1 - \bar{C}} \cdot \frac{1}{\tau_R} \tag{6.36}$$

where

$$\bar{C} = 1 - \frac{\bar{C}_A}{\bar{C}_{A,in}} \tag{6.37}$$

the average conversion. With the aid of Equations 6.36 and 6.37, Equation 6.34 can be written as:

$$[\tau_R(1 - \bar{C})s + 1]\delta C_A = \frac{\bar{C}_A \cdot \bar{C}}{\bar{\phi}}\delta\phi_{in} + (1 - \bar{C})\delta C_{A,in} - \frac{\bar{C}_A\bar{C}}{\bar{V}}\delta V \tag{6.38}$$

From Equations 6.38 and 6.33 the following transfer functions can be determined:

$$\left.\begin{aligned}\frac{\delta V}{\delta \phi_{out}} &= -\frac{1}{s}\\ \frac{\delta V}{\delta \phi_{in}} &= \frac{1}{s}\end{aligned}\right\} \qquad (6.39)$$

$$\frac{\delta C_A}{\delta C_{A,in}} = \frac{1 - \bar{C}}{1 + s\tau_R(1 - \bar{C})} \qquad (6.40)$$

$$\frac{\delta C_A}{\delta \phi_{in}} = \frac{\bar{C}_A}{\bar{\phi}} \frac{\bar{C}}{1 + s\tau_R(1 - \bar{C})} \qquad (6.41)$$

$$\frac{\delta C_A}{\delta V} = -\frac{\bar{C}_A}{\bar{V}} \cdot \frac{\bar{C}}{1 + s\tau_R(1 - \bar{C})} \qquad (6.42)$$

Figure 6.4 shows the information flow diagram.

CONTROL OF THE IDEALLY MIXED, ISOTHERMAL TANK REACTOR WITH VARIABLE CONTENT

In the reactor, two state variables can be controlled: the level, corresponding to the volume V, and the concentration C_A. If the inlet concentration $C_{A,in}$ is considered as a disturbance variable, the inlet flow ϕ_{in} and the outlet flow ϕ_{out} remain for control. First, proportional level control will be analyzed when the level is controlled by the inlet flow ϕ_{in}.

For simplicity, measurement lag and correction lag will be ignored. The information flow diagram for this control system is given in Figure 6.5.

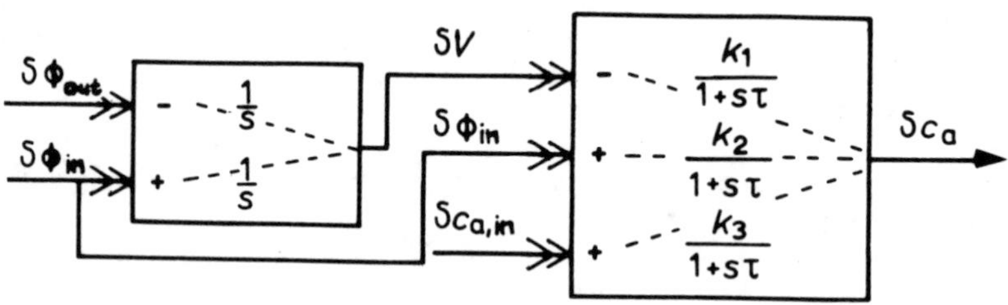

Figure 6.4 Information flow diagram for reactor with variable inventory.

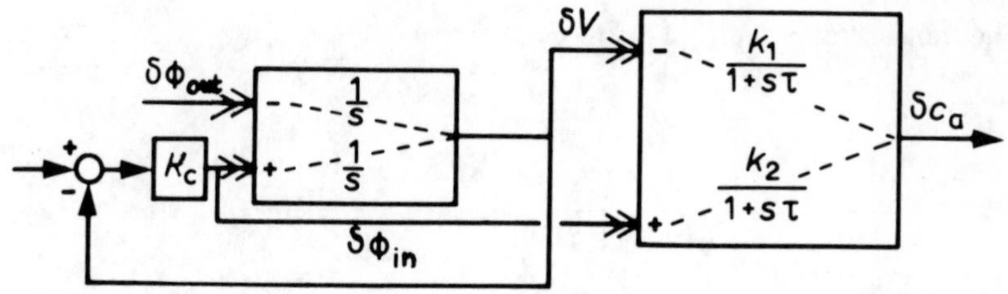

Figure 6.5 Information flow diagram for level control by outlet flow.

The model equations are now:

$$\delta C_A = \frac{\bar{C}_A}{\bar{\phi}} \frac{\bar{C}}{1 + s\tau_R(1 - \bar{C})} \delta\phi_{in} - \frac{\bar{C}_A}{\bar{V}} \frac{\bar{C}}{1 + s\tau_R(1 - \bar{C})} \delta V \tag{6.43}$$

$$\delta V = -\frac{1}{s} \delta\phi_{out} + \frac{1}{s} \delta\phi_{in} \tag{6.44}$$

$$\delta\phi_{in} = -K_{c1} \frac{\delta V}{\tau_R} \tag{6.45}$$

from which the transfer function from ϕ_{out} to C_A can be determined:

$$\frac{\delta C_A}{\delta\phi_{out}} = \frac{\bar{C}_A}{\bar{\phi}} \frac{\bar{C}}{1 + s\tau_R(1 - \bar{C})} \cdot \frac{1}{1 + \frac{\tau_R}{K_{c1}} s} \left(\frac{K_{c1} + 1}{K_{c1}}\right) \tag{6.46}$$

If the level in the reactor were constant (which means $\delta\phi_{in} = \delta\phi_{out}$), Equation 6.41 would represent the transfer function from ϕ_{out} to C_A.

If the level were not constant, the transfer function is given by Equation 6.46. It can be seen that level control leads to an extra lag. The time constant of this extra lag can be kept small, when the gain of the proportional level controller is made large.

The level can also be controlled by the outlet flow ϕ_{out}. The model is then given by Equations 6.43, 6.44 and:

$$\delta\phi_{out} = K_{c1} \frac{\delta V}{\tau_R} \tag{6.47}$$

The transfer function from ϕ_{in} to C_A then becomes:

$$\frac{\delta C_A}{\delta \bar{\phi}_{in}} = \frac{\bar{C}_A}{\phi} \frac{\bar{C}}{1 + s\tau_R(1 - \bar{C})} \cdot \frac{(1 - K_{cl}^{-1}) + \left(\frac{\tau_R}{K_{c1}} \cdot s\right)}{1 + \frac{\tau_R}{K_{c1}} s} \tag{6.48}$$

The first part in the right side of this equation corresponds again with the transfer function for constant level; the second part represents the nonideality of the level control.

Figure 6.6 shows the step response of the second part for different values of the controller gain K_c. Here δC_A^* represents the value that δC_A would have if level control were ideal. For K_{c1} equal to infinity, the result is just a step, which could be expected since level control is ideal in that case. For $1 < K_{c1} < \infty$, the response shows an overshoot followed by a slow decrease to a final steady-state value. For K_c equal to one, this static value is equal to zero, δC_A is insensitive to δC_A^* in the equilibrium situation and control is impossible because the power of control is zero.

The physical interpretation for this phenomenon is that an increase of ϕ_{in} increases the content of the reactor to such an extent that the residence time remains constant. For K_{c1} less than one, the static gain even becomes negative. The step response first goes to the one side and later to the other

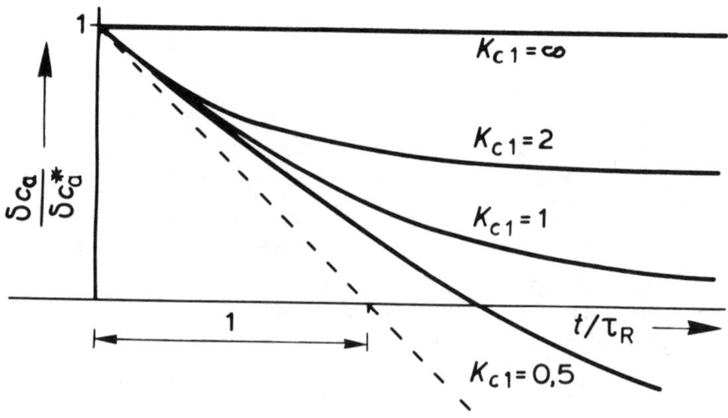

Figure 6.6 Step response of the second element in Equation 6.48.

side. This behavior represents nonminimum-phase behavior. To maintain the principle of negative feedback, here the conversion controller has to give reverse action: if C_A is too large (and consequently the conversion too small) then the inlet flow has to be increased.

A nonminimum-phase character is very disadvantageous for the speed of control, which can be seen from the limit of stability for the conversion control loop. The equation for a proportional conversion controller is:

$$\delta\phi_{in} = \pm K_{c2} \frac{\bar{\phi}}{\bar{C}_A} \delta C_A \tag{6.49}$$

where the sign has still to be determined. The characteristic equation for control can be obtained by combining Equations 6.48 and 6.49, resulting in:

$$(s\tau_R)^2 K_{c1}^{-1}(1 - \bar{C}) + s\tau_R [K_{c1}^{-1} + (1 - \bar{C}) \pm K_{c1}^{-1} K_{c2} \bar{C}] + [1 \pm K_{c2} \bar{C}(1 - K_{c1}^{-1})] = 0 \tag{6.50}$$

If K_{c1} is greater than one, the plus sign should be taken and the solution is always stable (all coefficients have the same sign). If K_{c1} is less than one, the minus sign should be taken and the system is only stable when:

$$K_{c1}^{-1} + (1 - \bar{C}) - K_{c1}^{-1} K_{c2} \bar{C} > 0 \tag{6.51}$$

or

$$K_{c2} < \frac{1 + (1 - \bar{C}) K_{c1}}{\bar{C}} \tag{6.52}$$

When this equation is considered as an equality, the controller gain on the limit of stability can be found. Substitution of this equality into Equation 6.50 results in:

$$(s\tau_R)^2 K_{c1}^{-1}(1 - \bar{C}) + [K_{c1}^{-1} - (1 - \bar{C}) K_{c1} + (1 - \bar{C})] = 0 \tag{6.53}$$

from which follows for the oscillation frequency on the limit of stability:

$$\frac{1}{\omega_n^2} = \frac{\tau_R^2(1 - \bar{C})}{1 + (1 - \bar{C})K_{c1} - (1 - \bar{C})K_{c1}^2} \tag{6.54}$$

resulting in an oscillation period:

$$P_n = \frac{2\pi}{\omega_n} = 2\pi\tau_R \sqrt{\frac{1 - \bar{C}}{1 + (1 - \bar{C})K_{c1} - (1 - \bar{C})K_{c1}^2}} \tag{6.55}$$

The oscillation period for values of K_{c1} less than one are mainly determined by the average conversion $\bar{C}$. For $\bar{C}$ equal to 0.95 the oscillation period P_n is approximately equal to 1.4 τ_R and for $\bar{C}$ equal to 0.5, the oscillation period is on the order of 4.2 to 4.4 τ_R.

ADIABATIC REACTOR WITH FIXED CATALYST BED

Until now, isothermal behavior was assumed, and consequently the energy balance was not taken into consideration. This is not realistic for many commercial-scale reactors; therefore, we must include temperature variations.

A common type of reactor has a fixed catalyst bed, through which reactants flow in liquid or gaseous form. The flow behavior is quite similar to plug flow; thus, the reactor can be approximated by a cascade of ideally mixed reactors, as discussed previously. One of these stages will be analyzed (Figure 6.7). It will be assumed that such a stage has a constant volume.

Model Equations

The partial balance has the following form:

$$\epsilon V \frac{dC_A}{dt} = \phi(C_{A,in} - C_A) - \epsilon V r_A \tag{6.56}$$

where
- ϵ = porosity of the catalyst bed
- V = reactor volume
- C_A = reactant concentration
- ϕ = volumetric flow
- r_A = reaction rate

If the heat exchange between gas or liquid and catalyst is ideal, the energy balance becomes:

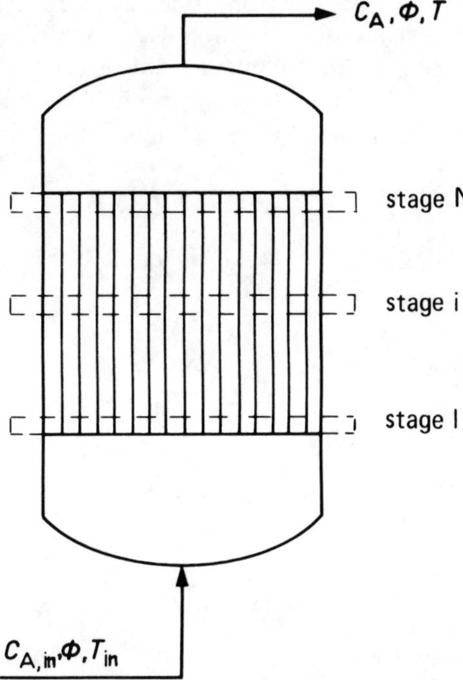

Figure 6.7 Division of a reactor with fixed catalyst bed into N stages.

$$(1 - \epsilon)V\rho_c\gamma_c \frac{dT}{dt} + \epsilon V\rho\gamma \frac{dT}{dt} = \phi\rho\gamma(T_{in} - T) + V\epsilon r_A \Delta H_r \tag{6.57}$$

where ρ = density
γ = specific heat
T = temperature
ΔH_r = heat of reaction, positive for exothermal reactions

The subscript c refers to catalyst. The rate of reaction is now also a function of the temperature. For a first order reaction we have the Arrhenius equation:

$$r_A = C_A k_0 e^{-E/RT} \tag{6.58}$$

where k_0 = a preexponential factor
R = gas constant
E = activation energy for the reaction.

After substitution of Equation 6.58 into Equations 6.56 and 6.57, the model consists of two simultaneous first-order differential equations, characterized by two state variables, the concentration and temperature. The behavior can therefore be presented in a state space representation with, for example, the conversion $C = 1 - C_A/C_{A,in}$, and the temperature T along the axes.

Static Behavior

First, static behavior can be presented in this state space. In the static case, Equations 6.56 and 6.57 become:

$$\bar{\phi}(\bar{C}_{A,in} - \bar{C}_A) - \epsilon V \bar{r}_A = 0 \tag{6.59}$$

$$\bar{\phi}\rho\gamma(\bar{T}_{in} - \bar{T}) + \epsilon V \bar{r}_A \Delta H_r = 0 \tag{6.60}$$

Adding Equation 6.60 to Equation 6.59 multiplied by Equation 6.59 results in:

$$\Delta H_r(\bar{C}_{A,in} - \bar{C}_A) = \rho\gamma(\bar{T} - \bar{T}_{in}) \tag{6.61}$$

from which

$$\bar{T} = \bar{T}_{in} + \frac{\Delta H_r \bar{C}_{A,in}}{\rho\gamma} \cdot \bar{C} \tag{6.62}$$

with $\bar{C}$ = the average conversion. In the state plane this is a straight line (Figure 6.8) representing the temperature accompanying an increase in conversion. As the second equation, we select Equation 6.59, which presents the increase in conversion as a result of an increase in temperature. With the aid of Equation 6.58, Equation 6.59 can be written as:

$$\bar{C} = \frac{k_0 \tau_R}{k_0 \tau_R + e^{E/RT}} \tag{6.63}$$

where

$$\tau_R = \frac{\epsilon V}{\phi} \tag{6.64}$$

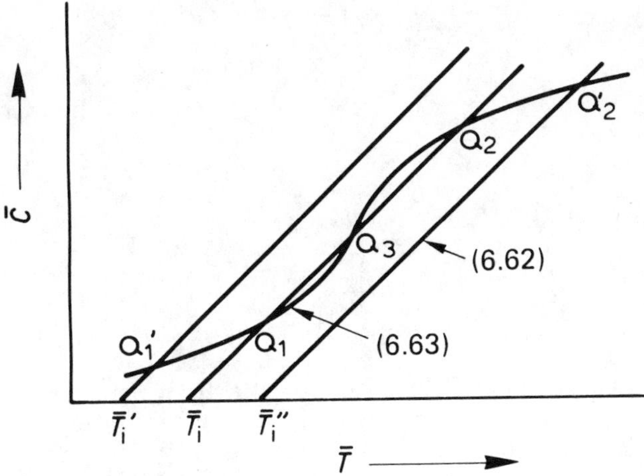

Figure 6.8 State plane for reactor with equilibrium lines.

If $\bar{T}$ increases from a low to a high value, $\bar{C}$ goes from 0 to 1 according to an S-shaped curve. The straight line and the S-shaped curve can have one or three intersections. In Figure 6.8 the situation is shown for three values of the inlet temperature. For low values of the inlet temperature (T_i') the reaction rate is small. The intersection Q_1' indicates a stable equilibrium (unless the dynamics create problems, as we will see later). At a high inlet temperature (T_i''), the reaction rate is large and the conversion is almost complete. Also, this intersection Q_2' indicates a stable equilibrium unless the dynamics again create problems in this case.

When the inlet temperature has an intermediate value (T_i) there can be three intersections (Q_1, Q_2 and Q_3). Q_1 and Q_2 are again intersections where a stable equilibrium exists; Q_3, however, indicates an unstable equilibrium. If initially the reactor is at Q_3 and an increase in temperature occurs, the heat of reaction (according to the S-shaped curve) increases more than can be removed by flow (according to the straight line). As a result, temperature will increase, until finally the state of the reactor can be characterized by Q_2. In a similar way, after an initial decrease of the temperature the reactor will go from Q_3 to Q_1.

Dynamic Behavior

Final conclusions about the stability can only be drawn on the basis of dynamic equations. When Equations 6.56 and 6.57 are written explicitly in terms of derivatives:

$$\frac{dc}{dt} = -\frac{C}{\tau_R} + k_0(1 - C)e^{-E/RT} \tag{6.65}$$

and

$$\frac{dT}{dt} = \frac{(T_i - T)/\tau_R + (1 - C)\dfrac{C_{A,in}\Delta H_r}{\rho\gamma}}{1 + \dfrac{1-\epsilon}{\epsilon} \cdot \dfrac{\rho_c\gamma_c}{\rho\gamma}} \tag{6.66}$$

the ratio of Equations 6.65 and 6.66 can be given as:

$$\frac{dc}{dT} = -\frac{\left[\dfrac{C}{\tau_R} - k_0(1 - C)e^{-E/RT}\right]\left[1 + \dfrac{1-\epsilon}{\epsilon} \cdot \dfrac{\rho_c\gamma_c}{\rho\gamma}\right]}{(T_i - T)/\tau_R + (1 - C)\dfrac{C_{Ai}\Delta H_r}{\rho\gamma}\,k_0e^{-E/RT}} \tag{6.67}$$

This ratio indicates a direction in every point of the state plane (Figure 6.9). The arrow points up on the right side of the S-curve because there the right side of Equation 6.65 is positive; the arrow points down for points on the left side of the S-curve.

The directions in various points of the state plane can now be connected into trajectories. Figure 6.9 shows that the reactor goes to one of the two stable equilibrium points, depending on the starting point. The state plane can be divided into two parts, separated by a separatrix. Figure 6.9 confirms the conclusions which already were drawn on the basis of the static behavior.

Information Flow Diagram

To determine the transfer functions, Equations 6.56 to 6.58 are linearized and written in operational notation:

$$\epsilon Vs\delta C_A = \delta\phi(\bar{C}_{A,in} - \bar{C}_A) + \bar{\phi}(\delta C_{A,in} - \delta C_A) - \epsilon V\delta r_A \tag{6.68}$$

This equation can be rewritten as:

$$\delta C_A = \frac{K_1}{\tau_R s + 1}\delta\phi + \frac{1}{\tau_R s + 1}\delta C_{A,in} - \frac{\tau_R}{\tau_R s + 1}\delta r_A \tag{6.69}$$

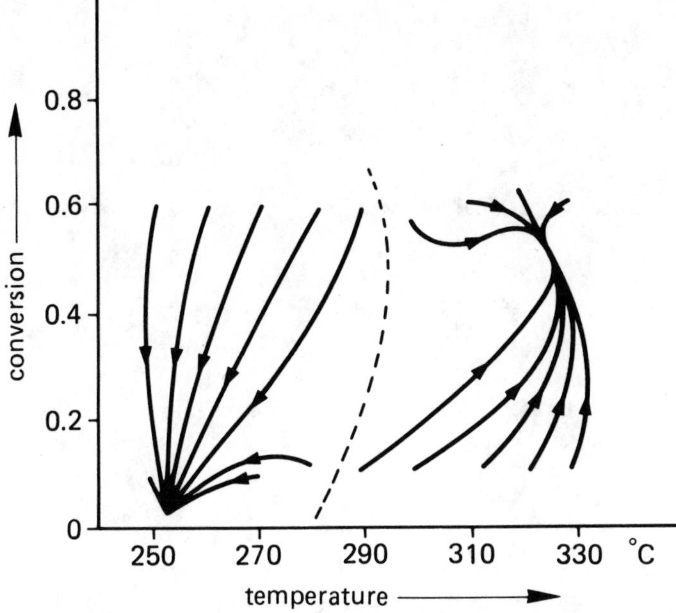

Figure 6.9 State plane for adiabatic ideally mixed reactor.

where

$$K_1 = \frac{\bar{C}_{A,in} - \bar{C}_A}{\bar{\phi}} \tag{6.70}$$

Linearization of Equation 6.57 gives:

$$\delta T = -\frac{K_2}{\tau_1 s + 1}\,\delta\phi + \frac{1}{\tau_1 s + 1}\,\delta T_{in} + \frac{K_3}{\tau_1 s + 1}\,\delta r_A \tag{6.71}$$

where

$$\tau_1 = \frac{(1-\epsilon)V\rho_c\gamma_c + \epsilon V\rho\gamma}{\bar{\phi}\rho\gamma} \tag{6.72}$$

$$K_2 = \frac{\bar{T} - \bar{T}_{in}}{\bar{\phi}} \tag{6.73}$$

$$K_3 = \frac{\epsilon V \Delta H_r}{\bar{\phi} \rho \gamma} \tag{6.74}$$

Linearization of Equation 6.58 gives:

$$\delta r_A = \frac{\bar{r}_A}{\bar{C}_A} \delta C_A + \frac{E \bar{r}_A}{R \bar{T}^2} \delta T = r_1 \delta C_A + r_2 \delta T \tag{6.75}$$

With the aid of Equations 6.69, 6.71 and 6.75, the information flow diagram can be determined (Figure 6.10). It is also possible to substitute Equation

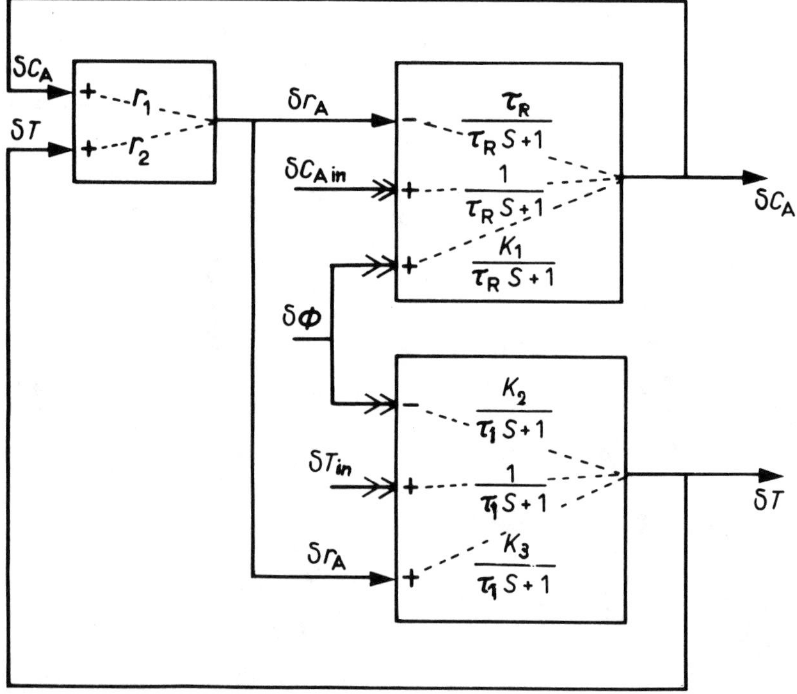

Figure 6.10 Information flow diagram for one stage of a reactor with fixed catalyst bed.

6.58 first into Equations 6.56 and 6.57 followed by linearization. This approach is not followed here because the former approach is somewhat clearer.

Conditions for Stability

If there are no external disturbances, thus $\delta\phi = 0$, $\delta C_{A,in} = 0$ and $\delta T_{in} = 0$, Equation 6.69 reduces to:

$$\delta C_A = -\frac{\tau_R}{\tau_R s + 1}\,\delta r_A \tag{6.76}$$

Combination with Equation 6.75 gives:

$$(\tau_R s + 1 + r_1\tau_R)\delta C_A = -r_2\tau_R\delta T \tag{6.77}$$

Without disturbances, Equation 6.71 reduces to:

$$\delta T = \frac{K_3}{\tau_1 s + 1}\,\delta r_A \tag{6.78}$$

Combination with Equation 6.75 gives:

$$(\tau_1 s + 1 - r_2K_3)\delta T = r_1K_3\delta C_A \tag{6.79}$$

Combination of Equations 6.77 and 6.79 results in the characteristic equation:

$$\tau_1\tau_R s^2 + (\tau_R - r_2\tau_R K_3 + \tau_1 + r_1\tau_1\tau_R)s + (r_1\tau_R - r_2K_3 + 1) = 0 \tag{6.80}$$

This equation can also be derived directly from Figure 6.10. The first condition for stability (to be called static condition) is:

$$r_1\tau_R - r_2K_3 + 1 > 0 \tag{6.81}$$

Substitution of parameters gives:

$$\frac{\bar{r}_A \epsilon V}{\bar{C}_A \bar{\phi}} - \frac{E \bar{r}_A \epsilon V \Delta H_r}{R \bar{T}^2 \bar{\phi} \rho \gamma} + 1 > 0 \tag{6.82}$$

or

$$\frac{\bar{C}_{A,in}}{\bar{C}_A} > a_T \tag{6.83}$$

where

$$a_T = \frac{E}{R\bar{T}^2} \frac{\Delta H_r}{\rho\gamma} (\bar{C}_{A,in} - \bar{C}_A) \tag{6.84}$$

The second condition is (for dynamic stability):

$$\tau_R - r_2 \tau_R K_3 + \tau_1 + r_1 \tau_1 \tau_R > 0 \tag{6.85}$$

from which:

$$\frac{\bar{C}_{A,in}}{\bar{C}_A} \left[1 + \frac{(1-\epsilon)}{\epsilon} \frac{\rho_c \gamma_c}{\rho\gamma}\right] + 1 > a_T \tag{6.86}$$

If, however, Expression 6.83 is satisfied, Expression 6.86 is also satisfied. If Expression 6.83 is not satisfied, the reactor is monotonically unstable. Its behavior is then characterized by an increasing exponential time function.

IDEALLY MIXED REACTOR WITH COOLING

Reactors in which exothermal reactions take place are often provided with cooling. Figure 6.11 shows schematically a cooling coil with inlet temperature $T_{c,in}$ and outlet temperature $T_{c,out}$. For simplicity the heat capacity of the cooling coil is ignored. We assume that the cooling liquid and cooling coil can be characterized by an ideal mixing temperature T_c. Furthermore we assume that the reactor is entirely filled with liquid. The energy balance now has the following form:

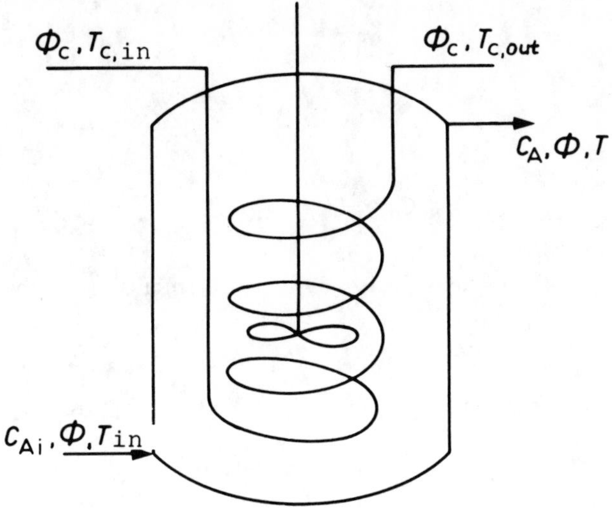

Figure 6.11 Reactor with cooling coil.

$$\rho V \gamma \frac{dT}{dt} = \phi \rho \gamma (T_{in} - T) + V \Delta H_r r_A - UA(T - T_c) \tag{6.87}$$

where U = overall heat transfer coefficient
A = heat transfer area.

The partial mass balance is:

$$V \frac{dC_A}{dt} = \phi(C_{A,in} - C_A) - V r_A \tag{6.88}$$

The mathematical model now consists of Equations 6.58, 6.87 and 6.88. For examination of the stability of the reactor, it is sufficient to consider variations in C_A and T. Linearization and introduction of the operational notation gives:

$$\tau_R s \delta T = -\delta T + K_3 \delta r_A - K_4 \delta T \tag{6.89}$$

with

$$K_4 = \frac{\bar{U}A}{\bar{\phi}\rho\gamma} \tag{6.90}$$

and

$$\tau_R s\delta C_A = -\delta C_A - \tau_R \delta r_A \tag{6.91}$$

Combination of Equations 6.75, 6.89 and 6.91 results in the characteristic equation:

$$\tau_R^2 s^2 + s\tau_R(2 + r_1\tau_R + K_4 - r_2K_3) + (1 + r_1\tau_R + K_4 + r_1\tau_R K_4 - r_2K_3) = 0 \tag{6.92}$$

The static condition is:

$$1 + r_1\tau_R + K_4 + r_1\tau_R K_4 > r_2K_3 \tag{6.93}$$

Substitution of parameters gives:

$$\frac{\bar{C}_{A,in}}{\bar{C}_A}\left(1 + \frac{UA}{\bar{\phi}\rho\gamma}\right) > a_T \tag{6.94}$$

If there is no cooling coil, $\bar{U}A = 0$ and Equation 6.94 reduces to 6.83. The dynamic condition pertains to the second term:

$$2 + r_1\tau_R + K_4 > r_2K_3 \tag{6.95}$$

or

$$\frac{\bar{C}_{A,in}}{\bar{C}_A} + 1 + \frac{\bar{U}A}{\bar{\phi}\rho\gamma} > a_T \tag{6.96}$$

We see that also here Equation 6.96 reduces to Equation 6.86 for $\bar{U}A$ equal to zero and ϵ equal to one (entirely liquid, as was assumed). It may happen, however, that Equation 6.94 is satisfied, but that Equation 6.96 is not satisfied. This indicates an oscillatory instability. If the condition in Equation 6.94 is not satisfied, this indicates a monotonic instability.

An interpretation of this behavior can be obtained with an information flow diagram for equations 6.75, 6.89 and 6.91 (Figure 6.12). A small increase in temperature δT results in an increased reaction rate $r_2 \delta T$. As a result, the temperature increases further with a first-order lag with time constant $\tau_R/(1 + \bar{U}A/\bar{\phi}\rho\gamma)$. This gives positive feedback.

Due to the increased reaction rate, the reactant concentration decreases, however, with a first-order lag with time constant τ_R. This gives a negative feedback. Evidently, negative feedback reacts more slowly than positive feedback. As a result, the temperature may temporarily increase, but then the effect of the negative feedback makes itself felt, and temperature will decrease, and so on. The result will be an oscillation. This situation is entirely different from the situation of the reactor with a fixed catalyst bed, where positive feedback comes with a larger lag ($\tau_1 > \tau_R$) and negative feedback with smaller lag τ_R.

CONTROL OF THE REACTOR BY COOLING WATER FLOW MANIPULATION

To determine the transfer function from cooling water flow ϕ_c to the reactor temperature T, Equations 6.87 and 6.88 are linearized, and d/dt is

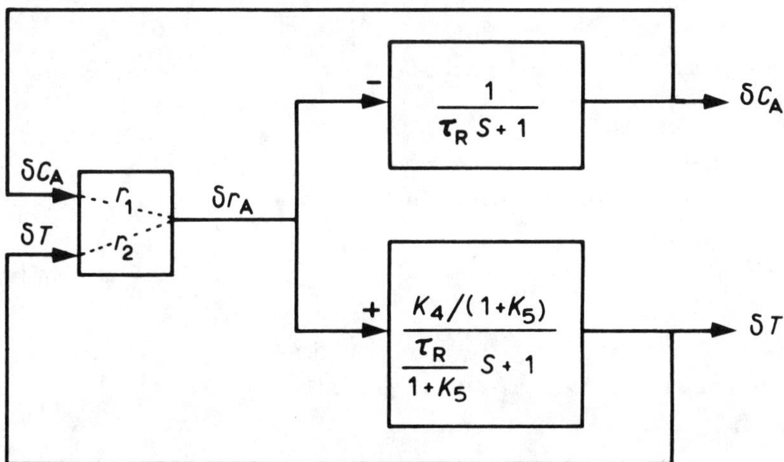

Figure 6.12 Information flow diagram for stability analysis of the reactor with cooling coil.

replaced by s. For the determination of this transfer function, only variations in T, C_A, r_A and U need to be considered. Equation 6.87 then becomes:

$$\rho V s \delta T = -\bar{\phi}\rho\gamma T + V\Delta H_r \delta r_A - \bar{U}A\delta T - A(\bar{T} - \bar{T}_c)\delta U \qquad (6.97)$$

and Equation 6.88 becomes:

$$V s \delta C_A = -\phi\delta C_A - V\delta r_A \qquad (6.98)$$

Let the overall heat transfer coefficient be determined by:

$$\frac{1}{U} = \frac{1}{\alpha_0} + \frac{1}{\alpha_i} \qquad (6.98)$$

where α_0 is the heat transfer coefficient at the water side and α_i the coefficient at the reactor side. From Equation 6.99 we can determine the variation δU. If α_i is assumed to be constant:

$$\delta U = \frac{\bar{\alpha}_i^2}{(\bar{\alpha}_i + \alpha_0)^2}\delta\alpha_0 \qquad (6.100)$$

When fouling exists inside the cooling coil and when the heat resistance of the wall cannot be ignored, Equation 6.99 has to be extended with corresponding terms. If we assume that the flow of the cooling water inside the coil is turbulent, and if the physical properties of the cooling water are constant, we may write for heat transfer coefficient α_0:

$$\alpha_0 = C_1^* \phi_c^{0.8} \qquad (6.101)$$

where C_1^* can be determined from the well known Nu-Re-Pr correlation. Linearization results in:

$$\delta\alpha_0 = \frac{0.8\bar{\alpha}_0}{\bar{\phi}_c}\delta\phi_c \qquad (6.102)$$

Combination of Equations 6.100 and 6.102 leads to:

$$\delta U = C_2^* \delta \phi_c \tag{6.103}$$

Combination of Equations 6.75, 6.97, 6.98 and 6.103 results in:

$$\{\tau_R^2 s^2 + \tau_R s[(1 + r_1\tau_R) + (1 + K_4) - r_2K_3] + [(1 + r_1\tau_R)(1 + K_4) - r_2K_3]\}\delta T$$
$$= -K_5(r_1\tau_R + 1 + \tau_R s)\delta\phi_c \tag{6.104}$$

where

$$K_5 = \frac{0.8\bar{\alpha}_0\bar{\alpha}_i^2 A(\bar{T} - \bar{T}_c)}{(\bar{\alpha}_i + \bar{\alpha}_0)^2\bar{\phi}_c\bar{\phi}\rho\gamma} \tag{6.105}$$

If Equation 6.104 is written as:

$$(a_2s^2 + a_1s + a_0)\delta T = -(b_1s + b_0)\delta\phi_c \tag{6.106}$$

then

$$\frac{\delta T}{\delta \phi_c} = -\frac{b_1s + b_0}{a_2s^2 + a_1s + a_0} \tag{6.107}$$

When ignoring measurement and correlation lag, the characteristic equation for control is:

$$1 + K_c\frac{b_1s + b_0}{a_2s^2 + a_1s + a_0} = 0 \tag{6.108}$$

or

$$a_2s^2 + (a_1 + b_1K_c)s + a_0 + b_0K_c = 0 \tag{6.109}$$

Note that a_2, b_1 and b_0 are all positive (see Equations 6.104 and 6.105). Hence, when a_1 and/or a_0 happen to be negative, the reactor can be stabilized by increasing K_c until the corresponding coefficient in Equation 6.109 is positive.

In practice, however, the result is often less favorable. This is caused by secondary effects, such as measurement and correcting device lags, and the heat capacity of the wall. The control valve also may have a certain hysteresis. Hence, in the previous analysis we have not tried to offer a complete model but we have presented the way of thinking which plays an important role in developing models.

The engineer himself has to determine if secondary effects may be ignored or not, and if they significantly affect process stability.

EXAMPLE: CONTROL OF REACTOR TEMPERATURE

Process Description

The reaction A + B → C is carried out in the liquid phase in an adiabatic reactor with fixed catalyst bed. The reaction is of first-order in component A; the fresh feed contains 10.6 mol of A/m^3. To restrict the temperature rise resulting from the reaction, two-thirds of the reactor effluent is recycled. Its temperature is reduced to the feed temperature by cooling.

Data:

- reactor volume = 0.34 m^3,
- bed porosity = 0.35,
- specific heat of the catalyst = 0.84 kJ/kg-°K,
- density catalyst = 2500 kg/m^3,
- specific heat of the feed = 2.1 kJ/kg-°K,
- density of the feed = 800 kg/m^3,
- activation energy = 8.0 kJ/mol of A,
- reaction rate constant = 0.01 sec^{-1},
- heat of reaction = 5 × 10^4 kJ/mol of A,
- fresh feed rate = 7.9 × 10^{-5} m^3/sec, and
- inlet temperature = 300°K.

Control

The outlet temperature of the reactor is controlled by adjustment of the recycle flow. The measurement lag is 50 sec; the correction lag may be ignored. Present the dynamic model in an information flow diagram with transfer functions. The reactor may be approximated by a cascade of two ideally mixed tank reactors.

Calculation of Static Behavior

The situation is shown in Figure 6.13, where the notation is also given. The partial mass balance around the mixing point upstream of the reactor is:

$$\phi_0 C_{A0} + \phi_R C_A = \phi_{in} C_{A,in} \tag{6.110}$$

with $\phi_R = \frac{2}{3}\phi_{in}$ we have: $\phi_0 = \frac{1}{3}\phi_{in}$, thus

$$\frac{1}{3} C_{A0} + \frac{2}{3} C_A = C_{A,in} \tag{6.111}$$

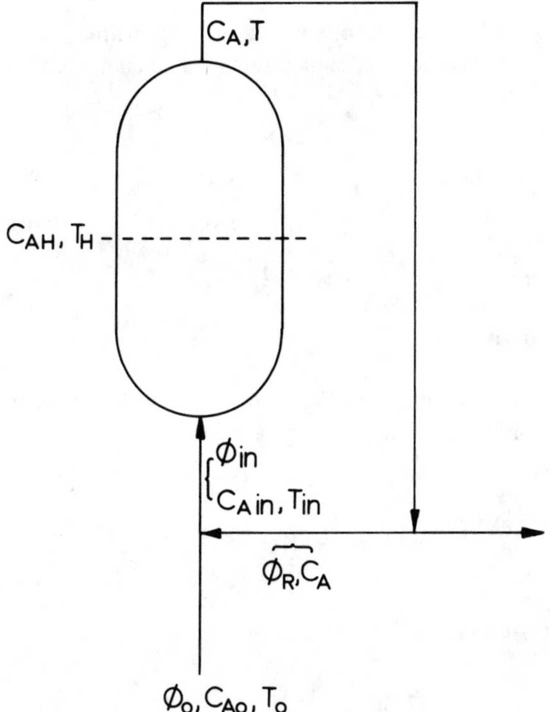

Figure 6.13 Flow diagram of reactor with recycle.

or

$$C_{A,in} = \frac{1}{3} \times 10.6 + \frac{2}{3} C_A = 3.53 + 0.667\ C_{A,in} \tag{6.112}$$

As the fresh feed equals 7.9×10^{-5} m^3/sec, the inlet flow is equal to:

$$\phi_{in} = 3\phi_0 = 23.7 \times 10^{-5}\ m^3/sec \tag{6.113}$$

The static partial mass balance and energy balance are:

$$\phi_{in}(C_{A,in} - C_A) = \epsilon V k_0 C_A e^{-E/RT} \tag{6.114}$$

$$\phi_{in}\rho_\ell \gamma_\ell (T - T_{in}) = \epsilon V \Delta H k_0 C_A e^{-E/RT} \tag{6.115}$$

When the reactor is divided into two sections, each with a volume of 0.17 m^3, the outlet concentrations and temperatures can be computed using Equations 6.114 and 6.115:

Reactor Section	C_A	T
1	7.51	330.5
2	6.38	364.5

Dynamic Analysis

Using the notation shown in Figure 6.13, the following balances hold for the first reactor section. The partial mass balance is:

$$\epsilon V_1 \frac{dc_{AH}}{dt} = \phi_{in}(C_{A,in} - C_{AH}) - \epsilon V_1 C_{AH} k_0 e^{-E/RT_H} \tag{6.116}$$

and the energy balance is:

$$[(1 - \epsilon)V_1\rho_c\gamma_c + \epsilon V_1\rho_\ell\gamma_\ell] \frac{dT_H}{dt} = \phi_{in}\rho_\ell\gamma_\ell(T_{in} - T_H) + \epsilon V_1 C_{AH} k_0 \Delta H e^{-E/RT_H} \tag{6.117}$$

At the mixing point in front of the reactor the following equations hold:

$$\phi_0 C_{A0} + \phi_R C_A = \phi_{in} C_{A,in} \tag{6.118}$$

$$\phi_{in} = \phi_0 + \phi_R \tag{6.119}$$

$$\phi_0 T_0 + \phi_R T_0 = \phi_{in} T_{in} \tag{6.120}$$

Equation 6.116 can be rewritten with the aid of Equations 6.118 and 6.119 as:

$$\epsilon V_1 \frac{dC_{AH}}{dt} = \phi_0 C_{A0} + \phi_R C_A - \phi_0 C_{AH} - \phi_R C_{AH} - \epsilon V_1 C_{AH} k_0 e^{-E/RT_H} \tag{6.121}$$

When ϕ_0 and C_{A0} are constant, Equation 6.121 may be linearized to:

$$[\epsilon V_1 s + \epsilon V_1 k_0 e^{-E/R\bar{T}_H} + \bar{\phi}_0 + \bar{\phi}_R]\delta C_{AH}$$

$$= \delta\phi_R(\bar{C}_A - \bar{C}_{AH}) + \bar{\phi}_R \delta C_A - \epsilon V_1 k_0 \bar{C}_{AH} e^{-E/R\bar{T}_H} . \frac{E}{R\bar{T}_H^2} \delta T_H \tag{6.122}$$

or

$$\delta C_{AH} = -\frac{K_1}{\tau_1 s + 1}\delta\phi_R + \frac{K_2}{\tau_1 s + 1}\delta C_A - \frac{K_3}{\tau_1 s + 1}\delta T_H \tag{6.123}$$

with

$$K_1 = \frac{-\bar{C}_A + \bar{C}_{AH}}{\epsilon V_1 k_0 e^{-E/R\bar{T}_H} + \bar{\phi}_0 + \bar{\phi}_R} = 4196 \tag{6.124}$$

$$K_2 = \frac{\bar{\phi}_R}{\epsilon V_1 k_0 e^{-E/R\bar{T}_H} + \bar{\phi}_0 + \bar{\phi}_R} = 0.59 \tag{6.125}$$

$$K_3 = \frac{\epsilon V_1 k_0 \bar{C}_{AH} e^{-E/R\bar{T}_H} . (E/R\bar{T}_H^2)}{\epsilon V_1 k_0 e^{-E/R\bar{T}_H} + \bar{\phi}_0 + \bar{\phi}_R} = 0.066 \tag{6.126}$$

and

$$\tau_1 = \frac{\epsilon V_1}{\epsilon V_1 k_0 e^{-E/R\bar{T}_H} + \bar{\phi}_0 + \bar{\phi}_R} = 221 \text{ sec} \tag{6.127}$$

Equation 6.117 can be rewritten with the aid of Equations 6.119 and 6.120 as:

$$[(1-\epsilon)V_1\rho_c\gamma_c + \epsilon V_1\rho_\ell\gamma_\ell]\frac{dT_H}{dt} = \phi_0\rho_\ell\gamma_\ell T_0 + \phi_R\rho_\ell\gamma_\ell T_0 - (\phi_0 + \phi_R)\rho_\ell\gamma_\ell T_H$$
$$+ \epsilon V_1 C_{AH} k_0 \Delta H e^{-E/RT_H} \tag{6.128}$$

When T_0 is also constant, this equation can be linearized to:

$$\Bigg\{[(1-\epsilon)V_1\rho_c\gamma_c + \epsilon V_1\rho_\ell\gamma_\ell]s + (\bar{\phi}_0 + \bar{\phi}_R)\rho_\ell\gamma_\ell$$
$$- \epsilon V_1 \bar{C}_{AH} k_0 \Delta H e^{-E/R\bar{T}_H} \cdot \frac{E}{R\bar{T}_H^2}\Bigg\}\delta T_H$$
$$= \rho_\ell\gamma_\ell(\bar{T}_0 - \bar{T}_H)\delta\phi_R + \epsilon V_1 k_0 \Delta H e^{-E/R\bar{T}_H}\delta C_{AH} \tag{6.129}$$

or

$$\delta T_H = -\frac{K_4}{\tau_2 s + 1}\delta\phi_R + \frac{K_5}{\tau_2 s + 1}\delta C_{AH} \tag{6.130}$$

with

$$K_4 = \frac{\rho_\ell\gamma_\ell(\bar{T}_H - \bar{T}_0)}{(\bar{\phi}_0 + \bar{\phi}_R)\rho_\ell\gamma_\ell - \epsilon V_1 \bar{C}_{AH} k_0 \Delta H e^{-E/R\bar{T}_H} \cdot \frac{E}{R\bar{T}_H^2}} = 1.76 \times 10^5 \tag{6.131}$$

$$K_5 = \frac{\epsilon V_1 k_0 \Delta H e^{-E/R\bar{T}_H} \cdot (E/R\bar{T}_H^2)}{(\bar{\phi}_0 + \bar{\phi}_R)\rho_\ell\gamma_\ell - \epsilon V_1 \bar{C}_{AH} k_0 \Delta H e^{-E/R\bar{T}_H} \cdot (E/R\bar{T}_H^2)} = 0.05 \tag{6.132}$$

and

$$\tau_2 = \frac{(1-\epsilon)V_1\rho_c\gamma_c + \epsilon V_1\rho_\ell\gamma_\ell}{(\bar{\phi}_0 + \bar{\phi}_R)\rho_\ell\gamma_\ell - \epsilon V_1\bar{C}_{AH}k_0\Delta H e^{-E/R\bar{T}_H}\cdot(E/R\bar{T}_H^2)} = 1140 \text{ sec} \qquad (6.133)$$

For the second section of the reactor, transfer functions can be determined in a similar way. The time constants and gains are now dependent on the average conditions in the second section. From the partial mass balance and energy balance the following linearized equations can be derived:

$$\delta C_A = \frac{K_6}{\tau_3 s + 1}\delta\phi_R + \frac{K_7}{\tau_3 s + 1}\delta C_{AH} - \frac{K_8}{\tau_3 s + 1}\delta T \qquad (6.134)$$

and

$$\delta T = -\frac{K_9}{\tau_4 s + 1}\delta\phi_R + \frac{K_{10}}{\tau_4 s + 1}\delta T_H + \frac{K_{11}}{\tau_4 s + 1}\delta C_A \qquad (6.135)$$

From the static situation the following parameters may be calculated:

- $K_6 = 4044$
- $K_7 = 0.85$
- $K_8 = 0.058$
- $K_9 = 2.02 \times 10^5$
- $K_{10} = 1.41$
- $K_{11} = 0.05$
- $\tau_3 = 213$ sec
- $\tau_4 = 1174$ sec

The information flow diagram is given in Figure 6.14.

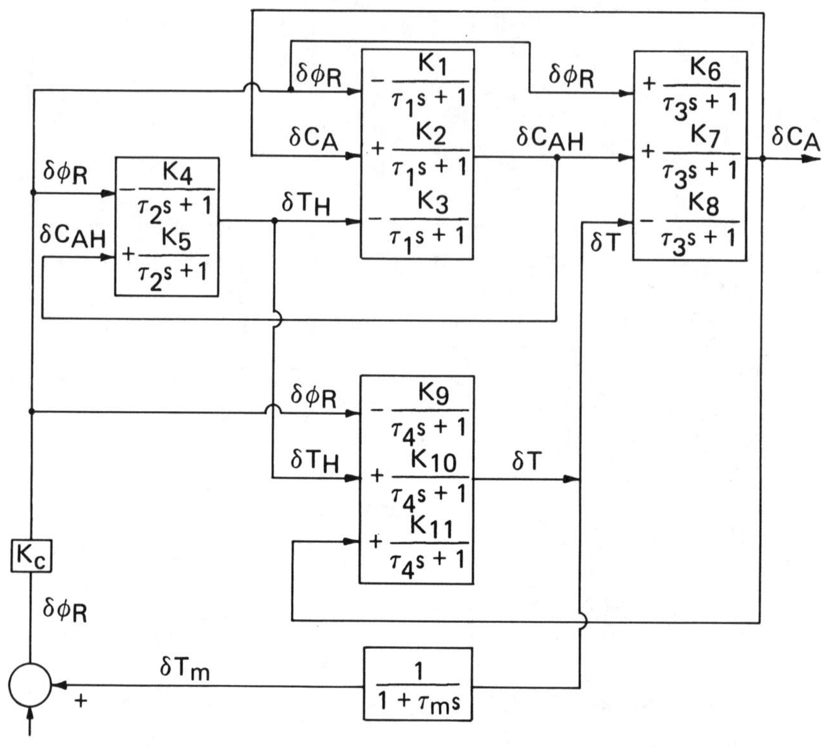

Figure 6.14 Information flow diagram for reactor with recycle.

CHAPTER 7

DYNAMICS AND CONTROL OF BIOLOGICAL REACTORS

PROCESS DESCRIPTION

Treatment by biological oxidation plays an important role in the total process of wastewater purification. A distinction is often made between municipal and industrial wastewater, although sometimes one treatment is possible for both. Wastewater usually contains the following substances:

- coarse solid substance;
- rapidly sedimenting substances, including sand;
- slowly sedimenting substances, e.g., sludge;
- floating substances;
- colloidal and dissolved organic substances; and
- inorganic dissolved substances.

The treatment of wastewater is strongly related to the abovementioned substances. The following phases in the treatment can be distinguished:

1. catching the solid substances with the aid of a grid and the processing of this waste material;
2. grinding the wastewater and thus reducing the size of the substances that have passed the grid;
3. removal of sand;
4. removal of a part of the slowly sedimenting substances in the primary sedimentation;
5. purification by biological oxidation, where two main types can be distinguished: (1) activated sludge installations; microorganisms are suspended in the liquid and air is blown through this suspension and (2) oxidation beds; microorganisms grow on a fixed surface and wastewater flows over the surface;
6. secondary sedimentation; and
7. drying of sludge or mechanical sludge processing.

Depending on the composition of the wastewater, the number of treatments may possibly be increased.

PURIFICATION BY BIOLOGICAL OXIDATION

Inorganic compounds are left undisturbed by biological processes. Organic compounds, however, are used as nutrition, and if conditions are favorable (enough oxygen, right pH, etc.) the number of microorganisms will increase.

We shall restrict ourselves to activated sludge installations and determine the dynamic behavior of these installations. The activated sludge process is schematically shown in Figure 7.1. The feed to the aeration tank consists of wastewater influent and concentrated sludge in which the concentration of living microorganisms is high. In the aeration tank or "biological reactor," the microorganisms utilize organic substances (substrate), resulting in an increase in the number of microorganisms. In the secondary sedimentation tank or clarifier, the sludge, consisting of microorganisms, some substrate and organic substances, is separated from the liquid. Part of the sludge is returned to the aeration tank; the other part is processed.

For simplicity we shall assume that the aeration tank can be described as an ideally mixed reactor. Furthermore, we assume that conditions in the reactor are aerobic, which means that there is enough oxygen, and microorganisms are not hindered in their growth. This assumption means that we can leave the change of oxygen concentration out of consideration; hence the model can be restricted to a substrate phase and a microorganism phase.

When developing mathematical models for biological reactors, there are a number of problems:

1. lack of a detailed kinetic model for biological systems;
2. the composition of the microorganism phase is not homogeneous but heterogeneous; and
3. lack of a model in which the effects can be described of heterogeneous composition of the substrate on growth and composition of the microorganism population.

Therefore we shall define one substrate and one microorganism concentration as if the compositions were homogeneous.

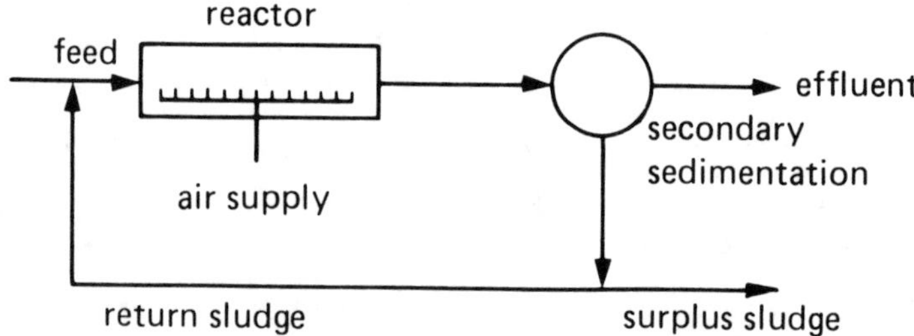

Figure 7.1. Diagram of the activated sludge process.

KINETICS OF BACTERIAL REACTIONS

Before going into detail about the dynamic model it is useful to pay some attention to microbial growth and decay. A pioneer in this field was Monod [11], although more recent publications also contribute to the understanding of kinetics [12,13].

While there is microbial growth, substrate will be consumed. The growth rate depends on the number of microorganisms present and on the amount of substrate. Most authors [12,14] assume that microbial growth is proportional to the number of living microorganisms. Microbial attrition, however, may not be ignored, especially in the case where part of the activated sludge is returned. It is usually assumed that the attrition rate is proportional to the concentration of living microorganisms, although sometimes also a dependence on the substrate concentration is assumed [14]. The widely accepted kinetic model for the net microbial growth is:

$$r_g = k_g C_B - k_d C_B \tag{7.1}$$

where k_g = growth rate constant (sec^{-1})
k_d = attrition rate constant (sec^{-1})
C_B = concentration of microorganisms (kg/m^3)

Monod [11] has shown that the growth rate constant k_g is a function of the concentration of substrate, according to:

$$k_g = \frac{k_{g0} C_s}{K_s + C_s} \tag{7.2}$$

where k_{g0} = the maximum specific growth rate (sec^{-1})
C_s = the concentration of substrate (kg/m^3)
K_s = a saturation constant, the concentration of substrate at which the observed specific growth rate is one-half the maximum value (kg/m^3)

Monod further proposed that the ratio of microbial growth to substrate consumption is constant:

$$-\frac{k_g C_B}{r_{s1}} = Y \tag{7.3}$$

where r_{s1} = the gross rate of substrate consumption (kg/m^3-sec)
Y = a conversion factor (<1).

Some authors have proposed that Y is a function of microbial growth rate. Others also distinguish a dispersed and coagulated phase, with related kinetics for each phase. However, to make the model not too complicated, we assume that there is only one microbial phase for which Y in Equation 7.3 is constant. The rate at which the substrate is consumed will also depend on the rate at which dead microorganisms dissolve. These dissolved microorganisms will partly contribute to an increase of the substrate concentration.

The net rate at which substrate is consumed can then be given by:

$$r_s = -\frac{k_{g0}C_sC_B}{Y(K_s + C_s)} + \beta k_d C_B \tag{7.4}$$

where β is a proportionality constant. Some authors say that the second term on the right side of Equation 7.4 will be proportional to the product of living and dead microorganism concentration [14]. Equation 7.4, however, is used by many authors. Other authors ignore the attrition of microorganisms entirely, which results in a very simple expression for the kinetics.

For the development of the dynamic model, however, we will use Equation 7.1, 7.2 and 7.4 as a description of the kinetics. We have to realize that these equations may show a certain degree of incompleteness.

MATHEMATICAL MODEL

Some assumptions were already discussed in the foregoing sections. It is further assumed that during secondary sedimentation microbial growth can be ignored and that there is no consumption of substrate. The concentration of microorganisms in the feed will be ignored. Since the return sludge flowrate ϕ_R is much smaller than the feed rate ϕ, and since the return substrate concentration C_{sR} is much smaller than the substrate concentration in the feed C_{si}, the term $\phi_R C_{sR}$ can be neglected compared to ϕC_{si}. For the biological reactor, the following partial mass balances hold. For the substrate (Figure 7.2):

$$V_R \frac{dC_{s1}}{dt} = (\phi + \phi_R)(C_{s0} - C_{s1}) + r_s V_R \tag{7.5}$$

and for the microbial phase:

$$V_R \frac{dC_{B1}}{dt} = (\phi + \phi_R)(C_{B0} - C_{B1}) + r_g V_R \tag{7.6}$$

The volume V is expressed in m^3, the flowrate ϕ in m^3/sec. At the point where the feed and return sludge are combined, the partial mass balances are:

$$\phi C_{si} = (\phi + \phi_R)C_{s0} \tag{7.7}$$

$$\phi_R C_{BR} = (\phi + \phi_R)C_{B0} \tag{7.8}$$

Most articles dealing with activated sludge process modeling emphasize the modeling of the reactor.

The behavior of the secondary sedimentation tank is often approximated by simple algebraic relationship, or the concentration of the substrate and microbial phase in the return sludge flow are even supposed to be constant. These concentrations, however, are not constant; they depend on a number of factors, such as the sedimentation properties of the sludge, the flow of the return sludge, the flow to the secondary sedimentation tank etc. The concentrations are a function of time and distance coordinate, hence the description of the dynamic behavior will result in partial differential equations.

As the particles in the tank both move horizontally and vertically, the concentration is a function of two (possibly three) position coordinates and a time coordinate. To make a reasonably simple approximation of the dynamic behavior, we assume that the particles mainly move vertically, and in this direction two ideal tanks are assumed, as shown in Figure 7.3 (this can be easily extended to n tanks).

The flows ϕ_1 and ϕ_2 indicate an internal circulation. A mass balance for the first volume gives:

$$\phi_1 - \phi_2 = \phi_R + \phi_s \tag{7.9}$$

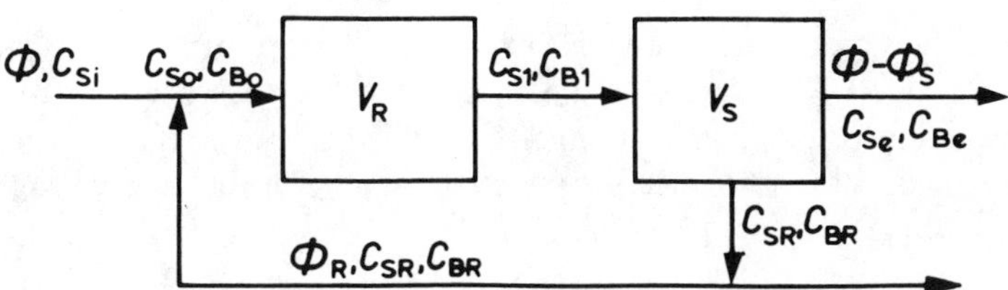

Figure 7.2. Symbols for the mathematical model.

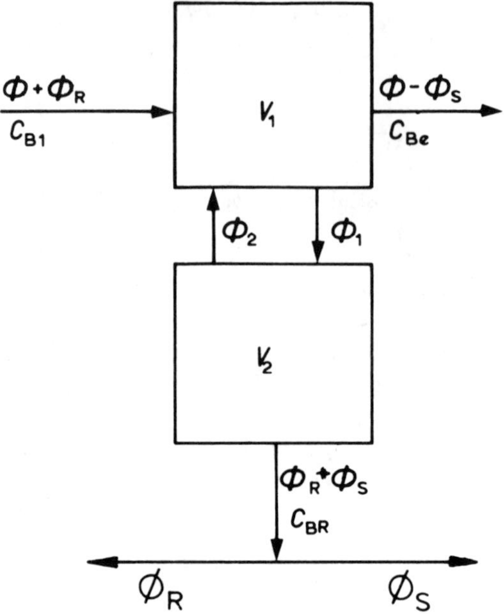

Figure 7.3. Model for the secondary sedimentation tank.

The internal flows will increase when the flow to the sedimentation tank increases. Assume that the following proportionality holds:

$$\phi_1 = K_e(\phi + \phi_R) \tag{7.10}$$

where K_e is a proportionality constant, depending on the degree of mixing.

In the static situation there will be a certain vertical concentration profile in the sedimentation tank (Figure 7.4). The solid line represents the real concentration profile; the dotted line the approximation. The approximation will be better when the tank is divided into more sections. In the first tank the average concentration is equal to C_{B2}. In the top of the first tank the concentration is equal to C^*_{B2} and in the bottom C^{**}_{B2}. Suppose that the concentration C^{**}_{B2} is equal to g times C_{B2} and that C^*_{B2} is equal to 1/g times C_{B2}. A similar situation holds for the second tank. The boundary conditions are in the top of the first tank:

$$C^*_{B2} = \frac{1}{g} \cdot C_{B2} = C_{Be} \tag{7.11}$$

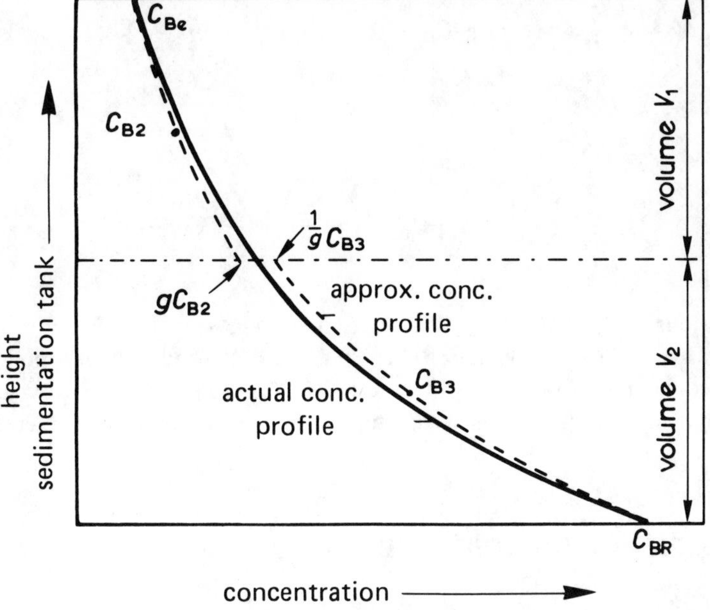

Figure 7.4. Concentration profile in the secondary sedimentation tank.

and in the bottom of the second tank:

$$C_{B3}^{**} = gC_{B3} = C_{BR} \tag{7.12}$$

The partial mass balance for both volumes are now:

$$V_1 \frac{dC_{B2}}{dt} = (\phi + \phi_R)C_{B1} - (\phi - \phi_s)C_{Be} - \phi_1 gC_{B2} + \phi_2 \frac{1}{g} \cdot C_{B3} \tag{7.13}$$

and

$$V_2 \frac{dC_{B3}}{dt} = \phi_1 gC_{B2} - \phi_2 \cdot \frac{1}{g} C_{B3} - (\phi_R + \phi_s)C_{BR} \tag{7.14}$$

With the aid of Equations 7.11 and 7.12, Equations 7.13 and 7.14 may be rewritten as:

$$gV_1 \frac{dC_{Be}}{dt} = (\phi + \phi_R)C_{B1} - (\phi - \phi_s + \phi_1 g^2)C_{Be} + \frac{\phi_2}{g^2} C_{BR} \tag{7.15}$$

and

$$\frac{V_2}{g} \frac{dC_{BR}}{dt} = \phi_1 g^2 C_{Be} - \left(\phi_R + \phi_s + \frac{\phi_2}{g^2}\right) C_{BR} \tag{7.16}$$

Equations 7.9, 7.10, 7.15 and 7.16 now represent the model for the secondary sedimentation tank. It should be emphasized that this model for the sedimentation tank only offers one possible description for the dynamics. Experimental verification is needed to determine the unknown parameters.

LINEARIZATION OF THE MATHEMATICAL MODEL

The model as given in Equations 7.1, 7.2, 7.4 to 7.10, 7.15 and 7.16 will now be linearized. Combination of Equations 7.4, 7.5 and 7.7, linearization and replacing d/dt by s results in:

$$\delta C_{s1} = \frac{K_{p1}}{\tau_1 s + 1} \delta\phi + \frac{K_{p2}}{\tau_1 s + 1} \delta C_{si} - \frac{K_{p3}}{\tau_1 s + 1} \delta C_{B1} - \frac{K_{p4}}{\tau_1 s + 1} \delta\phi_R \tag{7.17}$$

The parameters in this equation are given in Table 7.1. In a similar way, Equations 7.1, 7.2, 7.6 and 7.8 give:

$$\delta C_{B1} = \frac{K_{p5}}{\tau_2 s + 1} \delta C_{BR} - \frac{K_{p6}}{\tau_2 s + 1} \delta\phi + \frac{K_{p7}}{\tau_2 s + 1} \delta C_{s1} + \frac{K_{p8}}{\tau_2 s + 1} \delta\phi_R \tag{7.18}$$

An information flow diagram for Equations 7.17 and 7.18 is given in Figure 7.5. Combination of Equations 7.9, 7.10 and 7.16, linearization and replacing d/dt by s, gives:

$$\delta C_{BR} = \frac{K_1}{\tau_3 s + 1} \delta C_{Be} - \frac{K_2}{\tau_3 s + 1} \delta\phi - \frac{K_3}{\tau_3 s + 1} \delta\phi_R - \frac{K_4}{\tau_3 s + 1} \delta\phi_s \tag{7.19}$$

Table 7.1 Parameters in the Linearized Equations 7.17 and 7.18

$$N1 = \bar{\phi} + \bar{\phi}_R + \frac{k_{g0} V \bar{C}_{B1} K_s}{Y(K_s + \bar{C}_{s1})^2}$$

$$\tau_1 = V_R/N1$$

$$K_{p1} = (\bar{C}_{si} - \bar{C}_{s1})/N1$$

$$K_{p2} = \bar{\phi}/N1$$

$$K_{p3} = \left[\frac{k_{g0} V \bar{C}_{s1}}{Y(K_s + \bar{C}_{s1})} - \beta k_d V\right] \cdot N1^{-1}$$

$$K_{p4} = C_{s1}/N1$$

$$M_1 = \bar{\phi} + \bar{\phi}_R - \frac{k_{g0} V \bar{C}_{s1}}{K_s + \bar{C}_{s1}} + k_d V$$

$$\tau_2 = V_R/M1$$

$$K_{p5} = \bar{\phi}_R/M1$$

$$K_{p6} = \bar{C}_{B1}/M1$$

$$K_{p7} = \frac{k_{g0} V \bar{C}_{B1} K_s}{(K_s + \bar{C}_{s1})^2} \Big/ M1$$

$$K_{p8} = (\bar{C}_{BR} - \bar{C}_{B1})/M1$$

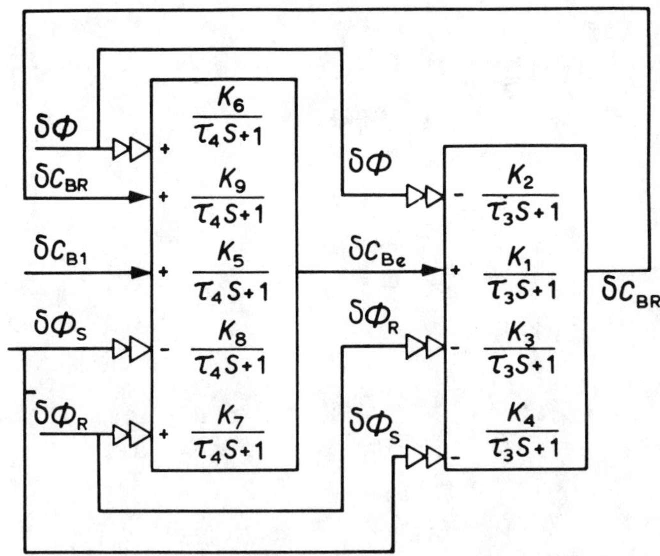

Figure 7.5. Information flow diagram for the biological reactor.

In a similar way, from Equations 7.9, 7.10 and 7.15

$$\delta C_{Be} = \frac{K_5}{\tau_4 s + 1}\delta C_{B1} + \frac{K_6}{\tau_4 s + 1}\delta\phi + \frac{K_7}{\tau_4 s + 1}\delta\phi_R - \frac{K_8}{\tau_4 s + 1}\delta\phi_s + \frac{K_9}{\tau_4 s + 1}\delta C_{BR} \tag{7.20}$$

The parameters of Equations 7.19 and 7.20 are given in Table 7.2. Depending on the values of K_e and g, Equations 7.19 and 7.20 can be simplified further. The information flow diagram for Equations 7.19 and 7.20 is given in Figure 7.6.

CONTROL OF THE ACTIVATED SLUDGE PROCESS

The return sludge flow from the sedimentation tank to the reactor is usually used as correcting condition. Theoretically, there are many possible ways to control the process, e.g., aeration, pH, temperature, residence time in the reactor and concentration of substrate.

Table 7.2 Parameters in Equations 7.19 and 7.20

$$N2 = \bar{\phi}_R + \bar{\phi}_s + \frac{K_e(\bar{\phi} + \bar{\phi}_R) - \bar{\phi}_R - \bar{\phi}_s}{g^2}$$

$$\tau_3 = \frac{1}{g} V_2/N2$$

$$M2 = \bar{\phi} - \bar{\phi}_s + g^2 K_e(\bar{\phi} + \bar{\phi}_R)$$

$$\tau_4 = gV_1/M2$$

$$K_1 = K_e(\bar{\phi} + \bar{\phi}_R)g^2/N2$$

$$K_2 = K_e(\bar{C}_{BR}/g^2 - g^2\bar{C}_{Be})/N2$$

$$K_3 = [K_e(\bar{C}_{BR}/g^2 - g^2\bar{C}_{Be}) + \bar{C}_{BR}(1 - 1/g^2)]/N2$$

$$K_4 = [\bar{C}_{BR}(1 - 1/g^2)]/N2$$

$$K_5 = (\bar{\phi} + \bar{\phi}_R)/M2$$

$$K_6 = [K_e(\bar{C}_{BR}/g^2 - g^2\bar{C}_{Be}) + \bar{C}_{B1} - \bar{C}_{Be}]/M2$$

$$K_7 = [K_e(\bar{C}_{BR}/g^2 - g^2\bar{C}_{Be}) + \bar{C}_{B1} - \bar{C}_{BR}/g^2]/M2$$

$$K_8 = (\bar{C}_{BR}/g^2 - \bar{C}_{Be})/M2$$

$$K_9 = [K_e(\bar{\phi} + \bar{\phi}_R) - \bar{\phi}_R - \bar{\phi}_s)/g^2]/M2$$

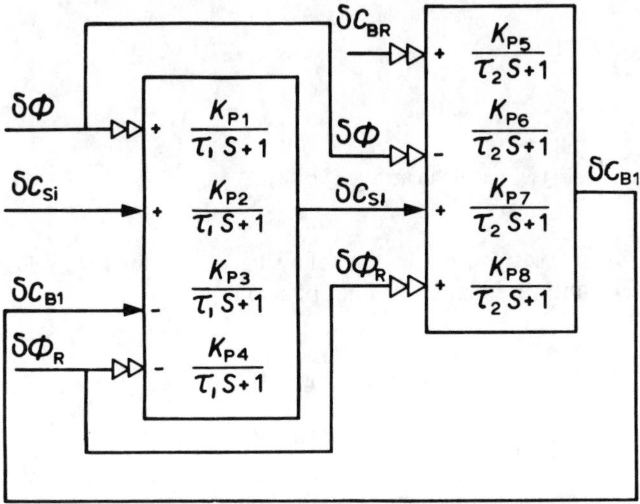

Figure 7.6. Information flow diagram for the sedimentation tank.

In practice, however, these factors cannot easily be influenced (concentration of substrate, residence time); are not economically controllable (temperature); have to be maintained at an optimal value (pH); or must be larger than a minimum value to guarantee growth and survival of microorganisms (aeration-flow). The controlled condition is usually the substrate concentration C_{s1}, which must be maintained as constant as possible. The equation for proportional control is:

$$\delta\phi_R = K_c \delta C_{s1} \tag{7.21}$$

The characteristic equation for control can easily be derived from Figures 7.5 and 7.6. The result is a fifth-order polynomial in s, of the form:

$$s^5 + a_4 s^4 + a_3 s^3 + a_2 s^2 + a_1 s + a_0 = 0 \tag{7.22}$$

where the coefficients are a function of the process gains, time constants and controller gain. The fifth-order process is stable if all coefficients in Equation 7.22 have the same sign and if the following conditions are satisfied [15]:

$$a_3a_4 - a_2 > 0 \tag{7.23}$$

and

$$(a_3a_4 - a_2)(a_1a_2 - a_0a_3) - (a_1a_4 - a_0)^2 > 0 \tag{7.24}$$

Dividing by $s^2 + \omega_n^2$ results in the following conditions for oscillation frequency and controller gain at the limit of stability:

$$\omega_n^4 - a_3\omega_n^2 + a_1 = 0 \tag{7.25}$$

$$a_4\omega_n^4 - a_2\omega_n^2 + a_0 = 0 \tag{7.26}$$

with ω_n = the oscillation frequency on the limit of stability.

Equations 7.25 and 7.26 cannot be solved easily because the coefficients are complicated functions of the process conditions. However, it is possible to derive an approximate expression. With the aid of Figures 7.5 and 7.6, δC_{BR} can be expressed in $\delta\phi_R$. The result is:

$$\delta C_{BR} = \left[-\frac{K_3}{\tau_3 s + 1} + \frac{K_1K_7}{(\tau_3 s + 1)(\tau_4 s + 1)} - \ldots + \ldots \right] \delta\phi_R \tag{7.27}$$

The speed of control and stability of the control loop are mainly determined by the first term in this expression. Therefore Equation 7.27 is approximated by:

$$\delta C_{BR} = -\frac{K_3}{\tau_3 s + 1}\,\delta\phi_R \tag{7.28}$$

It should be noted that this approximation is allowed only if the numerator in the right side of Equation 7.27 does not have a called nonminimum-phase character. This is the case when the numerator of Equation 7.27 contains a term like $1 - s\tau$. Combination of Equations 7.17, 7.18, 7.21 and 7.28 results, for $\delta\phi = 0$ and $\delta C_{si} = 0$, in the characteristic equation:

$$s^2 + b_2s^2 + b_1s + b_0 = 0 \tag{7.29}$$

Table 7.3 Coefficients of Equation 7.29

$b_0 = (1 + K_cK_{p4} - K_3K_{p3}K_cK_{p5} + K_{p3}K_{p8}K_c + K_{p3}K_{p7})/\tau_1\tau_2\tau_3$
$b_1 = (\tau_1 + \tau_2 + \tau_3 + K_cK_{p4}\tau_2 + K_cK_{p4}\tau_3 + K_{p3}K_{p7}\tau_3 + K_{p3}K_{p8}K_c\tau_3)/\tau_1\tau_2\tau_3$
$b_1 = (\tau_1\tau_2 + \tau_1\tau_3 + \tau_2\tau_3 + K_cK_{p4}\tau_2\tau_3)/\tau_1\tau_2\tau_3$

The coefficients b are given in Table 7.3. The limit of stability is determined by:

$$b_1b_2 = b_0 \tag{7.30}$$

which results in a second-order equation for the controller gain.

CHAPTER 8

DYNAMICS OF DISTILLATION COLUMNS

Much has been written about dynamic models for distillation columns. A survey is given by Rijnsdorp and co-workers [16,17].

In this chapter the dynamics of distillation columns will be illustrated by means of a simple distillation column (Figure 8.1). In the reboiler, the liquid is partially evaporated. On every tray the rising vapor flow is in intensive contact with the liquid (Figure 8.2). On its way to the top, the vapor flow will contain more and more of the most volatile components. The vapor flow leaving the top of the column is condensed in a condenser and collected in an accumulator. Part of the condensed vapor is fed back into the column as reflux; another part is withdrawn as top product. In Figure 8.3 a detail of a tray is given with the notation used.

Every tray can be associated with three lags, which determine different aspects of dynamic behavior. The lags for propagation of concentration variations are usually the largest. The next largest time constant appears in the propagation of liquid flow changes, as a result of the change in the amount of liquid on a tray due to flow variations. Smaller lags appear in the propagation of vapor flow variations. These variations are related to the change of the vapor rate through the trays due to a change in pressure.

MODEL EQUATIONS

Let us assume that both vapor and liquid on a tray are ideally mixed, and heat losses to the surroundings can be ignored. The mass balance for a certain component on tray i is:

$$\frac{d}{dt}(M_{\ell,i} + M_{v,i}) = L_{i+1} - L_i + V_{i-1} - V_i + F_i \tag{8.1}$$

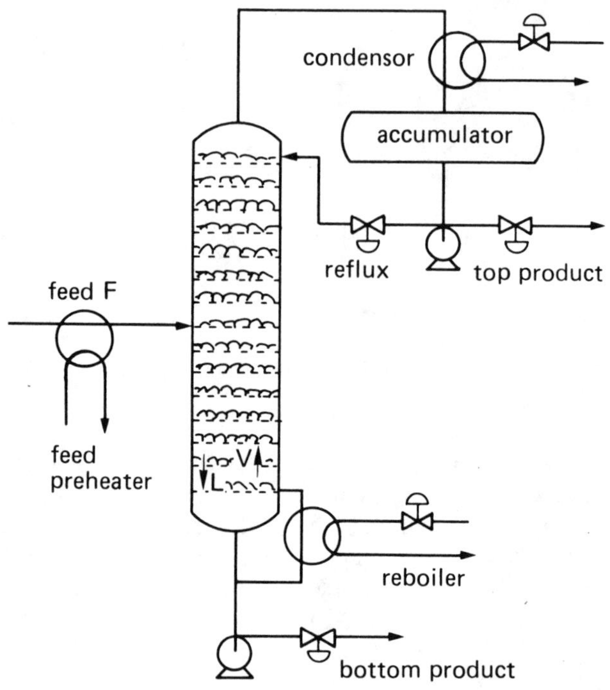

Figure 8.1 Schematic diagram of a distillation column.

where $M_{\ell,i}, M_{v,i}$ = mass of liquid or vapor, respectively (kg)
L = liquid flow (kg/sec)
V = vapor flow (kg/sec)
F = feed flow (kg/sec)

The partial mass balance is:

$$\frac{d}{dt}(M_{\ell,i}x_i + M_{v,i}y_i) = L_{i+1}x_{i+1} - L_i x_i + V_{i-1}y_{i-1} - V_i y_i + F_i z_i \qquad (8.2)$$

where x = mole fraction of component in the liquid
y = mole fraction of component in the vapor
z = mole fraction of component in the feed
$H_{f,i}$ = enthalpy of the feed (J/kg)

The energy balance is:

Figure 8.2 Vapor-liquid contact on the tray of a small distillation column.

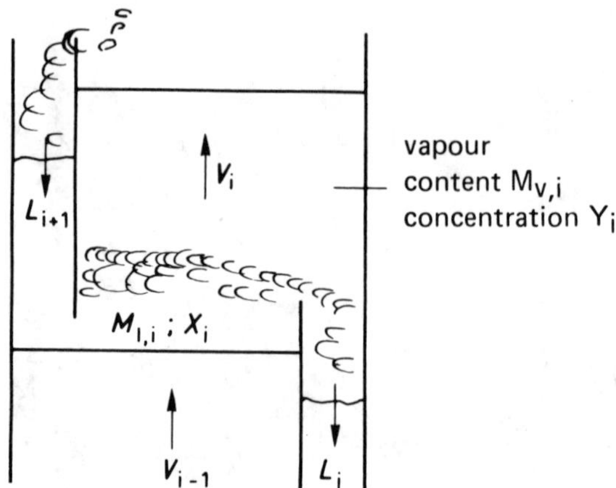

Figure 8.3 Detail of a tray.

$$\frac{d}{dt}(M_{\ell,i}H_{\ell,i} + M_{v,i}H_{v,i} + M_t\gamma_tT_i) - V_{t,i}\frac{dP_i}{dt}$$

$$= L_{i+1}H_{\ell,i+1} - L_iH_{\ell,i} + V_{i-1}H_{v,i-1} - V_iH_{v,i} + F_iH_{f,i} \tag{8.3}$$

where M_t = mass of the tray (kg)
H = enthalpy (J/kg)
V_t = tray volume (m^3)
P = pressure (N/m^2)
T = temperature (°K)
γ_t = specific heat of the tray material (J/kg-°K)

The term $V_{t,i}(dP_i/dt)$ appears in the equation because enthalpies are used in the first two terms in the left side of the equation instead of internal energies. This term can usually be ignored.

To complete the model, another seven equations are needed. The form of these equations will depend on further assumptions concerning the amount of liquid and vapor and the properties of the mixture on a tray. A frequently used assumption is that of equimolal overflow. The enthalpies of liquid and vapor can then be approximated by:

$$H_{\ell,i} = f(P_i) - a_1x_i \tag{8.4}$$

$$H_{v,i} = f(P_i) - a_1y_i + H_{e,i} \tag{8.5}$$

in which a_1 is a constant. The heat of evaporation $H_{e,i}$ will usually be a function of the pressure P_i:

$$H_{e,i} = f(P_i) \tag{8.6}$$

The amount of vapor on a tray is small; hence, it can approximately be considered as a function of pressure only:

$$\frac{dM_{v,i}}{dt} = \left(\frac{dM_{v,i}}{dP_i}\right)\frac{dP_i}{dt} \tag{8.7}$$

When it is further assumed that the mixture to be distilled is a binary one with constant relative volatility α, we have:

$$y_i = \frac{\alpha x_i}{1 + (\alpha - 1)x_i} \tag{8.8}$$

and

$$T_i = f(P_i, x_i) \tag{8.9}$$

For many types of trays the following equation can be written for the relationship between pressure drop and vapor flow:

$$P_{i-1} - P_i = a_2 M_{\ell,i} + a_3 V_{i-1}^2 + \text{constant} \tag{8.10}$$

where a_2 and a_3 are constants again. For liquid flow from the tray, the Francis weir formula is often used [18]:

$$L_i = a_4 \left[\frac{M_{\ell,i}}{\rho_{f,i} A_t} - h_w\right]^{3/2} \rho_{f,i} \tag{8.11}$$

where a_4 = a geometric constant ($m^{3/2}/sec$)
h_w = height of the overflow weir (m)
ρ_f = foam density (kg/m^3)
A_t = tray area (m^2)

The foam height can be assumed to be constant or can be corrected to other tray parameters. Although the proposed model gives a reasonably good and detailed description of the dynamic behavior, it will not be analyzed further. To promote clarity and to obtain a better insight in column dynamics, a strongly simplified approach will be followed.

Equations 8.1 to 8.11, however, can be solved only when correct parameter values are available.

LIQUID FLOW RESPONSES

When considering liquid flow responses, concentration fluctuations are usually left out of consideration. Furthermore, vapor flow variations propagate relatively rapidly, hence:

$$\delta V_{i-1} \simeq \delta V_i \simeq \delta V \tag{8.12}$$

Assuming that pressure is accurately controlled at the column top, and that feed conditions are constant, Equation 8.1 reduces to:

$$\frac{dM_{\ell,i}}{dt} = L_{i+1} - L_i \tag{8.13}$$

The amount of liquid flowing from tray i to i − 1 is usually a function of the amount of liquid on the tray $M_{\ell,i}$ and the vapor flow V to the tray; thus, a variation in L_i can be written as:

$$\delta L_i = \left(\frac{\partial L_i}{\partial M_{\ell,i}}\right)_V \delta M_{\ell,i} + \left(\frac{\partial L_i}{\partial V}\right)_{M_{\ell,i}} \delta V \tag{8.14}$$

Linearization of Equation 8.13 and combination with Equation 8.14 gives:

$$\delta L_i = \frac{1}{\tau_\ell}\left(\frac{\delta L_{i+1}}{s} - \frac{\delta L_i}{s}\right) + \lambda \delta V \tag{8.15}$$

or

$$\delta L_i = \frac{\delta L_{i+1}}{\tau_\ell s + 1} + \frac{\tau_\ell \lambda s}{\tau_\ell s + 1}\ \delta V = \frac{\delta L_{i+1}}{\tau_\ell s + 1} + \lambda\left(1 - \frac{1}{\tau_\ell s + 1}\right)\delta V \qquad (8.16)$$

where

$$\tau_\ell = \left(\frac{\partial M_{\ell,i}}{\partial L_i}\right)_V \qquad (8.17)$$

the hydraulic time constant of the tray, and

$$\lambda = \left(\frac{\partial L_i}{\partial V}\right)_{M_{\ell,i}} \qquad (8.18)$$

Repeated application of Equation 8.16 yields, for the overflow from the lowest tray:

$$\delta L_1 = \frac{1}{(\tau_\ell s + 1)^n}\ \delta L_{n+1} + \lambda\left[1 - \frac{1}{(\tau_\ell s + 1)^n}\right]\delta V \qquad (8.19)$$

which, for large values of n, can be approximated as:

$$\delta L_1 \simeq e^{-n\tau_\ell s}\delta L_{n+1} + \lambda(1 - e^{-n\tau_\ell s})\delta V \qquad (8.20)$$

where δL_{n+1} is the reflux flow variation in the top. Figure 8.4 shows the various elements in terms of unit step responses. Inputs are the reflux (δL_{n+1}) and the heating medium to the reboiler (U). The time constant for liquid flow propagation τ_ℓ has usually a magnitude on the order of some seconds to some tens of seconds. When there are many trays, the total lag for liquid flow variations from top to bottom is not inconsiderable, for example when $n = 60$ and $\tau_\ell = 10$ sec, the total lag will be 600 sec.

The parameter for vapor flow variations λ can either be positive or negative. If λ is negative or smaller than +0.5, no large difficulties are to be expected. If λ is larger than one, however, the effect (the temporary increase in liquid flow) exceeds the cause (the increase of the vapor flow). Then, for a step increase in the rate of evaporation, the bottom level will temporarily increase

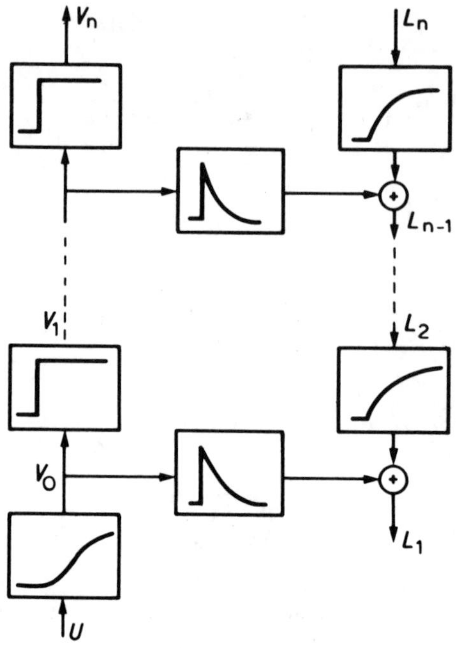

Figure 8.4 Liquid flow dynamics.

(go into the wrong direction). It will decrease (go into the right direction) only after about $n\tau_\ell$ time units. This is an example of nonminimum-phase behavior, which severely restricts the speed of control of the bottom level to reboiler heating. As, in practice, the bottom level should be maintained within a low and a high limit, slow control inevitably leads to a larger difference between these limits, hence to a taller column and to a higher capital investment.

For λ between 0.5 and 1, it can be shown that behavior is also nonminimum-phase, albeit of a different type than for $\lambda > 1$. Here the initial part of the bottom level response $(1 - \lambda)\delta V$ is too small compared to the delayed part $\lambda\delta V$, so the lag represented by the latter part significantly decreases the speed of bottom level control by reboiler heating.

VAPOR FLOW RESPONSES

If the pressure drop over the trays can be ignored compared to the absolute value of the pressure, vapor flow variations will propagate relatively rapidly.

However, even small lags have a strong influence on pressure control, as will be made clear in the final chapters of this book. The vapor flow responses can be found from the energy balance. Substitution of Equations 8.4 and 8.5 into 8.3, followed by addition of a_1 times Equation 8.2 yields:

$$\frac{dM_{\ell,i}f(P_i)}{dt} + \frac{dM_{v,i}f(P_i)}{dt} + \frac{dM_{v,i}H_{e,i}}{dt} + \frac{dM_t\gamma_t T_i}{dt} - V_t\frac{dP_i}{dt}$$

$$= L_{i+1}f(P_{i+1}) - L_i f(P_i) + V_{i-1}[f(P_{i-1}) + H_{e,i-1}] - V_i[f(P_i) + H_{e,i}] \qquad (8.21)$$

Subtraction of $f(P_i)$ times Equation 8.1 from this expression leads to:

$$(M_{\ell,i} + M_{v,i})\frac{df(P_i)}{dt} + M_{v,i}\frac{dH_{e,i}}{dt} + H_{e,i}\frac{dM_{v,i}}{dt} + M_t\gamma_t\frac{dT_i}{dt} - V_t\frac{dP_i}{dt}$$

$$= L_{i+1}[f(P_{i+1}) - f(P_i)] + V_{i-1}[f(P_{i-1}) + H_{e,i-1} - f(P_i) - H_{e,i}]$$

$$- (V_{i-1} - V_i)H_{e,i} \qquad (8.22)$$

The time derivatives of $f(P_i)$ and $f(P_i) + H_{e,i}$ can be written in terms of time derivatives of the pressure:

$$\frac{df(P_i)}{dt} = \gamma_\ell \frac{\partial T}{\partial P}\frac{dP_i}{dt} \qquad (8.23)$$

$$d/dt[f(P_i) + H_{e,i}] = \gamma_v \frac{\partial T}{\partial P}\frac{dP_i}{dt} \qquad (8.24)$$

where γ_ℓ = specific heat of liquid (J/kg-°K)
γ_v = specific heat of vapor (J/kg-°K)
$\frac{\partial T}{\partial P}$ = slope of equilibrium line (K/P_a)

The same can be done, with some approximation, for the time derivatives of $M_{v,i}$ and T_i:

$$\frac{dM_{v,i}}{dt} = \frac{\partial M_v}{\partial P}\frac{dP_i}{dt} \tag{8.25}$$

$$\frac{dT_i}{dt} = \frac{\partial T}{\partial P}\frac{dP_i}{dt} \tag{8.26}$$

Further, the first and second terms on the right side of Equation 8.22 are small, and more or less cancel. As a result, Equation 8.22 is now reduced to:

$$C_i \frac{dP_i}{dt} = V_{i-1} - V_i \tag{8.27}$$

with

$$C_i = [(\gamma_1 M_{1,i} + \gamma_v M_{v,i} + \gamma_t M_t)\frac{\partial T}{\partial P} + \frac{\partial M_v}{\partial P} - V_t]H_{e,i}^{-1} \tag{8.28}$$

Fortunately, these tedious conversions have resulted in a rather simple expression, which can also be interpreted in a physical way.

Equation 8.27 says that a pressure change is accompanied by a difference in vapor flow. This difference is due to tray capacitance C_i, which consists of a term for heating/cooling the tray and its contents by condensation/evaporation, a term for compression/expansion of the vapor holdup, and a small correction term for the difference between specific enthalpies and internal energies. Numerically, the most important element in C_i is that corresponding to the heating/cooling of the tray liquid hold-up.

Neglecting variations in $M_{\ell,i}$ in Equation 8.10 leads to a relation between vapor flow and pressure drop. After linearization:

$$\delta P_{i-1} - \delta P_i = R\delta V_{i-1} \tag{8.29}$$

with

$$R = 2a_3\overline{V} \tag{8.30}$$

Linearization of Equation 8.27 and substitution of Equations 8.29 and 8.30 yields:

$$sC\delta P_i = \frac{(\delta P_{i-1} - \delta P_i)}{R} - \frac{(\delta P_i - \delta P_{i+1})}{R} \tag{8.31}$$

Equations 8.29 and 8.31 can be represented by an electrical analog (Figure 8.5). The terms on the right side represent currents, the net difference of which is stored in capacitance C.

A similar analog can be set up for tray i + 1 (Figure 8.5a, etc.). All analogs connected together form the network for pressure variations, in which currents through the resistances represent vapor flow variations (Figure 8.5b).

CONCENTRATION RESPONSES

The response of the concentration of a component can be found with the partial mass balance for the tray (Equation 8.2): a change in the liquid flow δL_{i+1} initially transports an extra flow of component i from tray i + 1, equal to:

$$\delta L_{i+1}\,(\bar{x}_{i+1} - \bar{x}_i) \tag{8.32}$$

where $\bar{x}$ is the static value of the concentration in the liquid. In the same manner, the change of vapor flow to tray i is:

$$\delta V_{i-1}(\bar{y}_{i-1} - \bar{y}_i) \tag{8.33}$$

where $\bar{y}$ is the static value of the concentration in the vapor. The static partial mass balance of a tray (without feed):

$$\bar{L}\,(\bar{x}_{i+1} - x_i) + \bar{V}(\bar{y}_{i-1} - y_i) = 0 \tag{8.34}$$

Using this result, Equation 8.33 can be rewritten and combined with Equation 8.32 to:

$$\bar{L}(\bar{x}_{i+1} - \bar{x}_i)\left(\frac{\delta L_{i+1}}{\bar{L}} - \frac{\delta V_{i-1}}{\bar{V}}\right) \tag{8.35}$$

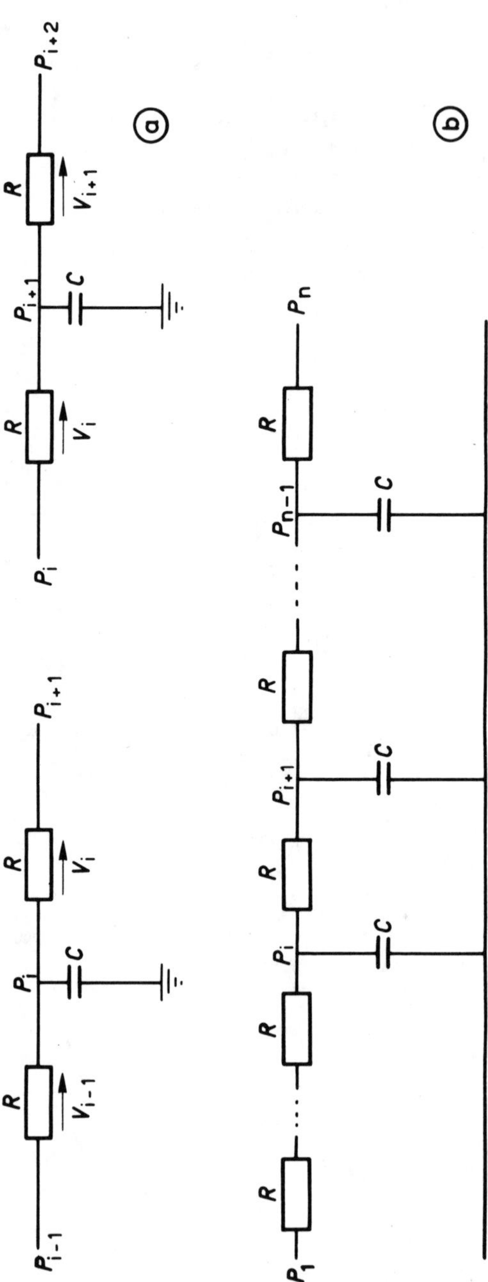

Figure 8.5 Electrical network for pressure and vapor flow variations.

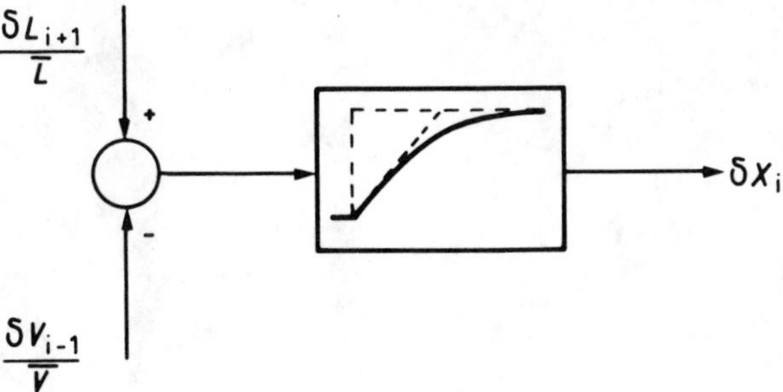

Figure 8.6 Concentration dynamics on tray i.

This extra flow of component is mixed with the liquid content of the tray $M_{\ell,i}$. Therefore, the initial response of the concentration is proportional to the integral of Equation 8.35:

$$\delta x_i = \int \frac{\overline{L}}{M_\ell} (\overline{x}_{i+1} - \overline{x}_i) \left(\frac{\delta L_{i+1}}{\overline{L}} - \frac{\delta V_{i-1}}{\overline{V}} \right) dt \tag{8.36}$$

Gradually, the concentration differences between the trays will change, however, and finally there will be a new equilibrium. This means that in Equation 8.36, the integral must be replaced by a first-order response with a very large time constant. The variation of the concentration on tray i can thus be approximated by the difference of the responses of the relative liquid and vapor flow variations, followed by a first-order response with a large time constant (Figure 8.6).

Here the parameter λ can have also an unfavorable influence, particularly for the concentrations on the trays near the bottom. Because relative flow variations are now important, a relative λ is relevant:

$$\overline{\lambda} = \lambda(\overline{V}/\overline{L}) = (\overline{V}/\overline{L}) \left(\frac{\partial L}{\partial V} \right)_{M_\ell} \tag{8.37}$$

CHAPTER 9

MODELING OF DISTRIBUTED PARAMETER SYSTEMS

In many actual cases, system variables are not only a function of time but also of one or more geometric dimensions. This is the case for many heat exchangers, tubular reactors, gas pipelines, pollution in a river, etc. One can approximate the actual behavior by dividing the system into a number of subsystems, each of which is ideally mixed. This approach was followed in the previous chapters and will be discussed again in a later chapter. Here, however, we shall maintain the distributed character of the system. The mathematical model, therefore, contains one or more partial differential equations. The solution of these equations can be simplified by linearizing, hence by restricting attention to small variations around the static values. Now we shall go into more detail about the method of dynamic analysis of distributed systems.

SELECTION OF INDEPENDENT GEOMETRIC VARIABLES

The construction of the mathematical model starts with the question: in how many geometric dimensions should the distributed character of the system be developed? The most general case is a description in three dimensions, resulting in partial derivatives to x, y, z and time. Evidently, the solution of such partial differential equations is not a simple matter. However, it is not certain that such a detailed model will also give better results. For example, the flow pattern in a river or a trickle-phase fixed-bed reactor is very complicated. One simply does not know enough of the system parameters to expect useful results, even when numerical problems can be overcome.

A considerable simplification is obtained by using only two geometric dimensions. This is possible in cylinder symmetrical systems, where the partial derivatives to z (axis direction of the cylinder) and r (distance to the axial) remain. Often one goes one step further: then there is only one independent geometric variable in the direction in which the changes are the largest.

Average values are used in other geometric dimensions. Himmelblau and Bischoff [9] call such an approach a maximum gradient description. In this and following chapters this approach will mainly be used.

SELECTION OF DISTRIBUTED (DEPENDENT) VARIABLES

The second step is the selection of the distributed state variables, which are a function of time and distance. Selection is strongly related to the assumptions on which the model is based. In fact, one can construct a number of models for a certain physical system or process that differ in the number of distributed dependent variables and therefore in complexity. For example, by not considering the density as a state variable in flow, incompressibility is assumed. Generally the choice of the model depends strongly on the type of questions to be answered. An example is the model of a distillation column. When its startup time would be of interest, only the slowest dynamic effects, the concentration changes, need be considered. For vapor and liquid flow variations, one can use algebraic equations, because hydraulic changes generally are an order of magnitude faster than composition changes [19].

When, however, we are studying the behavior of a control loop, say a pressure control on the reboiler, we are primarily interested in vapor flow variations, and hardly in concentration fluctuations; hence in this case a hydraulic model is constructed in which concentrations are assumed to be constant. Finally, if we are interested in optimal operation of a distillation column, a static model is sufficient [20]. Similarly, models for distributed systems are dependent on the application.

CONTINUITY EQUATIONS

The distributed dependent variables require an equal number of partial differential equations, which can usually be derived from conservation laws for mass, momentum and energy. This makes the distributed dependent variables equivalent to distributed state variables. Every balance can be expressed as:

Accumulation in a volume element

= Generation in the volume element

− Transport by flow through the system surface

− Transport by molecular motion through the system surface

− Loss through the system surface by other mechanisms (9.1)

In the maximum gradient description, only that direction is considered in which the changes are the largest (maximum gradient: in Fig. 9.1, the z direction). Changes in x and y directions do not play any role; hence, derivatives with respect to these coordinates do not appear in the differential equations. The different terms in Equation 9.1 will not be considered.

Mass Balance

The simplest equation which can be set up is the mass balance. Accumulation of mass in the volume element $A_c \Delta z$ can be represented by:

$$\frac{\partial M}{\partial t} = A_c \Delta z \frac{\partial \rho}{\partial t} \tag{9.2}$$

where $A_c = \Delta x \Delta y$, system area (m^2)
ρ = density (kg/m^3)

Transport by flow through the surface A_c at location z is equal to:

$$(\phi_m)_z = A_c(\rho v)_z \tag{9.3}$$

where ϕ_m = mass flow (kg/sec)
v = velocity (m/sec)

At location z + Δz:

$$(\phi_m)_{z+\Delta z} = A_c(\rho v)_{z+\Delta z} \tag{9.4}$$

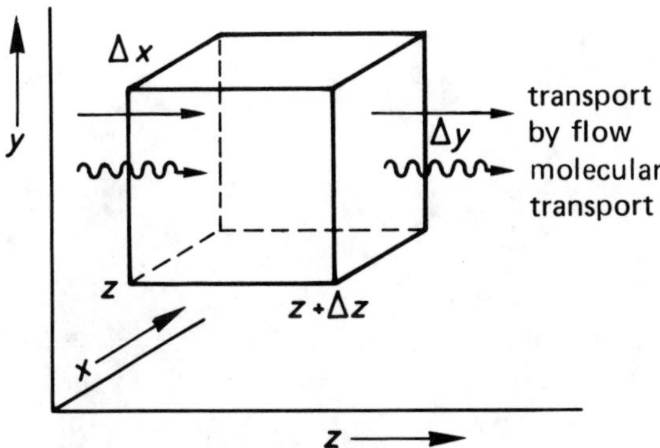

Figure 1. Volume element for the derivation of the model equations.

The net transport by flow is equal to the difference between Equations 9.3 and 9.4:

$$A_c[(\rho v)_z - (\rho v)_{z+\Delta z}] \qquad (9.5)$$

Using the well known rule for analytical functions:

$$\lim_{\Delta z \to 0} \frac{f(z + \Delta z) - f(z)}{\Delta z} = \frac{\partial f}{\partial z} \qquad (9.6)$$

Equation 9.5 can be written in the limit case, as:

$$\lim_{\Delta z \to 0} A_c \Delta z \left[\frac{(\rho v)_z - (\rho v)_{z+\Delta z}}{\Delta z} \right] = -A_c \Delta z \frac{\partial(\rho v)}{\partial z} \qquad (9.7)$$

As the accumulation term is equal to the flow term, the well known continuity equation results:

$$\frac{\partial \rho}{\partial t} = -\frac{\partial(\rho v)}{\partial z} \qquad (9.8)$$

Partial Mass Balance

When the fluid consists of different components, concentration variations may occur. For each component, a partial mass balance can be set up.

Accumulation

Accumulation represents the change of concentration of a certain component A per unit of time, multiplied by the volume of the element $A_c\Delta z$:

$$\frac{\partial}{\partial t}(Vc_A) = A_c \Delta z \frac{\partial c_A}{\partial t} \text{ kg/sec} \qquad (9.9)$$

with c_A = the concentration of component A in kg/m^3.

Production

In the volume element, component A may appear or disappear, e.g., as a result of chemical reaction. To keep the equation general, this term is represented as the rate of reaction R(z,t) times the volume of the element:

$$R(z,t)A_c\Delta z \text{ kg/sec} \qquad (9.10)$$

When the kinetics are given, the actual reaction rate in terms of concentrations may be substituted.

Flow

The flux of component A due to flow is equal to the velocity v multiplied by the concentration of component A. The rate of transport of A is then equal to the flux of A multiplied by the area perpendicular to the direction of flow. At location z the mass of A flowing through the area per unit of time is:

$$A_c (vc_A)_z \tag{9.11}$$

The mass flowing through the area at location $z + \Delta z$ is

$$A_c (vc_A)_{z+\Delta z} \tag{9.12}$$

The net transport by flow is equal to the difference between Equations 9.11 and 9.12:

$$A_c [(vc_A)_z - (vc_A)_{z+\Delta z}] \tag{9.13}$$

When Δz goes to zero, the limit of Equation 9.13 may be taken:

$$\lim_{\Delta z \to 0} A_c \Delta z \left[\frac{(vc_A)_z - (vc_A)_{z+\Delta z}}{\Delta z} \right] = -A_c \Delta z \frac{\partial (vc_A)}{\partial z} \tag{9.14}$$

Diffusion and Dispersion

The flux of component A due to molecular diffusion can be determined with the first law of Fick:

$$\phi_A = -D \frac{\partial c_A}{\partial z} \text{ kg/m}^2\text{-sec} \tag{9.15}$$

where D = diffusion coefficient (m^2/sec)

The minus sign indicates that the flux is positive in the direction of decreasing concentration. The transport of A is equal to the flux multiplied by the area perpendicular the direction of diffusion. The net transport is then:

$$\left[-\left(D_A \frac{\partial c_A}{\partial z} \right)_z + \left(D_A \frac{\partial c_A}{\partial z} \right)_{z+\Delta z} \right] A_c \tag{9.16}$$

In the limit:

$$\frac{\partial}{\partial z}\left(D_A \frac{\partial c_A}{\partial z}\right) A_c \Delta z \tag{9.17}$$

Combination of Equations 9.9, 9.10, 9.14 and 9.17 gives:

$$\frac{\partial c_A}{\partial t} = R(z,t) - \frac{\partial (v c_A)}{\partial z} + \frac{\partial}{\partial z}\left(D_A \frac{\partial c_A}{\partial z}\right) \tag{9.18}$$

For turbulent flow, D_A represents dispersion instead of diffusion.

Other Mechanisms

It could happen that mass transport occurs due to another mechanism, which has not yet been mentioned. When component A, for example, is produced not by reaction but as a result of mass transport between two phases, the term R(z,t) in Equation 9.18 has to be replaced by a mass transfer term:

$$k(c_A - c_A^*) \tag{9.19}$$

where k = mass transfer coefficient (m^2/sec)
c_A^* = equilibrium concentration of component A at the transfer surface (kg/m^3)

Energy Balance

The energy balance is rather similar to the partial mass balance. The accumulation term is:

$$\frac{\partial}{\partial t}(M c_p T) = A_c \Delta z \frac{\partial (\rho c_p T)}{\partial t} \tag{9.20}$$

where M = mass (kg)
c_p = specific heat (J/kg-°K)
T = temperature (°K)

and the flow term is:

$$-A_c \Delta z \frac{\partial (\rho v c_p T)}{\partial z} \tag{9.21}$$

The term which is a result of molecular heat transport can be found with Fourier's law:

$$\phi_H = -k_t \frac{\partial T}{\partial z} \ \text{W/m}^2 \tag{9.22}$$

where k_t = heat conduction coefficient (W/m-°K)
ϕ_H = heat flux (W/m^2)

Analogous to Equations 9.15, 9.16 and 9.17, the term due to molecular transport becomes:

$$\frac{\partial}{\partial z}\left(k_t \frac{\partial T}{\partial z}\right) A_c \Delta z \tag{9.23}$$

Under turbulent flow, k_t represents the turbulent heat transfer coefficient. Finally, heat can appear or disappear, e.g., as a result of chemical reaction. The corresponding term is:

$$Q(z,t) = R(z,t) A_c \Delta z \Delta H_R \tag{9.24}$$

where ΔH_R = heat of reaction (J/kg).

The total energy balance now becomes:

$$\frac{\partial(\rho c_p T)}{\partial t} = R(z,t)\Delta H_R - \frac{\partial(\rho v c_p T)}{\partial z} + \frac{\partial}{\partial z}\left(k_t \frac{\partial T}{\partial z}\right) \tag{9.25}$$

Momentum Balance

When there is a pressure gradient in the z direction, a force is acting on the volume element equal to:

$$A_c(P_z - P_{z+\Delta z}) \tag{9.26}$$

The expression in Equation 9.26 can be written for small Δz as:

$$-A_c \Delta z \frac{\partial P}{\partial z} \tag{9.27}$$

It is possible that gravity forces also act on the volume element. This term, which is equal to mass times acceleration, becomes:

$$\rho g_z A_c \Delta z \tag{9.28}$$

where g_z = acceleration of gravity in the z direction (m/sec^2)

Hence, the momentum production is equal to the sum of pressure and gravity forces:

$$A_c \Delta z \left(-\frac{\partial P}{\partial z} + \rho g_z\right) \tag{9.29}$$

The first term in this expression has a positive contribution since the pressure usually decreases in positive z direction. The mass in the volume element $A_c\Delta z$ is equal to $\rho A_c \Delta z$. The momentum, which is equal to mass times velocity, is:

$$\rho v A_c \Delta z \tag{9.30}$$

The momentum accumulation can now be written as:

$$A_c \Delta z \frac{\partial(\rho v)}{\partial t} \tag{9.31}$$

The term ρv is sometimes called the "momentum concentration." Analogous to Equations 9.9 and 9.14, the momentum transport due to flow is given by:

$$-A_c \Delta z \frac{\partial(\rho v^2)}{\partial z} \tag{9.32}$$

if the concentration is replaced by the product of density and velocity. In a maximum gradient model, friction is usually represented by an algebraic term, based on an empirical relationship:

$$\Delta p_f = -\mu \frac{\partial v}{\partial z} \; N/m^2 \tag{9.33}$$

where μ = viscosity (kg/m-sec)

The momentum transport in the volume element $A_c\Delta z$ can in the limit case be represented by:

$$\frac{\partial}{\partial z}\left(\mu \frac{\partial v}{\partial z}\right) A_c \Delta z \tag{9.34}$$

in a similar way as the mass transport in Equation 9.17 and the energy transport in 9.23. Combination of Equations 9.29, 9.31, 9.32 and 9.34 finally gives the momentum balance:

$$\frac{\partial(\rho v)}{\partial t} = -\frac{\partial P}{\partial z} + \rho g_z - \frac{\partial(\rho v^2)}{\partial z} + \frac{\partial}{\partial z}\left(\mu \frac{\partial v}{\partial z}\right) \tag{9.35}$$

Equations 9.8, 9.18, 9.25 and 9.35 are summarized in Table 9.1. Often D, λ and μ are assumed to be constant, resulting in a simplification of the equations.

ADDITIONAL EQUATIONS

At this stage it is useful to compare the number of dependent variables with the number of equations. Evidently one has to check if the equations are independent. Often the result is that the number of variables exceeds the number of equations. Then additional equations have to be formulated to construct a model which can be solved. One can think of physical laws, such as the gas law, empirical relations or kinetic expressions.

INITIAL AND BOUNDARY CONDITIONS

No mathematical model is complete without definition of initial and boundary conditions. Initial conditions have to be defined for the dependent

Table 9.1 Equations for the Change of Mass, Concentration, Thermal Energy and Momentum

Accumulation	Change as a Result of Production	Change as a Result of Flow	Change as a Result of Molecular Transport
$\frac{\partial \rho}{\partial t}$		$-\frac{\partial(\rho v)}{\partial z}$	
$\frac{\partial c_A}{\partial t}$	$R(z,t)$	$-\frac{\partial(v c_A)}{\partial z}$	$\frac{\partial}{\partial z}\left(D_A \frac{\partial c_A}{\partial z}\right)$
$\frac{\partial(\rho c_p T)}{\partial t}$	$R(z,t)\Delta H_R$	$-\frac{\partial(\rho v c_p T)}{\partial z}$	$\frac{\partial}{\partial z}\left(k_t \frac{\partial T}{\partial z}\right)$
$\frac{\partial(\rho v)}{\partial t}$	$-\frac{\partial P}{\partial z} + \rho g_z$	$-\frac{\partial(\rho v^2)}{\partial z}$	$\frac{\partial}{\partial z}\left(\mu \frac{\partial v}{\partial z}\right)$

variables as a function of the geometric independent variables. Boundary conditions are a function of time. This will be illustrated later by means of some examples. The order of the system of differential equations determines the number of initial and boundary conditions. For example, a partial differential equation with a second derivative with respect to the geometric coordinate together with a partial differential equation with a first derivative to the geometric coordinate requires three boundary conditions.

Example

Consider a tubular reactor in which a first-order reaction occurs under isothermal conditions:

$$A \xrightarrow{k} B \tag{9.36}$$

One may assume constant fluid velocity and negligible axial dispersion. The different terms of the partial mass balance from Table 9.1 become:

$$R(z,t) = -kc_A \tag{9.37}$$

$$\frac{\partial}{\partial z}(vc_A) = v\,\frac{\partial c_A}{\partial z} \tag{9.38}$$

$$\frac{\partial}{\partial z}\left(D_A\,\frac{\partial c_A}{\partial z}\right) = 0 \tag{9.39}$$

thus

$$\frac{\partial c_A}{\partial t} + v\,\frac{\partial c_A}{\partial z} + kc_A = 0 \tag{9.40}$$

The initial condition is:

$$t = 0:\quad c_A = c_{A0}(z) \tag{9.41}$$

and the boundary condition:

$$z = 0:\quad c_A = c_{Ai}(t) \tag{9.42}$$

SUBSTITUTION OF ALGEBRAIC EQUATIONS INTO PARTIAL DIFFERENTIAL EQUATIONS

In many cases it is possible to substitute the algebraic equations into the partial differential equations. These have the following form:

$$\phi\left[\frac{\partial^2 x(z,t)}{\partial z^2}, \frac{\partial x(z,t)}{\partial z}, \frac{\partial x(z,t)}{\partial t}, x(z,t), u(t), w(t)\right] = 0 \qquad (9.43)$$

where x = the dependent variable
z = the geometric coordinate
u = the correcting condition
w = the disturbance variable

When ϕ and x are interpreted as vectors, Equation 9.43 represents a set of partial differential equations. The second derivative usually indicates dispersion, the first derivative indicates flow, u and w are important for control; u can be used to eliminate or compensate the influence of w on x. In exceptional cases, u and w have a distributed character.

STATIC SOLUTION

Often the dynamic model can be made more convenient by substitution of the static model. The latter is found from the former by putting partial derivatives with respect to time equal to zero. The result is a (set of) ordinary differential equation(s):

$$\overline{\phi}\left[\frac{d^2\overline{x}(z)}{dz^2}, \frac{d\overline{x}(z)}{dz}, \overline{x}(z), \overline{u},\overline{w}\right] = 0 \qquad (9.44)$$

LINEARIZATION

When variations around the steady state are not too large, linearization is allowed. This results in the following differential equation(s):

$$B_1 \frac{\partial^2 \delta x(z,t)}{\partial z^2} + B_2 \frac{\partial \delta x(z,t)}{\partial z} + B_3 \frac{\partial \delta x(z,t)}{\partial t}$$

$$+ B_4\, \delta x(z,t) + B_5\, \delta u(t) + B_6\, \delta w(t) = 0 \qquad (9.45)$$

where x is a vector, B_1-B_4 are matrices, and B_5 and B_6 are column vectors (if u and w are scalar) or matrices too (if u and w are vectors).

OPERATIONAL NOTATION

Just as has been done in previous chapters, derivatives with respect to time will be replaced by the Laplace operator s. This is only correct when the initial conditions are equal to zero (the system is initially in equilibrium). In this way, time is eliminated as an independent variable, and the differential equations only contain derivatives with respect to the geometric coordinate.

After transformation, Equation 9.45 can be written as:

$$B_1 \frac{d^2\delta x(z,s)}{dz^2} + B_2 \frac{d\delta x(z,s)}{dz} + (B_3 s + B_4)\delta x(z,s) + B_5 \delta u(s) + B_6 \delta w(s) = 0 \tag{9.46}$$

If the general Laplace transformation is applied, the right side of this equation also contains the initial condition(s) $-B_3$ x $(z,0^+)$ (see Appendix I for more details).

SOLUTION OF THE DEPENDENT VARIABLE

The solution of Equation 9.46 with the aid of the boundary conditions results in the equation:

$$\delta x(z,s) = F_1(z,s)\delta x(0^+) + F_2(z,s)\delta u(s) + F_3(z,s)\delta w(s) \tag{9.47}$$

The solution is simple when the coefficients B_1 to B_6 are independent of z. However, if they are dependent on z, one can try to convert the equations to a certain standard expression (for example Bessel differential equations) for which tabulated solutions are available, or one can assume average values.

EXAMPLE

The previously discussed approach will be applied to heat conduction in a semiinfinite rod. Semiinfinite means that only one end is considered;

the other end is infinitely far away. Consequently, temperature fluctuations at the one end have no influence on the temperature at the other end.

Using Table 9.1 it is evident that there are no terms due to production and flow, because heat transport occurs only as a result of molecular transport. The model equation is then:

$$\frac{\partial(\rho c_p T)}{\partial t} = \frac{\partial}{\partial z}\left(k_t \frac{\partial T}{\partial z}\right) \tag{9.49}$$

When it is assumed that ρ, c_p and k_t are constant, Equation 9.49 can be written as:

$$\frac{\partial T}{\partial t} = a \frac{\partial^2 T}{\partial z^2} \tag{9.50}$$

where $a = k_t/c_p\rho$, a temperature adjustment coefficient

The boundary and initial conditions are:

$$\left.\begin{aligned} &t = 0 &&: \quad T(z) = T_0 \ , \quad \text{the initial temperature} \\ &z = 0 \ , \quad t > 0 &&: \quad T = T_1 \ , \quad \text{a step change} \\ &z = L \ , \quad t > 0 &&: \quad T = T_0 \end{aligned}\right\} \tag{9.51}$$

Note that the model is already linear.

By substracting T_0 from all temperatures, the initial condition corresponds to:

$$T^* = T - T_0 = 0 \tag{9.52}$$

Introduction of operator s transforms Equation 9.50 into:

$$a \frac{d^2 T^*}{dz^2} - sT^* = 0 \tag{9.53}$$

Similarly, the boundary conditions become:

$$\left.\begin{aligned} &z = 0 &&: \quad T^* = T_1^*(s) \\ &z = L \to \infty &&: \quad T^* = 0 \end{aligned}\right\} \tag{9.54}$$

where $T_1^*(s)$ = the transform of $T_1 - T_0$

The general solution of Equation 9.53 is:

$$T^* = Ae^{-pz} + Be^{pz} \tag{9.55}$$

where

$$p = \sqrt{s/a} \tag{9.56}$$

The general solution can easily be verified by substitution into Equation 9.53. Substitution of the boundary conditions (Equation 9.54) into Equation 9.55 yields:

$$T_1^* = A + B \tag{9.57}$$

$$0 = Ae^{-pL} + Be^{pL} \tag{9.58}$$

Since Equation 9.58 must apply for L approaching infinity; it can only be satisfied if B = 0. Equation 9.57 yields:

$$A = T_1^* \tag{9.59}$$

The final solution then becomes:

$$\frac{T^*(s)}{T_1^*(s)} = e^{-\sqrt{s/a}\cdot z} \tag{9.60}$$

When a step is introduced at z = 0, the Laplace transform of $T_1 - T_0$ will be $T_1 - T_0/s$, thus Equation 9.60 becomes:

$$T - T_0 = \frac{T_1 - T_0}{s}\, e^{-\sqrt{s/a}\cdot z} \tag{9.61}$$

With a Laplace transform table, Equation 9.61 can be transformed to the time domain:

$$\begin{aligned} T &= T_0 + (T_1 - T_0)\mathrm{erfc}(z/2\sqrt{at}) \\ &= T_1 - (T_1 - T_0)\mathrm{erf}(z/2\sqrt{at}) \end{aligned} \tag{9.62}$$

where erf is the error function, tabulated in Table 9.2. In Figure 9.2 the temperature profiles are given for a steel rod with $a = 1.42 \times 10^{-5}\ m^2/sec$, at times equal to 10.50 and 200 sec, respectively.

Table 9.2 Tabulated Error Function

z	erf (z)	z	erf (z)
0	0	0.60	0.604
0.05	0.057	0.70	0.678
0.10	0.113	0.80	0.742
0.15	0.168	0.90	0.797
0.20	0.223	1.0	0.843
0.25	0.276	1.2	0.910
0.30	0.329	1.4	0.952
0.35	0.380	1.6	0.976
0.40	0.428	1.8	0.989
0.45	0.478	2.0	0.995
0.50	0.521	3.0	0.999
		∞	1.000

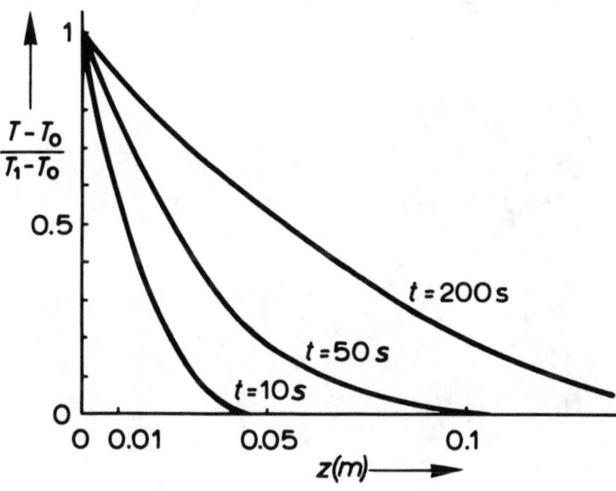

Figure 2. Temperature profiles in a steel rod.

CHAPTER 10

DISTRIBUTED MODEL OF A HEAT EXCHANGER

A heat exchanger is used to heat or cool a process stream. Many different types of heat exchangers exist. One type that is frequently used in industry is the one where liquid is heated and where condensing steam supplies the required heat. This chapter will be restricted to that type (Figure 10.1). The liquid flows through some tubes in one direction and through other tubes in the reverse direction. Steam condenses at the outer side of the tubes.

ASSUMPTIONS

If we were to make a detailed description of the process, the mathematical model would become very complicated. The disadvantage of very complicated models is that they can only be solved with the aid of a computer,

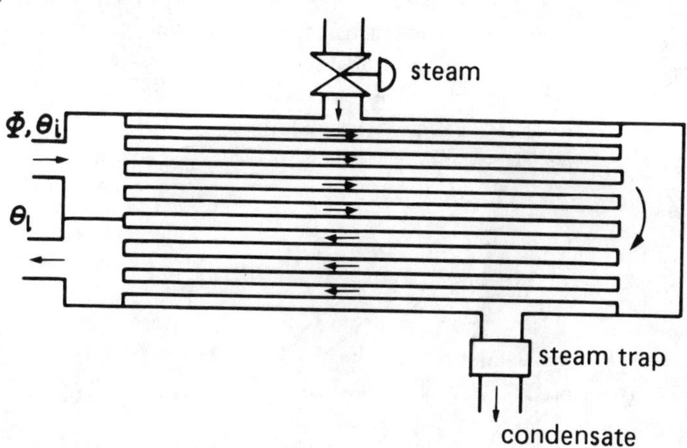

Figure 10.1 Heat exchanger.

and the results are often hard to interpret. On the other hand, the model must not be too simple, because an incomplete description of the process would be obtained. An intelligent engineer would keep his model as simple as possible, and compare the results with the experiments which have been done in practice. Differences between practice and model results can be an indication to which direction the model must be extended. The following assumptions will be made:

1. The dynamics on the steam side will be ignored, which means that the temperature of the condensing steam reacts instantaneously to changes in the steam supply. It is also assumed that the temperature of the condensing steam is the same in all locations.
2. The heat transfer coefficient of the steam to the tube wall is constant.
3. The radial heat conduction through the tube wall is ideal; the axial heat conduction can be ignored.
4. The heat transfer coefficient from the tube wall to the liquid depends only on the liquid velocity.
5. The liquid flow is plug flow with ideal mixing over the cross-sectional area. This means that the liquid temperature is the same over the cross-sectional area of the tube. This assumption is necessary to arrive at a maximum gradient description.
6. The specific heat and density of the liquid are constant.
7. The heat capacities of the shell and headings of the heat exchanger are not taken into account. An indication of the effect of this assumption on the dynamic behavior is given by Harriott [23].
8. The time delay of the liquid in the heat exchanger heading can be ignored.
9. There is no evaporation of the liquid.

MODEL EQUATIONS

When analyzing a process in which heat effects play a role, energy balances must be set up. As it is assumed that the liquid is incompressible, the mass balance gives irrelevant information. A partial mass balance does not play a role because no concentration differences exist. We are not interested in the momentum balance because we are not interested in the propagation of pressure and velocity fluctuations. This means that the model only consists of an energy balance for the tube wall and one for the liquid. For a section with length Δz, the energy balance for the tube wall is:

$$M_w c_w \Delta z \frac{\partial \theta_w}{\partial t} = h_s A_s \Delta z(\theta_s - \theta_w) - h_f A_f \Delta z(\theta_w - \theta) \qquad (10.1)$$

where M_w = mass of the tubes per unit of length (kg/m)
c_w = specific heat of the wall (J/kg-°K)
θ_w = wall temperature (°K)
h_s = heat transfer coefficient at the steam side (W/m^2-°K)

A_s = outside area of the tubes per unit of length (m^2/m)
h_f = heat transfer coefficient at the liquid side (W/m^2-°K)
A_f = inside area of the tubes per unit of length (m^2/m)
θ = liquid temperature (°K)

The term in the left side of Equation 10.1 represents the accumulation of energy; the first term in the right side is the heat transfer of the condensing steam to the tube wall; and the second term is the heat transfer from the wall to the flowing liquid.

The energy balance for the liquid is:

$$M_f c_f \Delta z \frac{\partial \theta}{\partial t} + \phi c_f \Delta z \frac{\partial \theta}{\partial z} = h_f A_f(\theta_w - \theta)\Delta z \qquad (10.2)$$

where M_f = liquid inventory per unit of length (kg/m)
c_f = specific heat of the liquid (J/kg-°K)
ϕ = liquid flow (kg/sec)

the first term represents the accumulation of energy; the second term, the transport of energy due to flow; and the third term, the flow of energy from tube wall to liquid.

The following time constants are defined:

$$T_{ff} = \frac{M_f c_f}{h_f A_f} \qquad (10.3)$$

$$T_{ws} = \frac{M_w c_w}{h_s A_s} \qquad (10.4)$$

$$T_{wf} = \frac{M_w c_w}{h_f A_f} \qquad (10.5)$$

The velocity of the liquid follows from:

$$v = \frac{\phi}{M_f} \qquad (10.6)$$

By using Equations 10.3 to 10.6, Equations 10.1 and 10.2 can be simplified to:

$$T_{ws} \frac{\partial \theta_w}{\partial t} = \theta_s - \theta_w - \frac{T_{ws}}{T_{wf}} (\theta_w - \theta) \qquad (10.7)$$

and

$$T_{ff}\frac{\partial\theta}{\partial t} + vT_{ff}\frac{\partial\theta}{\partial z} = \theta_w - \theta \tag{10.8}$$

To complete the model, two initial conditions are needed: one for θ and one for θ_w, and a boundary condition for θ. This will be discussed later.

STATIC MODEL

The static model can be found by setting the derivatives with respect to time equal to zero. Equation 10.7 then becomes:

$$\bar{\theta}_s - \bar{\theta}_w = \frac{\bar{T}_{ws}}{\bar{T}_{wf}}(\bar{\theta}_w - \bar{\theta}) \tag{10.9}$$

from which:

$$\bar{\theta}_w = \frac{\bar{T}_{wf}\bar{\theta}_s + \bar{T}_{ws}\bar{\theta}}{\bar{T}_{wf} + \bar{T}_{ws}} \tag{10.10}$$

The bar over a variable indicates the static value. In Equation 10.10, $\bar{\theta}_w$ is a weighted average between $\bar{\theta}_s$ and $\bar{\theta}$, in which the heat transfer conductances h_fA_f and h_sA_s are factors. This is clear after substitution of Equation 10.4 and 10.5 into Equation 10.10:

$$\bar{\theta}_w = \frac{\bar{h}_sA_s\theta_s + \bar{h}_fA_f\bar{\theta}}{\bar{h}_sA_s + \bar{h}_fA_f} \tag{10.11}$$

In the static situation, Equation 10.8 is:

$$\overline{vT}_{ff}\frac{d\bar{\theta}}{dz} = \bar{\theta}_w - \bar{\theta} \tag{10.12}$$

Substitution of Equation 10.11 into Equation 10.12 gives:

$$\overline{vT}_f\frac{d\bar{\theta}}{dz} = \bar{\theta}_s - \bar{\theta} \tag{10.13}$$

in which

$$\overline{T}_f = M_f c_f [(\overline{h}_f A_f)^{-1} + (\overline{h}_s A_s)^{-1}] \qquad (10.14)$$

Equation 10.13 is a first-order linear differential equation which can easily be solved. When Equation 10.13 is written as:

$$\overline{vT}_f \frac{d\overline{\theta}}{dz} + \overline{\theta} = \overline{\theta}_s \qquad (10.15)$$

it can be seen that the solution is:

$$\overline{\theta} = c_1 e^{-c_2 z} + \overline{\theta}_s \qquad (10.16)$$

with

$$c_2 = \frac{1}{\overline{vT}_f} \qquad (10.17)$$

The value of the constant c_1 can be found from the boundary condition for $z = 0$. Let the inlet liquid temperature be equal to $\overline{\theta}_i$:

$$\overline{\theta}(0) = \overline{\theta}_i \ , \quad z = 0 \qquad (10.18)$$

resulting in:

$$c_1 = \overline{\theta}_i - \overline{\theta}_s \qquad (10.19)$$

from combination of Equations 10.16 and 10.18, Equation 10.16 finally becomes:

$$\overline{\theta} = \overline{\theta}_s - (\overline{\theta}_s - \overline{\theta}_i) e^{-z/\overline{vT}_f} \qquad (10.20)$$

Substitution of Equation 10.20 into Equation 10.11 gives for the tube temperature $\overline{\theta}_w$:

$$\overline{\theta}_w = \overline{\theta}_s - \frac{\overline{h}_f A_f}{\overline{h}_s A_s + \overline{h}_f A_f} (\overline{\theta}_s - \overline{\theta}_i) e^{-z/\overline{vT}_f} \qquad (10.21)$$

The outlet temperature of the liquid $\overline{\theta}_u$ for $z = \ell$ can be obtained from Equation 10.20:

$$\overline{\theta}_u = \overline{\theta}_s - (\overline{\theta}_s - \overline{\theta}_i) e^{-L/\overline{T}_f} \qquad (10.22)$$

in which $L = \ell/\bar{v}$, the time delay of the liquid in the tubes. In Figure 10.2, Equation 10.20 is given for $\bar{v} = 1$ m/sec, $\ell = 10$ m, $\bar{T}_f = 8.33$, a steam temperature of 400°K and an inlet temperature of 300°K.

LINEARIZATION

It is assumed that variations are present in the inlet temperature of the liquid, steam temperature and liquid flow. The most important output variation is the one in the exit liquid temperature. Figure 10.3 shows the information flow diagram. In the block, both state variables $\delta\theta$ and $\delta\theta_w$ are indicated.

The heat exchanger can be controlled automatically by adjustment of ϕ or θ_s (via the control valve in the steam supply). Figure 10.4 shows the

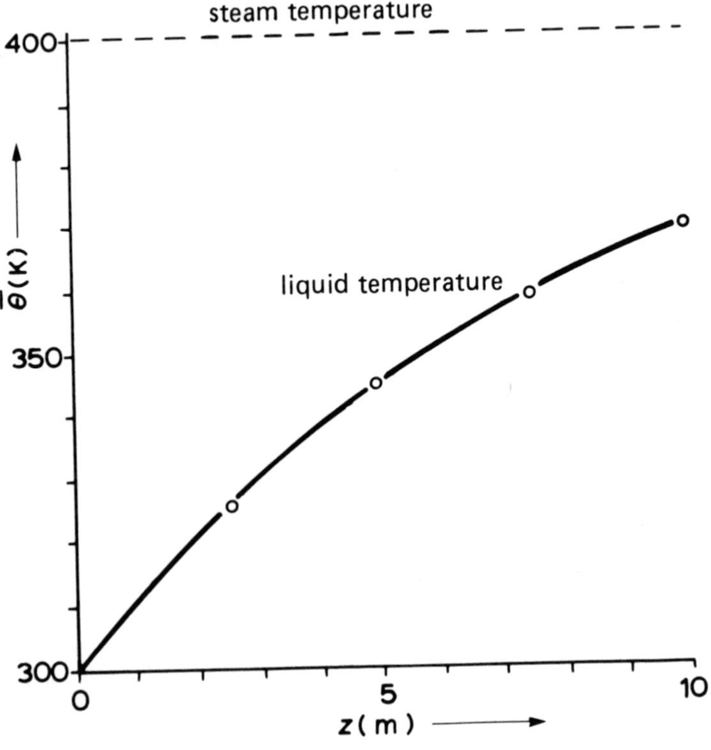

Figure 10.2 Static temperature profile of the liquid.

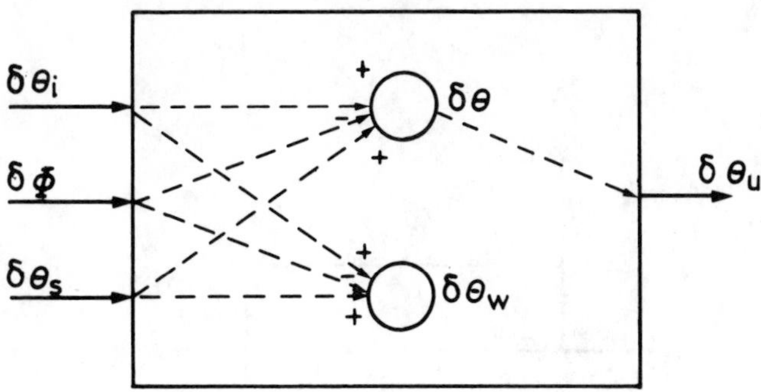

Figure 10.3 Information flow diagram of the heat exchanger.

latter possibility. Here $(\delta\theta_u)_m$ is the variation in the measured value of the outlet temperature, $(\delta\theta_u)_i$ is the variation in the set value (for control without systematic errors this corresponds to the desired value of the outlet temperature) and δP_s is the variation in the supply pressure of the steam. The block "correcting unit" contains the dynamics of the control valve and of the steam volume of the heat exchanger. The pressure drop across the control valve is influenced by the steam flow and, therefore, by the heat transfer Q (this is roughly a first-order system with a time constant that is determined by the resistance of the steam control valve and the total heat capacity of the steam side of the heat exchanger). We shall not go into further detail on control, but only analyze the dynamics of the heat exchanger. First, the model will be simplified by neglecting the heat capacity of the tube wall. Later the influence of this simplification on the temperature profile will be analyzed. Equation 10.7 now reduces to an algebraic equation:

$$\theta_s - \theta_w - \frac{T_{ws}}{T_{wf}} (\theta_w - \theta) = 0 \qquad (10.23)$$

Substitution of Equation 10.23 into Equation 10.8 results in:

$$T_f \frac{\partial\theta}{\partial t} + vT_f \frac{\partial\theta}{\partial z} = \theta_s - \theta \qquad (10.24)$$

which resembles the static Equation 10.13.

The process variables are now represented by the sum of their average values and a small variation:

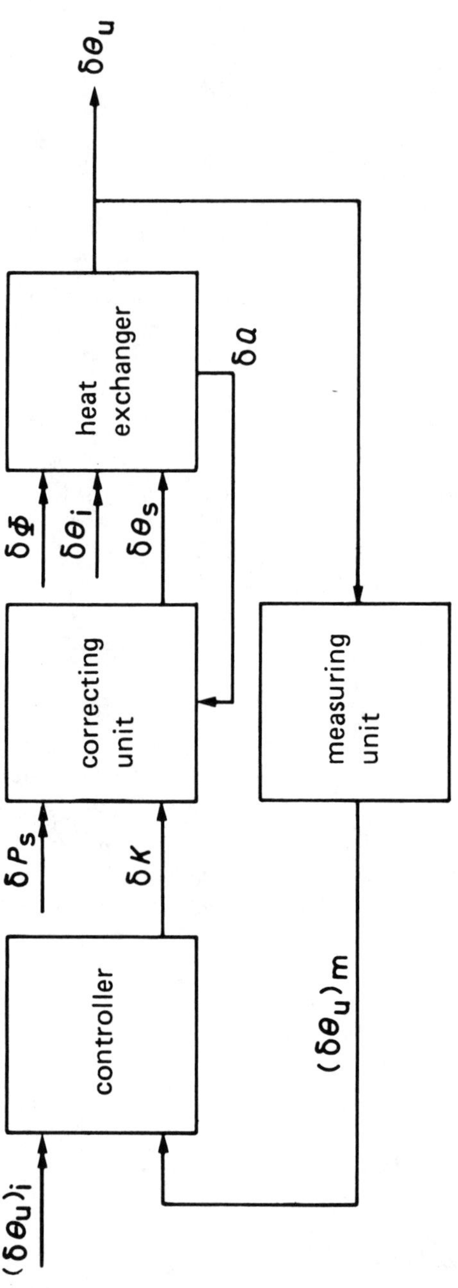

Figure 10.4 Information flow diagram of the control loop.

$$\theta = \bar{\theta} + \delta\theta$$

$$v = \bar{v} + \delta v \tag{10.25}$$

$$h_f = \bar{h}_f + \delta h_f$$

Substitution of Equation 10.25 into Equation 10.24 gives:

$$\frac{\partial(\bar{\theta} + \delta\theta)}{\partial t} + (\bar{v} + \delta v)\frac{\partial(\bar{\theta} + \delta\theta)}{\partial z} = (\bar{T}_f^{-1} + \delta T_f^{-1})(\bar{\theta}_s + \delta\theta_s - \bar{\theta} - \delta\theta) \tag{10.26}$$

in which

$$\bar{T}_f^{-1} + \delta T_f^{-1} = \frac{1}{M_f c_f}\,\frac{(\bar{h}_f + \delta h_f)A_f h_s A_s}{(\bar{h}_f + \delta h_f)A_f + h_s A_s} \tag{10.27}$$

The variation in the heat transfer coefficient h_f is the result of a variation in the fluid velocity v. According to the Nusselt equation, the film coefficient depends on the Reynolds number and hence on the velocity [24]:

$$h = c_1 v^{0.8} \tag{10.28}$$

For small variations, Equation 10.28 becomes:

$$\delta h/\bar{h} = 0.8\,\delta v/\bar{v} \tag{10.29}$$

The film coefficient h_f, however, is smaller than h due to the presence of a deposit on the tube wall:

$$h_f^{-1} = h^{-1} + h_{deposit}^{-1} \tag{10.30}$$

Linearization of this equation gives:

$$-\frac{\delta h_f}{\bar{h}_f^2} = -\frac{\delta h}{\bar{h}^2} \tag{10.31}$$

or

$$\frac{\delta h_f}{\bar{h}_f} = \left(\frac{\bar{h}_f}{\bar{h}}\right)\frac{\delta h}{\bar{h}} \tag{10.32}$$

Combination with Equation 10.29 gives:

$$\frac{\delta h_f}{\bar{h}_f} = 0.8 \cdot \frac{\bar{h}_f}{\bar{h}} \cdot \frac{\delta v}{\bar{v}} \tag{10.33}$$

After expanding Equation 10.26, neglecting higher-order terms in δ, subtracting the static Equation 10.3 and substituting Equation 10.33, we find:

$$\frac{\partial(\delta\theta)}{\partial t} + \bar{v}\,\frac{\partial(\delta\theta)}{\partial z} + \delta v\,\frac{\partial\bar{\theta}}{\partial z} = \bar{T}_f^{-1}(\delta\theta_s - \delta\theta) + \beta\bar{T}_f^{-1}(\bar{\theta}_s - \bar{\theta})\,\frac{\delta v}{\bar{v}} \tag{10.34}$$

in which

$$\beta = \frac{A_s h_s}{A_s h_s + A_f \bar{h}_f}\left(0.8\,\frac{\bar{h}_f}{\bar{h}}\right) \tag{10.35}$$

Equation 10.35 can also be written as:

$$\beta = 0.8\,\frac{(A_f\bar{h})^{-1}}{(A_f\bar{h}_f)^{-1} + (A_s h_s)^{-1}} \tag{10.36}$$

This can be interpreted as the ratio of the velocity-sensitive heat transfer resistance $A_f h$ to the total heat transfer resistance, multiplied by the exponent in the Nusselt equation. Substitution of Equations 10.13 and 10.20 into Equation 10.34 gives:

$$\frac{\partial(\delta\theta)}{\partial t} + \bar{v}\,\frac{\partial(\delta\theta)}{\partial z} + \bar{T}_f^{-1}\delta\theta = \bar{T}_f^{-1}\delta\theta_s - (1-\beta)\bar{T}_f^{-1}(\bar{\theta}_s - \bar{\theta}_i)e^{-z/v\bar{T}_f}\,\frac{\delta v}{\bar{v}} \tag{10.37}$$

OPERATIONAL NOTATION

Introducing the operator s into Equation 10.37 yields:

$$s\delta\theta + \bar{v}\,\frac{d(\delta\theta)}{dz} + \bar{T}_f^{-1}\delta\theta = \bar{T}_f^{-1}\delta\theta_s - (1-\beta)\bar{T}_f^{-1}(\bar{\theta}_s - \bar{\theta}_i)e^{-z/\overline{vT}_f}\,\frac{\delta v}{\bar{v}} \tag{10.38}$$

Equation 10.38 is an ordinary first-order differential equation for $\delta\theta$ and can be solved in the usual way. The general solution of Equation 10.38 can be found by setting the left side equal to zero. This solution is:

$$\delta\theta_{general} = A_1 e^{-(s-\bar{T}_f^{-1})z/\bar{v}} \tag{10.39}$$

The particular solution belonging to the right side of Equation 10.38 is of the form:

$$\delta\theta_{particular} = A_2\delta\theta_s + A_3 e^{-z/\bar{v}\bar{T}_f} \cdot \frac{\delta v}{\bar{v}} \tag{10.40}$$

Substitution of Equation 10.40 into Equation 10.38 gives the following conditions for A_2 and A_3:

$$A_2 = \frac{\bar{T}_f^{-1}}{s + \bar{T}_f^{-1}} \tag{10.41}$$

$$A_3 = -\frac{(1-\beta)\bar{T}_f^{-1}}{s}(\bar{\theta}_s - \bar{\theta}_i) \tag{10.42}$$

Combination of Equations 10.39 to 10.42 gives for the total solution:

$$\delta\theta(z,s) = A_1 e^{-(s+\bar{T}_f^{-1})z/\bar{v}} + \frac{\bar{T}_f^{-1}}{s + \bar{T}_f^{-1}}\delta\theta_s - \frac{(1-\beta)\bar{T}_f^{-1}}{s}(\bar{\theta}_s - \bar{\theta}_i)e^{-z/\bar{v}\bar{T}_f} \cdot \frac{\delta v}{\bar{v}} \tag{10.43}$$

The paramater A_1 can be determined from the boundary condition at $z = 0$:

$$\delta\theta = \delta\theta(0,s) \tag{10.44}$$

Substitution of $z = 0$ into Equation 10.43 gives:

$$\delta\theta = \delta\theta(0,s) = A_1 + \frac{\bar{T}_f^{-1}}{s + \bar{T}_f^{-1}}\delta\theta_s - \frac{(1-\beta)\bar{T}_f^{-1}}{s}(\bar{\theta}_s - \bar{\theta}_i)\frac{\delta v}{\bar{v}} \tag{10.45}$$

from which

$$A_1 = \delta\theta(0,s) - \frac{1}{1 + s\bar{T}_f}\delta\theta_s + \frac{(1-\beta)}{s\bar{T}_f}(\bar{\theta}_s - \bar{\theta}_i)\frac{\delta v}{\bar{v}} \tag{10.46}$$

Substitution of Equation 10.46 into Equation 10.43 results in:

$$\delta\theta(z,s) = \delta\theta(0,s)e^{-z/\bar{v}\bar{T}_f}\,e^{-sz/\bar{v}} + \frac{\delta\theta_s}{1 + s\bar{T}_f}\left[1 - e^{-z/\bar{v}\bar{T}_f}e^{-sz/\bar{v}}\right]$$

$$- \frac{\delta v}{\bar{v}}\frac{(1-\beta)}{s\bar{T}_f}(\bar{\theta}_s - \bar{\theta}_i)e^{-z/\bar{v}\bar{T}_f}(1 - e^{-sz/\bar{v}}) \tag{10.47}$$

This equation shows the three most important transfer functions of the heat exchanger: from the inlet and steam temperatures, and the liquid velocity to the outlet temperature.

SUBSTITUTION OF THE STATIC EQUATION

The result in Equation 10.47 can be made more suitable for practice. The term $e^{-z/\bar{v}\bar{T}_f}$ in this equation can be eliminated with the aid of Equation 10.22 for $z = \ell$. The outlet responses are:

$$\delta\theta_u = \left(\frac{\delta\theta_u}{\delta\theta_i}\right)\delta\theta_i + \left(\frac{\delta\theta_u}{\delta\theta_s}\right)\delta\theta_s + \left(\frac{\delta\theta_u}{\delta v/\bar{v}}\right)\frac{\delta v}{\bar{v}} \tag{10.48}$$

with

$$\left(\frac{\delta\theta_u}{\delta\theta_s}\right) = \frac{\bar{\theta}_s - \bar{\theta}_u}{\bar{\theta}_s - \bar{\theta}_i}\, e^{-sL} \tag{10.49}$$

$$\left(\frac{\delta\theta_u}{\delta\theta_s}\right) = \frac{1}{1 + s\bar{T}_f}\left(1 - \frac{\bar{\theta}_s - \bar{\theta}_u}{\bar{\theta}_s - \bar{\theta}_i}\, e^{-sL}\right) \tag{10.50}$$

$$\left(\frac{\delta\theta_u}{\delta v/\bar{v}}\right) = -(1 - \beta)\frac{L}{\bar{T}_f}\,(\bar{\theta}_s - \bar{\theta}_u)\,\frac{1 - e^{-sL}}{sL} \tag{10.51}$$

in which the ratio $L/\bar{T}_f$ can be determined from the static value of the temperatures with the aid of Equation 10.22:

$$\frac{L}{\bar{T}_f} = ln\,\frac{\bar{\theta}_s - \bar{\theta}_i}{\bar{\theta}_s - \bar{\theta}_u} \tag{10.52}$$

and

$$\delta\theta_i = \delta\theta(0,s) \;\; ; \;\; \delta\theta_u = \delta\theta(\ell,s) \tag{10.53}$$

These transfer functions can be calculated directly from inlet and outlet temperatures, the residence time of the liquid in the tubes, and an estimated value of β. Note that the steam temperature corresponds to the dew point temperature as the effects of overheating have been neglected.

INTERPRETATION

It is worthwhile to pay some attention to the physical meaning of these formulas. Equation 10.49 has a dead time character; variations in the liquid temperature propagate through the tubes while being attenuated. This attenuation can be determined directly from average temperatures.

The response to variations of the steam temperature (see Equation 10.50) contains a first-order in cascade with the difference between a direct and a delayed transfer function (Figure 10.5). The step response of the right part of Figure 10.5 is shown in Figure 10.6. If the liquid in the tubes would be ideally mixed, the total response would have a first-order character. Actually, the response looks very much like a first-order despite the distributed character. This is related to the fact that variations in the steam temperature influence the liquid temperature profile over the whole length of the tube wall at the same time.

The time delay effect is caused by the inlet temperature of the liquid, which does not follow the steam temperature. The transfer function for variations in the velocity (see Equation 10.51) can be considered as an integrator with the difference between a direct and delayed response (Figure 10.7). The step response has the form shown in Figure 10.8.

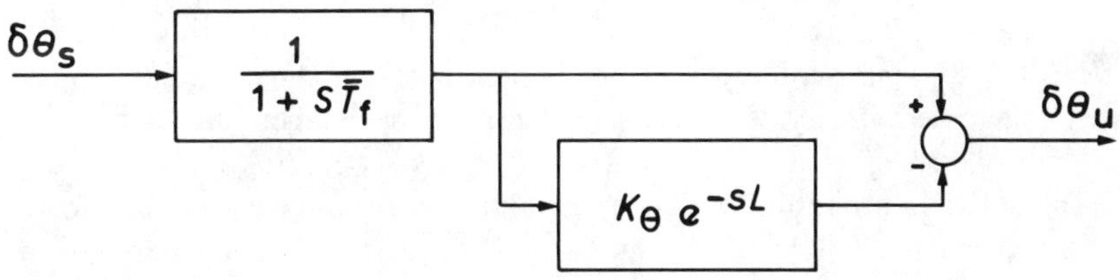

Figure 10.5 Subdivision of the transfer function for temperature variations.

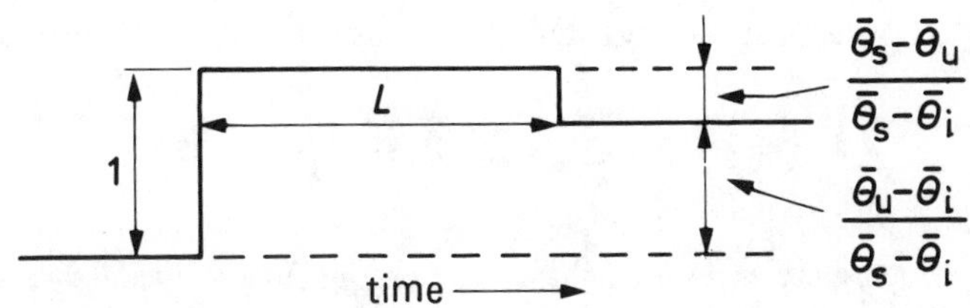

Figure 10.6 Step response of the right part of Figure 10.5.

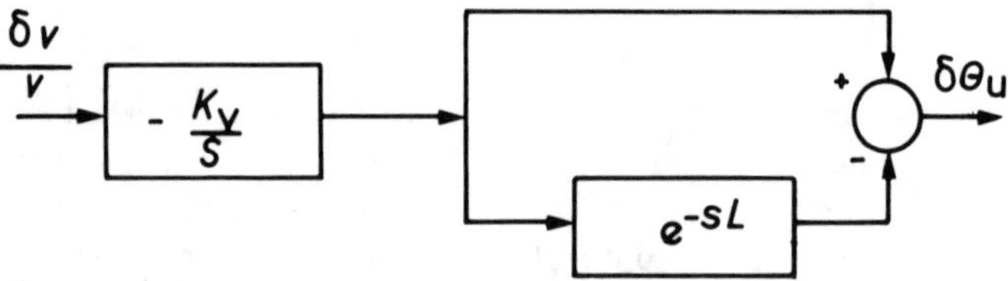

Figure 10.7 Subdivision of the transfer function for velocity variations.

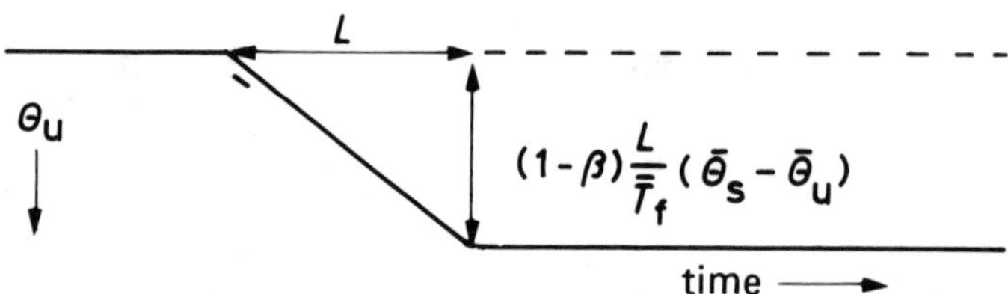

Figure 10.8 Response of the outlet temperature to a step in the liquid velocity.

It can be seen that there is a certain similarity between the transfer functions for velocity variations and for steam temperature variations; in both cases the temperature profile is influenced at the same time over the total length. The influence of the increase of the heat transfer when increasing the velocity is purely proportional: the temperature decrease is $1 - \beta$ smaller.

The static value of the outlet temperature to a step change in the liquid velocity can be determined with the final value theorem, as discussed in the Appendix.

$$\lim_{t\to\infty} \delta\theta(\ell,t) = \delta\theta_u(\infty) = \lim_{s\to 0} s\delta\theta_u(\ell, s) \tag{10.54}$$

The Laplace transform of a unit step is 1/s, thus Equation 10.51 becomes:

$$\delta\theta_u = -\frac{1}{s}(1-\beta)\frac{L}{\bar{T}_f}(\bar{\theta}_s - \bar{\theta}_u)\frac{1-e^{-sL}}{sL} \tag{10.55}$$

When developing the exponential power in a series for small values of s:

$$e^{-sL} \approx 1 - sL \tag{10.56}$$

Equation 10.54 can be written with the aid of Equations 10.55 and 10.56 as:

$$\lim_{t\to\infty} \delta\theta(\ell,t) = \lim_{s\to 0}\left[-\frac{1}{s}\cdot(1-\beta)\frac{L}{\bar{T}_f}(\bar{\theta}_s-\bar{\theta}_u)\frac{1-(1-sL)}{sL}\cdot s\right]$$

$$= -(1-\beta)\frac{L}{\bar{T}_f}(\bar{\theta}_s-\bar{\theta}_u) \qquad (10.57)$$

INFLUENCE OF THE HEAT CAPACITY OF THE TUBE WALL

Taking the heat capacity of the tube wall into account does not give any particular problems, although the formulas become more complicated. Linearization of Equations 10.5 and 10.7 and combination with Equations 10.9, 10.20, 10.21 and 10.33 gives:

$$\frac{\partial(\delta\theta_w)}{\partial t} + (\bar{T}_{ws}^{-1} + \bar{T}_{wf}^{-1})\delta\theta_w - \bar{T}_{wf}^{-1}\delta\theta$$

$$= \bar{T}_{ws}^{-1}\delta\theta_s - 0.8\frac{\bar{h}_f}{\bar{h}}(\bar{\theta}_s-\bar{\theta}_i)\bar{T}_w^{-1}e^{-z/v\bar{T}_f}\frac{\delta v}{\bar{v}} \qquad (10.58)$$

in which $\bar{T}_w$ is given by:

$$\bar{T}_w = M_w c_w\,[(\bar{h}_s A_s)^{-1} + (\bar{h}_f A_f)^{-1}] \qquad (10.59)$$

Linearization of Equations 10.3 and 10.8 and combination with Equations 10.12, 10.13, 10.20 and 10.33 gives:

$$\bar{v}\frac{\partial(\delta\theta)}{\partial z} + \frac{\partial(\delta\theta)}{\partial t} + \bar{T}_{ff}^{-1}\delta\theta - \bar{T}_{ff}^{-1}\delta\theta_w = -\left(1-0.8\frac{\bar{h}_f}{\bar{h}}\right)(\bar{\theta}_s-\bar{\theta}_i)\bar{T}_f^{-1}e^{-z/v\bar{T}_f}\frac{\delta v}{\bar{v}}$$

$$(10.60)$$

When Equations 10.58 and 10.60 are written with the operator s, and combined the result is:

$$\bar{v}\frac{d(\delta\theta)}{dz} + g_1(s)\delta\theta = g_2(s)\delta\theta_s - g_3(s)(\bar{\theta}_s-\bar{\theta}_i)\cdot e^{-z/v\bar{T}_f}\frac{\delta v}{\bar{v}} \qquad (10.61)$$

in which:

$$g_1(s) = s + \bar{T}_{ff}^{-1} \frac{1 + sT_{ws}}{1 + T_{ws}\bar{T}_{wf}^{-1} + sT_{ws}} \tag{10.62}$$

$$g_2(s) = \frac{\bar{T}_{ff}^{-1}}{1 + T_{ws}\bar{T}_{wf}^{-1} + sT_{ws}} \tag{10.63}$$

$$g_3(s) = \bar{T}_f^{-1}\left(1 - 0.8\frac{\bar{h}_f}{\bar{h}} \frac{1 + sT_{ws}}{1 + T_{ws}T_{wf}^{-1} + sT_{ws}}\right) \tag{10.64}$$

Solution of Equation 10.64 gives:

$$\delta\theta = e^{-g_1 z/\bar{v}}\delta\theta_i + \frac{g_2}{g_1}(1 - e^{-g_1 z/\bar{v}})\delta\theta_s - \frac{g_3}{g_1 - \bar{T}_f^{-1}}$$

$$\cdot(\bar{\theta}_s - \bar{\theta}_i)(e^{-z/\bar{v}\bar{T}_f} - e^{-g_1 z/\bar{v}})\frac{\delta v}{\bar{v}} \tag{10.65}$$

Equation 10.62 can be written in the following way:

$$g_1(s) = s + \bar{T}_f^{-1} + \frac{sT_{ws}\dfrac{T_{ws}}{T_{ws} + \bar{T}_{wf}}\bar{T}_{ff}^{-1}}{1 + s\dfrac{T_{ws}\bar{T}_{wf}}{T_{ws} + \bar{T}_{wf}}} \tag{10.66}$$

In many applications the third term in the right side of Equation 10.66 has only little importance: for low-frequency disturbances the second term exceeds the third term, for high-frequency disturbances the first term dominates. Therefore, the response in Equation 10.65 may be approximated by:

$$\frac{\delta\theta_u}{\delta\theta_i} \simeq e^{-L/\bar{T}_f}e^{-sL} \tag{10.67}$$

which corresponds to Equation 10.49. The second term becomes by approximation:

$$\frac{\delta\theta_u}{\delta\theta_s} \simeq \frac{1 - [(\bar{\theta}_s - \bar{\theta}_u)/(\bar{\theta}_s - \bar{\theta}_i)]\,e^{-sL}}{1 + s[\bar{T}_{ff} + T_{ws} + T_{ws}\bar{T}_{wf}^{-1}\bar{T}_{ff}] + s^2T_{ws}\bar{T}_{ff}} \tag{10.68}$$

The most important difference with Equation 10.50 is the second-order factor in the denominator. This can be written as the product of the first-order factors:

$$(1 + sT_1)(1 + sT_2) \tag{10.69}$$

The third term of Equation (10.65) is by approximation:

$$\frac{\delta\theta_u}{(\delta v/\bar{v})} \simeq -\frac{L}{\bar{T}_f} \frac{\left(1 - 0.8\dfrac{\bar{h}_f}{\bar{h}} + T_{ws}\bar{T}_{wf}^{-1}\right) + sT_{ws}\left(1 - 0.8\dfrac{\bar{h}_f}{\bar{h}}\right)}{1 + T_{ws}\bar{T}_{wf}^{-1}(1 + T_{ws}\bar{T}_{ff}^{-1}) + sT_{ws}}$$

$$\cdot (\bar{\theta}_s - \bar{\theta}_u)\left(\frac{1 - e^{-sL}}{sL}\right)\frac{\delta v}{\bar{v}} \tag{10.70}$$

The second factor is not much different from $1 - \beta$; thus, the result is rather similar to Equation (10.51). Apparently the effect of the heat capacity of the tube wall is not very strong, only an extra first-order is present in the response to steam temperature variations.

CHAPTER 11

POLLUTION IN A RIVER

The concentration of a pollutant in a river is interesting and of practical importance. When a plant drains off a certain amount of chemical pollutant into a river, one is interested in the pollutant concentration as a function of time and location in the river. With the aid of a good model of a river section, it is possible to determine the amount of pollutant that has been drained off, from measurements of the pollutant concentration as a function of time at a certain location along the river. These measurements and calculations may be useful when the maximal allowable emissions have to be checked.

ASSUMPTIONS

The description of the dynamic behavior of the pollutant concentration in a river would be impossible without any assumptions. A complicated model usually makes no sense [25]. For the development of a simple model, the following assumptions are made:

1. The river has everywhere the same cross-sectional area and flow. It is evident that the water level in the river is not the same everywhere, particularly not after long local rainfall. But here we shall work with an average flow.
2. Mixing over the cross-sectional area is ideal. This assumption is necessary to develop a maximum gradient model. Measurements in large rivers, however, have shown that emissions at one side of the river were not smoothed across the width of the river, even after several kilometers. However, when the measuring station is sufficiently far downstream, we can still maintain the assumption.
3. The pollution can be characterized by one concentration [e.g., biochemical oxygen demand (BOD) or chemical oxygen demand (COD)].
4. The pollution disappears via a first-order reaction.
5. Axial mixing can be characterized by a constant dispersion coefficient D.

MODEL DESCRIPTION

A river section with a length Δz is considered. The following mass balance holds:

$$A\Delta z \frac{\partial c}{\partial t} = Q(c)_z - Q(c)_{z+\Delta z} - AD\left(\frac{\partial c}{\partial z}\right)_z + (AD)_{z+\Delta z} - kA\Delta z(c)_z \qquad (11.1)$$

where
A = cross-sectional area of the river (m^2)
c = pollutant concentration (kg/m^3)
Q = flow (m^3/sec)
D = dispersion coefficient (m/sec^2)
k = reaction rate constant (sec^{-1})

The first term in Equation 11.1 is an accumulation term; the second and third term represent transport by flow; the following two terms stand for dispersion; and the last term represents the disappearance of pollutant by reaction. When Δz approaches to zero, Equation 11.1 gives in the limit case:

$$A \frac{\partial c}{\partial t} = -Q \frac{\partial c}{\partial z} + AD \frac{\partial^2 c}{\partial z^2} - Akc \qquad (11.2)$$

After division by A and substitution, the flow velocity v

$$v = \frac{Q}{A} \qquad (11.3)$$

Equation 11.2 becomes:

$$\frac{\partial c}{\partial t} = -v \frac{\partial c}{\partial z} + D \frac{\partial^2 c}{\partial z^2} - kc \qquad (11.4)$$

Linearization is not necessary here, because Equation 11.4 is already linear in c.

GENERAL SOLUTION

Equation 11.4 can easily be written with the operator s, or can be Laplace-transformed. The latter operation gives:

$$sc(z,s) - c(z,0^+) = -v\frac{dc(z,s)}{dz} + D\frac{d^2c(z,s)}{dz^2} - kc(z,s) \tag{11.5}$$

This equation is an ordinary differential equation in c and can be rewritten:

$$D\frac{d^2c}{dz^2} - v\frac{dc}{dz} - (k + s)c = -c(z,0^+) \tag{11.6}$$

The term $c(z,0^+)$ represents an initial condition. The general solution of Equation 11.6 can be determined from the characteristic equation:

$$DJ^2 - vJ - (k + s) = 0 \tag{11.7}$$

from which:

$$J_{1,2} = \frac{v \pm \sqrt{v^2 + 4D(k + s)}}{2D} \tag{11.8}$$

The general solution has then the following form:

$$c(z,s) = A_1e^{J_1z} + A_2e^{J_2z} \tag{11.9}$$

If the real part of s is positive (which is required for convergence of the transformation), the real part of J_1 is positive, and that of J_2 negative.

The real part of $J_{1,2}$ is determined by the real part of the denominator in Equation 11.8:

$$\begin{aligned} \text{real part } [J_{1,2}] &= \text{real part } \left[v \pm \sqrt{v^2 + 4D(k + s)}\right] \\ &= \text{real part } \left[1 \pm \sqrt{1 + \frac{4Dk}{v^2} + \frac{4Ds}{v^2}}\right] \end{aligned} \tag{11.10}$$

The square root form can be split into a real and imaginary part:

$$\sqrt{1 + \frac{4Dk}{v^2} + \frac{4Ds}{v^2}} = \sqrt{1 + \alpha + j\beta} = A + jB \tag{11.11}$$

Squaring gives:

$$1 + \alpha + j\beta = A^2 + 2jAB - B^2 \tag{11.12}$$

thus

$$\left.\begin{aligned} 1 + \alpha &= A^2 - B^2 \\ \beta &= 2AB \end{aligned}\right\} \tag{11.13}$$

Squaring and summing Equation (11.13) results in:

$$(1 + \alpha)^2 + \beta^2 = (A^2 + B^2)^2 \tag{11.14}$$

thus

$$A^2 + B^2 = \sqrt{(1 + \alpha)^2 + \beta^2} \tag{11.15}$$

In Equation 11.13, an expression was already found for $A^2 - B^2$. Combination with Equation 11.15 gives:

$$A = \sqrt{\frac{1 + \alpha + \sqrt{(1 + \alpha)^2 + \beta^2}}{2}} = \sqrt{\frac{1 + \alpha + 1 + \gamma}{2}} = 1 + \epsilon \tag{11.16}$$

As the real part of s is positive, α is positive; hence, ϵ is positive. Consequently, the real part of the square root form is greater than one

$$\text{real part}\left[1 \pm \sqrt{1 + \frac{4Dk}{v^2} + \frac{4Ds}{v^2}}\right] = 1 \pm A \gtrless 0 \tag{11.17}$$

thus

$$\text{real part } [J_1] > 0 \text{ and real part } [J_2] < 0 \tag{11.18}$$

BOUNDARY CONDITIONS

If a river is considered from the point $z = 0$ to a large distance from this point, the general solution has to converge for z approaching infinity. This means that A_1, the coefficient of the term with J_1, must be equal to zero. If, however, the river is very long in both directions, the general solution must converge for $z = \infty$ and $z = -\infty$. The first requirement means that $A_1 = 0$, the second one means that $A_2 = 0$. In this way nothing remains from the general solution. Therefore, we are forced to divide the river in two sections, each with its own general solution. Both solutions are:

$$c^+ = A_2\, e^{J_2 z} \text{ for } 0^+ < z < \infty \tag{11.19}$$

and

$$c^- = A_1 e^{J_1 z} \text{ for } -\infty < z < 0^- \tag{11.20}$$

This approach is necessary if a certain amount of pollutant is introduced at $z = 0$. If the initial concentration at time $t = 0$ is equal to zero everywhere, the concentration just after the pollutant introduction at time $t = 0$ will be infinite at the location $z = 0$ and zero everywhere else. This is described by a delta function:

$$c(z,0^+) = I\delta(z) \tag{11.21}$$

I is the intensity of the delta function, the amount of drained pollutant divided by the cross-sectional area of the river. The properties of the delta function can be presented as follows:

$$\left.\begin{aligned} \delta(z) &= \infty \text{ for } z = 0 \\ \delta(z) &= 0 \text{ for } z \neq 0 \\ \int_{-\infty}^{+\infty} \delta(z)dz &= 1 \end{aligned}\right\} \tag{11.22}$$

From this last equation it is clear that $\delta(z)$ has the dimension m^{-1}, thus the right side of Equation 11.21 has the dimension of unit mass per volume.

At the moment of pollutant introduction we have a two-dimensional delta function as shown in Figure 11.1. It should be noticed that it is impossible to draw the delta function, because the value at t = 0 and z = 0 is infinite. However, to give an impression, the function in Figure 11.1 is drawn in a somewhat smoothed form. From Figure 11.1 it is immediately clear that at time t = 0, we have to do with a delta function in the z direction $\delta(z)$, as already represented in Equation 11.21.

The boundary conditions follow from the condition of continuity for the concentration profile at z = 0:

$$c(0^+,t) = c(0^-,t) \tag{11.23}$$

and from the partial mass balance at z = 0. To set up this mass balance a small section $0^- < z < 0^+$ is considered. When the length of this section is equal to Δz, the partial mass balance is according to Equation 11.1:

$$A\Delta z \frac{\partial c(0^+,t)}{\partial t} = Avc(0^-,t) - Avc(0^+,t) - AD \frac{\partial c(0^-,t)}{\partial z} + AD \frac{\partial c(0^+,t)}{\partial z} - kA\Delta z c(0^+,t) + AI\delta(t) \tag{11.24}$$

In this equation the last term represents the amount of pollutant which is introduced at z = 0. From Figure 11.1 it is clear that one has to deal then

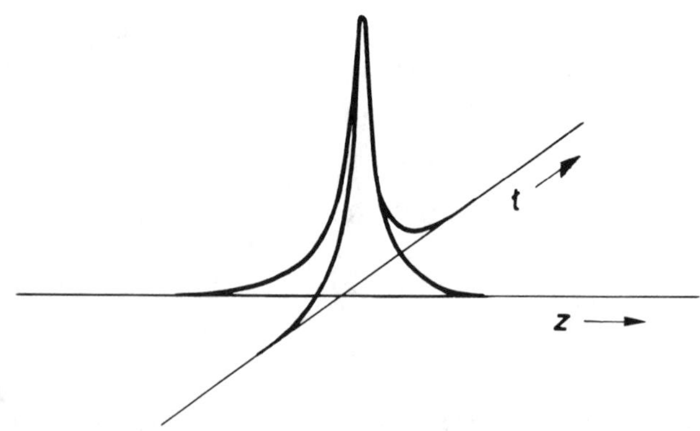

Figure 11.1 Representation of the delta function after pollutant introduction.

with the delta function in the t direction. In this limit case when Δz goes to zero, Equation 11.24 together with Equation 11.23 results in:

$$D\frac{\partial c(0^-,t)}{\partial z} - D\frac{\partial c(0^+,t)}{\partial z} = I\delta(t) \tag{11.25}$$

FINAL SOLUTION

When Equations 11.23 and 11.25 are Laplace-transformed, the result is:

$$c(0^+,s) = c(0^-,s) \tag{11.26}$$

and

$$D\frac{dc(0^-,s)}{dz} - D\frac{dc(0^+,s)}{dz} = I \tag{11.27}$$

Then from Equations 11.19, 11.20, 11.26 and 11.27, it follows:

$$A_1 = A_2 = \frac{I}{D(J_1 - J_2)} \tag{11.28}$$

thus

$$c^+ = \frac{I}{D(J_1 - J_2)}\, e^{J_2 z} \tag{11.29}$$

and

$$c^- = \frac{I}{D(J_1 - J_2)}\, e^{J_1 z} \tag{11.30}$$

Equations 11.29 and 11.30 can be combined and, after substitution of Equation 11.8, the result is:

$$c^{\pm} = \frac{1}{\sqrt{v^2 + 4D(k + s)}} e^{vz/2D} e^{\pm z/2D\sqrt{v^2 + 4D(k + s)}} \quad (11.31)$$

Equation 11.31 can be transformed from the s domain to the s* domain, for which:

$$s^* = s + k + \frac{v^2}{4D} \quad (11.32)$$

Equation 11.31 then becomes:

$$c^{\pm}(z,s^*) = \frac{1}{2\sqrt{D}\sqrt{s^*}} e^{vz/2D} e^{\mp(z/\sqrt{D})\sqrt{s^*}} \quad (11.33)$$

With tables for Laplace transforms [26], Equation 11.33 can easily be transformed to the time domain. However we have to take into account that the transformation is not from s to t but from s* to t. Translation of the Laplace transform s to s* means multiplication with

$$e^{(k+v^2/4D)t} \quad (11.34)$$

in the time domain, which can be seen from the following equations:

$$c(z,s) = \int_0^\infty c(z,t)e^{-st}dt \quad (11.35)$$

Equation 11.35 gives, with the aid of Equation 11.32:

$$c(z,s^*) = \int_0^\infty c(z,t)e^{(k+v^2/4D)t}\, e^{-s^*t}\, dt \quad (11.36)$$

The inverse transformation of Equation 11.33 now gives:

$$c^{\pm}(z,t)e^{(k+v^2/4D)t} = \frac{I}{2\sqrt{D}} e^{vz/2D} \frac{e^{-z^2/4Dt}}{\sqrt{\pi t}} \quad (11.37)$$

After some rearrangement, Equation 11.37 can be written as:

$$c^{\pm}(z,t) = Ie^{-kt}\left[\frac{1}{\sigma_z\sqrt{2\pi}}\, e^{-(z-vt)^2/2\sigma_z^2}\right] \tag{11.38}$$

in which

$$\sigma_z = \sqrt{2Dt} \tag{11.39}$$

This result is a pulse which has the form of a normal (gaussian) probability density function. The peak value moves with the flow, and the standard deviation increases with the square root of time; the intensity, which is the total area of the pulse, decreases exponentially with time. It is interesting to consider the limit of the part between brackets in Equation 11.38:

$$\lim_{t\to 0}\left[\frac{1}{2\sqrt{\pi Dt}}\, e^{-(z-vt)^2/4Dt}\right] \tag{11.40}$$

This limit satisfies all conditions of a delta function; thus, introduction of a delta function in the model gives a consistent picture.

DETERMINATION OF DISPERSION WITH THE AID OF MOMENTS

The solution found in the previous section can also be used as a starting point for experimental determination of the dispersion D. One could take samples at a certain time and at different locations after introduction of a "tracer," to determine the pulse shaped concentration change. However, it is easier experimentally to take samples at a fixed location at different times, thus determining the pulse while it passes. Due to increasing variance, the pulse as a function of time has not a normal distribution but an oblique distribution with a steep front and a longer tail as can also be seen from Figure 11.2. Commonly the tracer is an inert material, for which k is equal to zero. Then Equation 11.38 can, for a measurement station at $z = L$, be written as:

$$c(L,t) = \frac{I}{2\sqrt{\pi Dt}}\, e^{-(L-vt)^2/4Dt} \tag{11.41}$$

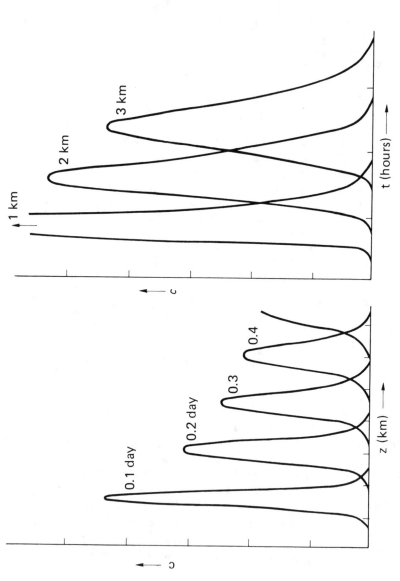

Figure 11.2 Concentration profiles at certain times and at certain locations, $D = 1\ km^2/day$, $k = 0.2\ day^{-1}$ and $v = 30\ km/day$.

The intensity I is equal to the amount of added material W divided by the cross-sectional area A of the river:

$$I = \frac{W}{A} = \frac{WL}{V} \tag{11.42}$$

with V the volume, Equation 11.41 becomes:

$$c = \frac{WL}{2V\sqrt{\pi Dt}} e^{-(L-vt)^2/4Dt} \tag{11.43}$$

This equation can also be written as:

$$\frac{cV}{W} = \frac{1}{2\sqrt{\pi\left(\frac{vt}{L}\right)\left(\frac{D}{vL}\right)}} \exp\left(\frac{-\left(1-\frac{vt}{L}\right)^2}{4\left(\frac{vt}{L}\right)\left(\frac{D}{vL}\right)}\right) \tag{11.44}$$

Plotting Equation 11.44 with cV/W and vt/L along the axes, results in a number of curves with D/vL as parameter. The latter parameter is the inverse Péclet number:

$$Pe^{-1} = \frac{D}{vL} \tag{11.45}$$

In Figure 11.3 these so called c curves are plotted for high values of the Péclet number. They approximately show a normal distribution. In Figure 11.4 the curves are plotted for low values of the Péclet number. The front of the curve becomes steeper and the tail smoother as the Péclet number increases.

In Figures 11.2 to 11.4 every curve can be characterized by an average value and a standard deviation. How can these parameters be determined from an arbitrary curve? As is clear from Figures 11.3 and 11.4, the average value and the standard deviation of a distribution depend on the velocity and the dispersion. The relationship can be easily found by calculating the moments of the distribution with the aid of the Laplace transform. According to the definition of the Laplace transform:

$$f(s) = \int_0^\infty e^{-st} f(t)dt \tag{11.46}$$

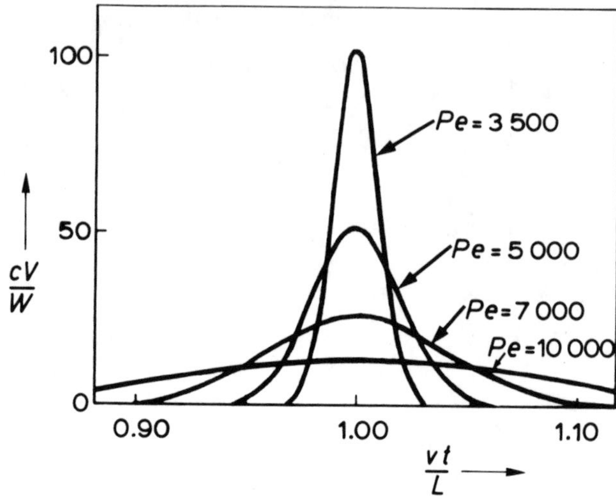

Figure 11.3 c curves for high values of the Péclet number.

With the aid of a series expansion for e^{-st}, Equation 11.46 can be written as:

$$f(s) = \int_0^\infty \left[1 - \frac{st}{1!} + \frac{(st)^2}{2!} - \frac{(st)^3}{3!} \ldots\right] f(t)dt$$

$$= \int_0^\infty f(t)dt - s\int_0^\infty tf(t)dt + \frac{1}{2}s^2 \int_0^\infty t^2 f(t)dt \ldots$$

$$= \mu_0 - s\mu_1 + \frac{1}{2}s^2\mu_2 - \ldots \tag{11.47}$$

where μ_i is the ith moment with respect to t = 0. The moments can thus be calculated by repeated differentiation of f(s):

$$\mu_i = (-1)^i \left[\frac{d^i f(s)}{ds^i}\right]_{s=0} \tag{11.48}$$

It can easily be seen that application of Equation 11.48 on Equation 11.46 results in Equation 11.47. When applying Equation 11.48 to Equation 11.33, the result is:

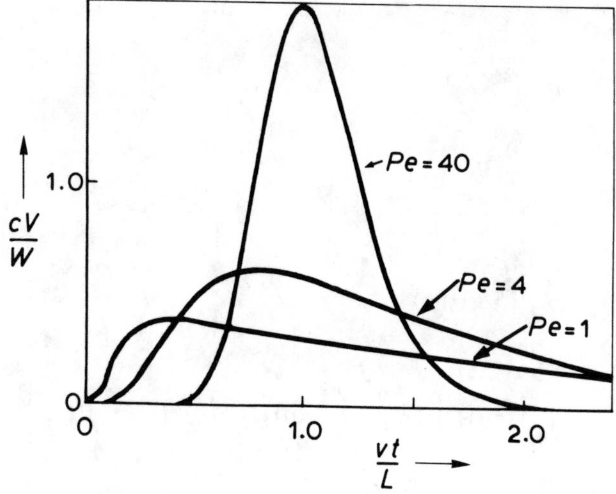

Figure 11.4 c curves for low values of the Péclet number.

$$\mu_0 = [c(z,s^*)]_{s=0}$$

$$\mu_1 = -\left[\frac{dc(z,s^*)}{ds^*}\right]_{s=0} \tag{11.49}$$

$$\mu_2 = \left[\frac{d^2c(z,s^*)}{ds^{*2}}\right]_{s=0}$$

Application of Equation 11.49 results in:

$$\frac{\mu_1}{\mu_0} = \frac{1}{2}(s^*)^{-1} + \frac{1}{2}\frac{z}{\sqrt{D}}(s^*)^{-1/2} \tag{11.50}$$

and

$$\frac{\mu_2}{\mu_0} = \frac{3}{4}(s^*)^{-2} + \frac{3}{4}\frac{z}{\sqrt{D}}(s^*)^{-3/2} + \frac{1}{4}\frac{z^2}{D}(s^*)^{-1} \tag{11.51}$$

Now the average μ is according to the definition:

$$\mu = \frac{\mu_1}{\mu_0} \tag{11.52}$$

and the variance:

$$\begin{aligned}
\sigma^2 &= \frac{1}{\mu_0}\int_0^\infty (t-\mu)^2 f(t)dt \\
&= \frac{1}{\mu_0}\left[\int_0^\infty t^2 f(t)dt - 2\frac{\mu_1}{\mu_0}\int_0^\infty tf(t)dt + \left(\frac{\mu_1}{\mu_0}\right)^2\int_0^\infty f(t)dt\right] \\
&= \frac{1}{\mu_0}\left[\mu_2 - 2\frac{\mu_1}{\mu_0}\cdot\mu_1 + \left(\frac{\mu_1}{\mu_0}\right)^2\cdot\mu_0\right] \\
&= \frac{\mu_2}{\mu_0} - \left(\frac{\mu_1}{\mu_0}\right)^2
\end{aligned} \tag{11.53}$$

It should be noticed that in a large part of the literature μ_0 is assumed to be equal to one, thus the average $\mu = \mu_1$. When the average $\mu_1 = 0$, the variance $\sigma^2 = \mu_2$, the second moment.

When Equations 11.50 and 11.51 are substituted into Equations 11.52 and 11.53, the result is:

$$\mu = \frac{1}{2}(s^*)^{-1} + \frac{1}{2}\frac{z}{\sqrt{D}}(s^*)^{-1/2} \tag{11.54}$$

and

$$\sigma^2 = \frac{1}{2}(s^*)^{-2} + \frac{1}{4}\frac{z}{\sqrt{D}}(s^*)^{-3/2} \tag{11.55}$$

with

$$(s^*)_{s=0} = k + \frac{v^2}{4D} \tag{11.56}$$

Substitution of Equation 11.56 into Equations 11.54 and 11.55 finally results in the average value and standard deviation of the distribution:

$$\mu = \frac{z}{\sqrt{v^2 + 4kD}} \left[1 + \frac{2}{Pe^*}\right] \tag{11.57}$$

and

$$\sigma^2 = \frac{z^2}{v^2 + 4kD} \left[\frac{2}{Pe^*} + \frac{8}{(Pe^*)^2}\right] \tag{11.58}$$

in which

$$Pe^* = \frac{z\sqrt{v^2 + 4kD}}{D} \tag{11.59}$$

If k = 0 these equations become:

$$\mu = \frac{z}{v} \left[1 + \frac{2}{Pe}\right] \tag{11.60}$$

and

$$\sigma^2 = \left(\frac{z}{v}\right)^2 \left[\frac{2}{Pe} + \frac{8}{Pe^2}\right] \tag{11.61}$$

in which

$$Pe = \frac{zv}{D}, \text{ a Péclet number} \tag{11.62}$$

APPLICATION

In an open pipe water flows with a velocity of 0.4 m/sec. A solution of a well conducting substance is injected into the water and the conductivity is measured 3 m downstream. The results are given in Table 11.1. The first three moments are approximated by:

$$\left.\begin{aligned} \mu_0 &= \int_0^\infty f(t)dt \simeq \Sigma\kappa \\ \mu_1 &= \int_0^\infty tf(t)dt \simeq \Sigma\kappa t \\ \mu_2 &= \int_0^\infty t^2 f(t)dt \simeq \Sigma\kappa t^2 \end{aligned}\right\} \quad (11.63)$$

in which κ is the conductivity in mmho. The calculations are also shown in Table 11.1. The average becomes:

$$\mu = \frac{\mu_1}{\mu_0} = \frac{2758}{255} = 10.82 \text{ sec}$$

Table 11.1 Conductivity Measurements in an Open Pipe

time t (sec)	conductivity κ (mmho)	κt	κt^2
0	0	0	0
2	7	14	28
4	35	140	560
6	43	258	1548
8	39	312	2496
10	32	320	3200
12	26	312	3744
14	19	266	3724
16	15	240	3840
18	10	180	3240
20	7	140	2800
22	6	132	2904
24	5	120	2880
26	3	78	2028
28	3	84	2352
30	2	60	1800
32	1	32	1024
34	1	34	1156
36	1	36	1296
38	0	0	0
40	0	0	0
	$\mu_0 = \Sigma\kappa = 255$	$\mu_1 = \Sigma\kappa t = 2758$	$\mu_2 = \Sigma\kappa t^2 = 40620$

and the variance:

$$\sigma^2 = \left(\frac{\mu_2}{\mu_0}\right) - \left(\frac{\mu_1}{\mu_0}\right)^2$$

$$= \frac{40{,}620}{255} - \left(\frac{2758}{255}\right)^2 = 42.31 \text{ sec}^2$$

As there is no reaction, k is equal to zero and Equation 11.60 can be applied.

Substitution of μ = 10.82 sec, z = 3 m and v = 0.4 m/sec results in a Péclet number of 4.52, substitution of σ^2, z and v in Equation 11.61 gives Pe = 4.85. The deviation between the calculated Péclet numbers is due to deviations between the model and reality, and to the approximation made in Equation 11.63. When using an average value of Pe = 4.68, the dispersion coefficient according to Equation 11.62 is:

$$D = \frac{vL}{Pe} = \frac{0.4 \times 3}{4.68} = 0.26 \text{ m}^2/\text{sec}$$

CHAPTER 12

DYNAMICS OF A TUBULAR FLOW REACTOR

In the previous chapter, the dynamics of pollutant concentration in a river were analyzed. The model description was represented by a second-order differential equation with initial and boundary conditions. Now an identical differential equation can be applied for describing the dynamic behavior of an isothermal tubular flow reactor with longitudinal dispersion and first-order reaction. The boundary conditions, however, are different, resulting in a different solution.

In a comment on an article by Levenspiel and Smith, Van der Laan [27] gave a survey of the calculation of the unknown dispersion coefficient by determining the average and variance from dynamic measurements. For each case the differential equation is the same; the boundary conditions are different. Van der Laan analyzed eight different cases. From the results it is clear that the boundary conditions have a large influence on the final solution. Although the dynamics of the isothermal tubular flow reactor are simple, the dynamics of the nonisothermal case are more important for practice. In many cases, exo- or endothermic reactions occur in the reactor; as a result, the temperature is not constant. In the case of varying temperature, the model description has to be extended with energy balances for the fluid and the catalyst, if present. The analysis of this situation, however, is much more complicated than the isothermal case.

Crider and Foss [28] analyzed the dynamics of a nonisothermal tubular flow reactor, filled with inert packing material. This system has many of the dynamic properties of packed catalytic reactors, although several experimental and analytical difficulties which occur in catalytic systems are avoided. When analyzing the dynamics of the nonisothermal tubular flow reactor, the work of Crider and Foss [28] and Douglas and Eagleton [29] will be used.

ISOTHERMAL TUBULAR FLOW REACTOR

The assumptions made in the previous chapter are relevant here. It is assumed that the cross-sectional area of the reactor is constant, that mixing over this cross-sectional area is ideal and that the axial mixing can be characterized by a constant dispersion coefficient D. The reactor temperature and flow of the fluid through the reactor are assumed constant. Finally, it is assumed that the component of interest disappears according to a first-order reaction. The dynamic behavior of the tubular flow reactor can be described by the partial differential equation:

$$\frac{\partial c}{\partial t} = -v\frac{\partial c}{\partial z} + D\frac{\partial^2 c}{\partial z^2} - kc \tag{12.1}$$

where
c = concentration (mol/m^3)
v = velocity (m/sec)
D = dispersion coefficient (m^2/sec)
k = reaction rate constant (sec^{-1})

When there is no dispersion in the section upstream of the reactor, the boundary condition at the entrance can be found from a partial material balance around the entrance (from $z = 0^-$ to $z = 0^+$ over a distance Δz):

$$A_c\Delta z\left(\frac{\partial c}{\partial t}\right)_{0^+} = A_c vc(0^-,t) - A_c vc(0^+,t) + A_c D\left(\frac{\partial c}{\partial z}\right)_{0^+} - A_c\Delta z kc(0^+,t) \tag{12.2}$$

where
A_c = cross-sectional area
$c(0^-,t)$ = concentration at the entrance of the element with length Δz, the inlet concentration
$c(0^+,t)$ = concentration at the outlet of the element with length Δz.

It is assumed that there is no dispersion in the section upstream of the reactor. The first term in Equation 12.2 represents the accumulation; the following two terms represent the transport due to flow; the next term is the dispersion term; and the last term represents the disappearance of component by reaction. In the limit case where Δz approaches to zero, Equation 12.2 reduces to:

$$vc(0^-,t) + D\left(\frac{\partial c}{\partial z}\right)_{0^+} - vc(0^+,t) = 0 \tag{12.3}$$

In a similar way at the end of the reactor:

$$vc(\ell^-,t) + D\left(\frac{\partial c}{\partial z}\right)_{\ell^-} - vc(\ell^+,t) = 0 \qquad (12.4)$$

Also here the dispersion outside the reactor is ignored. However, since:

$$c(\ell^-,t) = c(\ell^+,t) \qquad (12.5)$$

because there is no introduction of other materials at the end of the reactor, Equation 12.4 may be reduced to:

$$D\left(\frac{\partial c}{\partial z}\right)_{\ell} = 0 \qquad (12.6)$$

where ℓ^- is replaced by ℓ.

The model now consists of Equations 12.1, 12.3 and 12.6. The boundary conditions for the isothermal tubular flow reactor have been the subject of many discussions in literature during many years [30,31].

GENERAL SOLUTION AND SUBSTITUTION OF THE BOUNDARY CONDITIONS

After Laplace transformation, Equation 12.1 becomes:

$$sc(z,s) - c(z,0^+) = -v\frac{dc(z,s)}{dz} + D\frac{d^2c(z,s)}{dz^2} - kc(z,s) \qquad (12.7)$$

The general solution is again determined by the characteristic equation:

$$DJ^2 - vJ - (k+s) = 0 \qquad (12.8)$$

from which:

$$J_{1,2} = \frac{v \pm \sqrt{v^2 + 4D(k+s)}}{2D} \qquad (12.9)$$

The general solution then has the form:

$$c(z,s) = A_1 e^{J_1 z} + A_2 e^{J_2 z} \tag{12.10}$$

The coefficients A_1 and A_2 are determined by the boundary conditions. After Laplace transformation, Equation 12.3 becomes:

$$c(0^-,s) = c(0^+,s) - \frac{D}{v}\frac{dc(0^+,s)}{dz} \tag{12.11}$$

Equation 12.10 gives, for $z = 0^+$:

$$c(0^+,s) = A_1 + A_2 \tag{12.12}$$

Differentiation of Equation 12.10 and substitution of $z = 0^+$ results in:

$$\frac{dc(0^+,s)}{dz} = A_1 J_1 + A_2 J_2 \tag{12.13}$$

With the aid of Equations 12.12 and 12.13, Equation 12.11 can be written as:

$$c(0^-,s) = A_1\left(1 - J_1\frac{D}{v}\right) + A_2\left(1 - J_2\frac{D}{v}\right) \tag{12.14}$$

Differentiation of Equation 12.10 at $z = \ell$ and combination with the Laplace transformed Equation 12.6 gives:

$$\frac{dc(\ell,s)}{dz} = A_1 J_1 e^{J_1 \ell} + A_2 J_2 e^{J_2 \ell} = 0 \tag{12.15}$$

From Equations 12.14 and 12.15, A_1 and A_2 can be solved. The result is:

$$A_1 = \frac{-c(0^-,s)J_2 e^{J_2 \ell}}{J_1\left(1 - J_2\frac{D}{v}\right)e^{J_1 \ell} - J_2\left(1 - J_1\frac{D}{v}\right)e^{J_2 \ell}} \tag{12.16}$$

and

$$A_2 = \frac{c(0^-,s)J_1 e^{J_1 \ell}}{J_1\left(1 - J_2 \frac{D}{v}\right)e^{J_1\ell} - J_2\left(1 - J_1\frac{D}{v}\right)e^{J_2\ell}} \tag{12.17}$$

Substitution of Equations 12.16 and 12.17 in Equation 12.10 gives, after some rearrangement:

$$\frac{c(z,s)}{c(0^-,s)} = \frac{J_2 e^{-J_1(\ell - z)} - J_1 e^{-J_2(\ell - z)}}{J_2\left(1 - J_1\frac{D}{v}\right)e^{-J_1\ell} - J_1\left(1 - J_2\frac{D}{v}\right)e^{-J_2\ell}} \tag{12.18}$$

STATIC ANALYSIS

The static behavior can easily be determined by setting s equal to zero in Equations 12.9 and 12.18. When defining the Péclet number

$$Pe = \frac{v\ell}{D} \tag{12.19}$$

and parameter a:

$$a = \sqrt{1 + \frac{4kD}{v^2}} \tag{12.20}$$

Equation 12.18 can be rewritten as:

$$\frac{c(z)}{c(0^-)} = 2\exp\left(Pe \cdot \frac{z}{2\ell}\right)\frac{(1+a)\exp\left[Pe\frac{a}{2}\left(1 - \frac{z}{\ell}\right)\right] - (1-a)\exp\left[Pe\frac{a}{2}\left(\frac{z}{\ell} - 1\right)\right]}{(1+a)^2\exp\left[Pe\frac{a}{2}\right] - (1-a)^2\exp\left[-Pe\frac{a}{2}\right]} \tag{12.21}$$

For a = 1.2 and values of the Péclet number equal to 0.1, 1 and 10, $c(z)/c(0^-)$ is calculated as a function of z and given in Figure 12.1. For other values of a and the Péclet number, $c(z)/c(0^-)$ can easily be calculated with the aid

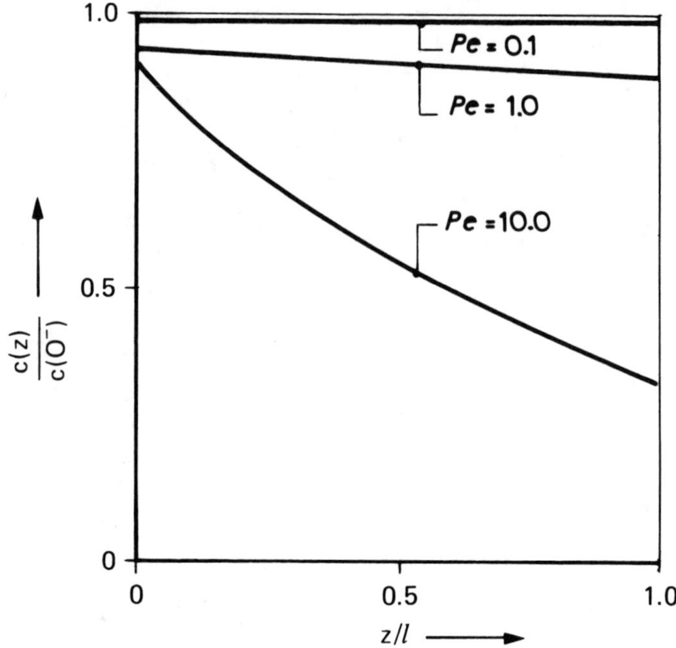

Figure 12.1 Concentration profile in the reactor for different values of the Péclet number and a = 1.2.

of Equation 12.21. From Figure 12.1 it can also be seen that the outlet concentration decreases, when the Péclet number increases. Hence the conversion defined by

$$C = 1 - \frac{c(\ell)}{c(0^-)} \tag{12.22}$$

increases for increasing Péclet number, which is well known from reactor engineering.

SPECIAL CASES

First, we shall analyze the case in which D approaches to zero. According to Equation 12.20, a is then approximately equal to 1, thus $1 - a \simeq 0$. Equation 12.21 can then be rewritten as:

$$\frac{c(z)}{c(0^-)} \simeq \exp\left(\mathrm{Pe}\,\frac{z}{2\ell}\right)\frac{\exp\left[\mathrm{Pe}\,\frac{a}{2}\left(1-\frac{z}{\ell}\right)\right]}{\exp\left[\mathrm{Pe}\,\frac{a}{2}\right]} \simeq \exp\left[\mathrm{Pe}\,\frac{z}{2\ell}\,(1-a)\right] \qquad (12.23)$$

When D is very small the following approximation can be made:

$$a = \sqrt{1+\frac{4kD}{v^2}} \simeq 1+\frac{2kD}{v^2} \qquad (12.24)$$

thus

$$1-a = -\frac{2kD}{v^2} \qquad (12.25)$$

after which Equation 12.22 can be written as:

$$\frac{c(z)}{c(0^-)} = \exp\left(-\frac{kz}{v}\right) \qquad (12.26)$$

From Equation 12.26 for $z = \ell$ in combination with Equation 12.22 follows:

$$C = 1-\exp\left(-\frac{k\ell}{v}\right) \qquad (12.27)$$

This is the well known equation for the conversion in a tubular flow reactor with plug flow.

Second, we shall analyze the opposite case, where D approaches infinity. For large values of D, a is proportional to $D^{1/2}$ and because Pe is proportional to D^{-1}, $\mathrm{Pe}(a/2)$ is proportional to $D^{-1/2}$ and therefore approaches zero. For small values of x the following approximation may be made:

$$\left.\begin{aligned} \exp x &\simeq 1+x \\ \exp(-x) &\simeq 1-x \end{aligned}\right\} \qquad (12.28)$$

Application of Equation 12.28 to Equation 12.21 results in:

$$\frac{c(z)}{c(0^-)} \simeq \frac{4}{4 + Pe(a^2 + 1)} \simeq \frac{4v^2D}{4v^2D + 2v^3\ell + 4v\ell kD} \tag{12.29}$$

As D approaches infinity, the second term in the denominator of Equation 12.29 can be neglected, resulting in:

$$\frac{c(z)}{c(0^-)} \simeq \frac{v}{v + k\ell} \tag{12.30}$$

The position coordinate does not appear in this equation, because infinite dispersion results in ideal mixing. Equation 12.30 therefore also holds for $z = \ell$, and the conversion now becomes:

$$C = \frac{k\ell}{v + k\ell} \tag{12.31}$$

which is the well known expression for conversion in an ideally mixed reactor.

DYNAMICS IN THE CASE OF NEGLIGIBLE DISPERSION

When the dispersion is negligible, Equation 12.1 can be written as:

$$\frac{\partial c}{\partial t} + v\frac{\partial c}{\partial z} + kc = 0 \tag{12.32}$$

Introduction of operator s gives:

$$sc + v\frac{dc}{dz} + kc = 0 \tag{12.33}$$

or:

$$\frac{dc}{dz} = -\left(\frac{(k + s)}{v}\right)c \tag{12.34}$$

When we start from an equilibrium situation with $c(z,0^+) = 0$ and use the initial condition:

$$c = c(0^-) \text{ at } z = 0 \tag{12.35}$$

the solution of Equation 12.34 is:

$$c = c(0^-) \exp\left[-\frac{(k+s)z}{v}\right] \tag{12.36}$$

thus:

$$\frac{c}{c(0^-)} = \exp\left(-\frac{zs}{v}\right) \exp\left(-\frac{kz}{v}\right) \tag{12.37}$$

in which $c(0^-)$ is the Laplace transformed inlet concentration. The response to a step in $c(0^-)$ can easily be analyzed. This step is delayed by the residence time z/v and attenuated by the last factor of Equation 12.37 (Figure 12.2). Hence the response in c equals:

$$c(z,t) = A \exp\left(-\frac{kz}{v}\right) \quad \text{for} \quad t > \frac{z}{v} \tag{12.38}$$

$$= 0 \quad \text{for} \quad 0 \leqslant t < \frac{z}{v} \tag{12.39}$$

where A = step size

DYNAMICS OF THE ADIABATIC TUBULAR FLOW REACTOR

As already discussed in the introduction, the isothermal tubular flow is less important in practice than the adiabatic tubular flow reactor. Heat effects due to reaction occur often, resulting in a nonuniform temperature.

Assume that the reactor is filled with catalyst material with a temperature T_s (Figure 12.3). The porosity or fraction of the volume occupied by the fluid is denoted by ϵ. Furthermore it is assumed that there are no radial gradients, the reaction is again first-order and the axial dispersion is negligible. At constant fluid velocity, the following partial mass balance holds:

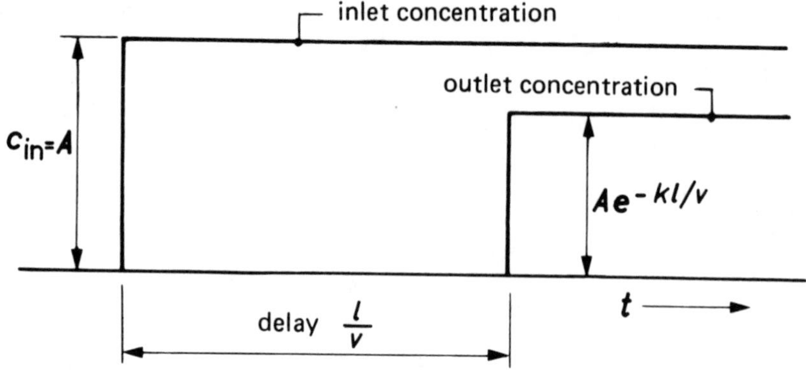

Figure 12.2 Step response of an isothermal tubular flow reactor without dispersion.

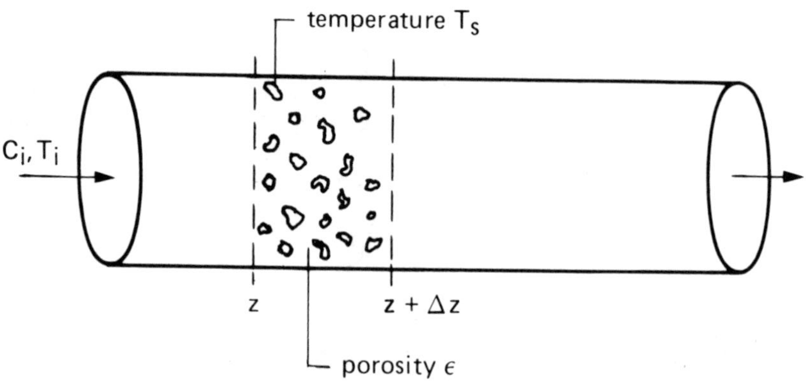

Figure 12.3 Packed tubular flow reactor.

$$A_c \Delta z \epsilon \frac{\partial c}{\partial t} = -A_c \Delta z \epsilon v \frac{\partial c}{\partial z} - A_c \Delta z \epsilon k c \tag{12.40}$$

The first term is an accumulation term, the second term represents the transport of component by flow and the third term the disappearance by reaction.

Energy balances can be formulated for the flowing medium and the catalyst. The form is determined by assumptions about the various heat transfer rates. One can often assume a uniform temperature distribution for the fluid throughout a cross section. However, this does not necessarily hold for the temperature inside the catalyst particles, particularly if they have a low heat conductivity. Here we shall use an equivalent catalyst temperature which interacts with the fluid temperature via an equivalent heat transfer coefficient (encompassing some distance of heat conduction in the catalyst). When it is assumed that heat development due to reaction occurs in the fluid, the energy generation term will appear in the energy balance of the fluid:

$$A_c \Delta z \epsilon \rho_f c_f \frac{\partial T}{\partial t} = -A_c \Delta z \epsilon \rho_f c_f v \frac{\partial T}{\partial z} + h_s A_s \Delta z (T_s - T) + A_c \Delta z \epsilon \Delta H k c \quad (12.41)$$

where ρ_f = fluid density (kg/m^3)
c_f = specific heat of fluid (J/kg-°K)
T = temperature of fluid (°K)
h_s = heat transfer coefficient (W/m^2-°K)
A_s = catalyst area (m^2)
T_s = catalyst temperature (°K)
ΔH = heat of reaction (J/mol)

The energy balance for the packing material is now:

$$A_c \Delta z (1 - \epsilon) \rho_s c_s \frac{\partial T_s}{\partial t} = h_s A_s \Delta z (T - T_s) \quad (12.42)$$

where ρ_s = catalyst density (kg/m^3)
c_s = specific heat of the catalyst (J/kg-°K)

The boundary and initial conditions can be defined as:

$$\left.\begin{array}{lll} t = 0 \quad c = c_0(z), & T = T_0(z), & T_s = T_{s0}(z) \\ z = 0 \quad c = c_i(t), & T = T_i(t) & \end{array}\right\} \quad (12.43)$$

The static behavior can be determined by setting the derivatives with respect to time equal to zero. Then Equation 12.40 gives:

$$\bar{v} \frac{d\bar{c}}{dz} = -\overline{kc} \quad (12.44)$$

where the bar over a variable denotes the value in the static situation. Similarly, for Equations 12.41 and 12.42:

$$\rho_f c_f \bar{v} \frac{d\bar{T}}{dz} = \Delta H \overline{kc} \tag{12.45}$$

and

$$\bar{T} = \bar{T}_s \tag{12.46}$$

Linearization of Equation 12.40 results in:

$$\frac{\partial(\delta c)}{\partial t} = -\bar{v}\frac{\partial(\delta c)}{\partial z} - \bar{k}\delta c - \bar{c}\delta k \tag{12.47}$$

The reaction rate constant is usually a function of the temperature according to:

$$k = k_0 e^{-E/RT} \tag{12.48}$$

Thus a variation in k can be written as:

$$\delta k = k_0 e^{-E/R\bar{T}} \cdot \frac{E}{R\bar{T}^2}\delta T = \frac{E\bar{k}}{R\bar{T}^2}\delta T \tag{12.49}$$

Substitution of Equation 12.49 into Equation 12.47 results in:

$$\frac{\partial(\delta c)}{\partial t} = -\bar{v}\frac{\partial(\delta c)}{\partial z} - \bar{k}\delta c - \frac{E\overline{kc}}{R\bar{T}^2}\delta T \tag{12.50}$$

Now the following dimensionless variables are defined:

$$\left.\begin{aligned} x_1 &= \frac{\delta c}{\bar{c}} \\ x_2 &= \frac{\rho_f c_f \delta T}{\Delta H \bar{c}} \\ \tau &= \frac{\bar{v} t}{\ell} \\ \zeta &= \frac{z}{\ell} \end{aligned}\right\} \tag{12.51}$$

Equation 12.50 can then be transformed to:

$$\frac{\partial x_1}{\partial \tau} + \frac{\partial x_1}{\partial \zeta} = -bx_1 - dx_2 \tag{12.52}$$

where the parameters b and d are functions of ζ and are defined by:

$$\left.\begin{aligned} b(\zeta) &= \frac{\bar{k}\ell}{\bar{v}} \\ d(\zeta) &= \frac{E\bar{k}\ell\Delta H\bar{c}}{\bar{v}R\bar{T}^2\rho_f c_f} \end{aligned}\right\} \tag{12.53}$$

Linearization of Equation 12.41 results in:

$$\rho_f c_f \frac{\partial(\delta T)}{\partial t} = -\rho_f c_f \bar{v} \frac{\partial(\delta T)}{\partial z} + \frac{h_s A_s}{A_c \epsilon}(\delta T_s - \delta T) + \Delta H\bar{k}\delta c + \frac{\Delta H\overline{kc}E}{R\bar{T}^2}\delta T \tag{12.54}$$

When in addition the following dimensionless parameters are introduced:

$$\left.\begin{aligned} x_3 &= \frac{\rho_f c_f \delta T_s}{\Delta H\bar{c}} \\ e &= \frac{h_s A_s \ell}{A_c \epsilon \bar{v} \rho_f c_f} \end{aligned}\right\} \tag{12.55}$$

Equation 12.54 can be written as:

$$\frac{\partial x_2}{\partial \tau} + \frac{\partial x_2}{\partial \zeta} = e(x_3 - x_2) + bx_1 + dx_2 \tag{12.56}$$

Equation 12.42 can be transformed with the aid of Equations 12.51 and 12.55 to:

$$\frac{\partial x_3}{\partial \tau} = ef(x_2 - x_3) \tag{12.57}$$

with

$$f = \frac{\epsilon}{(1-\epsilon)} \cdot \frac{\rho_f c_f}{\rho_s c_s} \tag{12.58}$$

The model, written in dimensionless parameters, is now formed by Equations 12.52, 12.56 and 12.57. Introduction of operator s yields:

$$\frac{dx_1}{d\zeta} = -(b+s)x_1 - dx_2 \tag{12.59}$$

$$\frac{dx_2}{d\zeta} = bx_1 + (d-e-s)x_2 + ex_3 \tag{12.60}$$

$$0 = efx_2 - (ef+s)x_3 \tag{12.61}$$

From the last equation, x_3 can be eliminated and substituted into Equation 12.60. Finally, a set of linear ordinary differential equations with position dependent coefficients results:

$$\begin{aligned} \frac{dx_1}{d\zeta} + \alpha(\zeta)x_1 &= -dx_2 \\ \frac{dx_2}{d\zeta} + \beta(\zeta)x_2 &= bx_1 \end{aligned} \tag{12.62}$$

in which

$$\begin{aligned} \alpha(\zeta) &= b+s \\ \beta(\zeta) &= \frac{es}{ef+s} - d + s \end{aligned} \tag{12.63}$$

The boundary conditions are:

$$\zeta = 0\!: \; x_1 = x_1(0,s) \;, \quad x_2 = x_2(0,s) \tag{12.64}$$

The set of equations can be solved analytically using the method of reduction of the order of the differential equation. This will not be discussed here; the interested reader is referred to the literature [28]. The final solution can be written in the form:

$$\left.\begin{aligned} x_1(\zeta,s) &= G_1(\zeta,s)x_1(0,s) + G_2(\zeta,s)x_2(0,s) \\ x_2(\zeta,s) &= G_3(\zeta,s)x_1(0,s) + G_4(\zeta,s)x_2(0,s) \end{aligned}\right\} \tag{12.65}$$

In every transfer function there is an influence of the heat capacity of the packing material and the fluid-solid heat transfer resistance, in all cases in the form $es/(s + ef)$. The transfer functions are very complicated expressions.

The transfer functions can be simplified only for extreme values of reactor parameters. For unpacked reactors, e is equal to infinity and the expression $es/(s + ef)$ becomes equal to zero. When there is no resistance to heat transfer between fluid and packing material d is equal to infinity, and the expression $es/(s + ef)$ becomes equal to s/f.

CHAPTER 13

CATALYST DEACTIVATION

In the previous chapters, linear distributed systems with or without variable coefficients were analyzed. In a number of cases, an analytical solution is possible; a large class of linear problems, however, can only be solved analytically by approximation. Nonlinear distributed systems are even more difficult to handle.

Nonlinear partial differential equations apply to processes in which slow changes occur. Examples are ion exchange, in which a slow saturation occurs as a result of the ion exchange process, and catalyst deactivation, in which, as a result of a cracking reaction, cokes deposits on the catalyst. In both cases it is not allowed to linearize the differential equation, because we cannot talk about a static situation. The problem of catalyst deactivation in a fixed bed reactor will be dealt with in somewhat more detail.

ASSUMPTIONS

Here we shall analyze a packed-bed tubular reactor with a gaseous mixture. To make the problem not too complex, the following assumptions are made:

1. dispersion is ignored;
2. gasflow through the reactor is independent of the geometric coordinate;
3. conditions are isothermal; and
4. gradients in radial and tangential direction can be ignored compared to gradients in axial direction.

MODEL EQUATIONS

When a component A is converted in the reactor, a partial mass balance for this component is:

$$\epsilon A_c \Delta z \frac{\partial c_a}{\partial t} + \frac{\partial}{\partial z}(Qc_a)\Delta z = -\epsilon A_c \Delta z r_a \qquad (13.1)$$

where ϵ = porosity
A_c = cross-sectional area of the reactor (m^2)
c_a = reactant concentration (kg/m^3)
Q = volumetric flow (m^3/sec)
r_a = rate of reaction (kg/m^3-sec)
Δz = length of the reactor section (m)

The first term represents the accumulation of component A, the second term the transport by flow and the third term the disappearance of A by reaction. The gasflow can be written as:

$$Q = v\epsilon A_c \qquad (13.2)$$

where v = the gas velocity (m/sec)

When the gasflow is constant, substitution of Equation 13.1 into Equation 13.2 yields:

$$\frac{\partial c_a}{\partial t} + v\frac{\partial c_a}{\partial z} = -r_a \qquad (13.3)$$

Now it is assumed that cokes deposits on the catalyst with a rate proportional to the conversion rate of component A, resulting in the following partial mass balance:

$$\rho_c(1-\epsilon)A_c\Delta z \frac{\partial c_c}{\partial t} = \alpha r_a \epsilon A_c \Delta z \qquad (13.4)$$

where ρ_c = catalyst density (kg/m^3)
c_c = coke concentration (kg/kg catalyst)
α = a proportionality factor

When component A is converted via a first-order reaction:

$$r_a = k' c_a \qquad (13.5)$$

Here, k' is a function of coke concentration. When the coke concentration increases, the rate of reaction r_a will decrease, because less catalyst area is available for reaction. Therefore k' will decrease as c_c increases. In the literature [32] two equations are represented for the relation between k' and the coke concentration. In the first equation an exponential relationship is assumed between k' and c_c according to:

$$k' = k_0' e^{-\beta c_c} \tag{13.6}$$

In the second equation a hyperbolic dependence is assumed:

$$k' = \frac{k_0'}{1 + k_1' c_c} \tag{13.7}$$

In both cases we see that if the coke concentration is equal to zero, k' is equal k_0' and Equation 13.5 gives the usual expression for the conversion rate of a first-order reaction. In Figure 13.1 k'/k_0' is given as a function of the coke concentration for $\beta = 100$ and $k_1' = 200$.

QUASISTATIC ANALYSIS

The rate at which the catalyst is deactivated is usually very small compared to all other changes. This means that concentration variations in component A may be assumed to adjust themselves instantaneously; hence the time variation in the partial mass balance of component A may be ignored. Equation 13.3 then reduces to:

$$v \frac{\partial c_a}{\partial z} = -r_a \tag{13.8}$$

To transform the model to a set of equations which can be solved analytically, the following substitutions are introduced:

$$\psi_a = c_a/\bar{c}_{ai}$$

$$\zeta = z/\ell$$

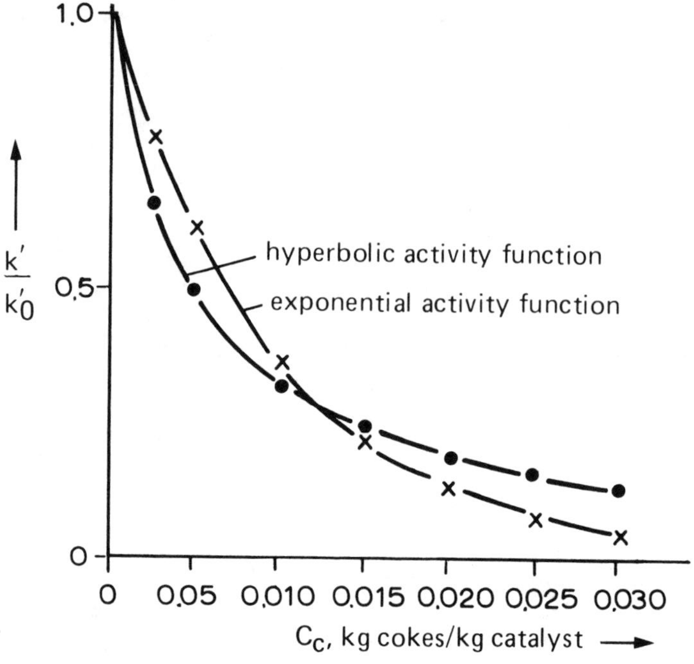

Figure 13.1 Hyperbolic and exponential activity function for $\beta = 100$ and $k_1' = 200$.

$$\tau = \alpha\bar{v}t/\ell$$

$$\psi_c = \frac{1-\epsilon}{\epsilon} \cdot \frac{c_c\rho_c}{\bar{c}_{ai}}$$

$$k_0 = k_0'\ell/\bar{v}$$

$$k_1 = \frac{k_1'}{\rho_c}\bar{c}_{ai}\frac{\epsilon}{1-\epsilon}$$

$$\gamma = \frac{\beta\bar{c}_{ai}}{\rho_c}\frac{\epsilon}{1-\epsilon}$$

$$\phi = v/\bar{v} \qquad (13.9)$$

in which $\bar{c}_{ai}$ is the average value of the inlet concentration of component A and where the bar over a variable indicates the value in the static situation. Equations 13.4 to 13.8 can now be written as:

$$\phi \frac{\partial \psi_a}{\partial \zeta} = -g(\psi_c)\psi_a \tag{13.10}$$

$$\frac{\partial \psi_c}{\partial \tau} = g(\psi_c)\psi_a \tag{13.11}$$

where, for exponential deactivation:

$$g(\psi_c) = k_0 e^{-\gamma \psi_c} \tag{13.12}$$

and, for hyperbolic deactivation:

$$g(\psi_c) = \frac{k_0}{1 + k_1 \psi_c} \tag{13.13}$$

As the model contains a derivative with respect to the location ζ and another with respect to time τ, the model description must be completed by defining a boundary and initial condition. When the concentration of component A at the reactor entrance is time-dependent, the following boundary condition holds:

$$\psi_a(0,\tau) = \psi_{a0}(\tau) \tag{13.14}$$

At time $\tau = 0$ a certain coke concentration will be present on the catalyst. As the concentration and reaction rate of component A are position-dependent, the coke concentration will also be position-dependent. For reasons of simplicity, however, it is assumed that the coke concentration at time $\tau = 0$ is constant, thus:

$$\psi_c(\zeta,0) = \psi_{c0} \tag{13.15}$$

When studying catalyst deactivation, we usually start with a clean catalyst, thus $\psi_{c0} = 0$ usually. In the previous chapters the next step was linearization of the mathematical model. This, however, is not possible here because there is a sustained dynamic situation in which the coke concentration increases and, therefore, the catalyst activity and the conversion gradually decrease. The model equations will be solved for constant fluid velocity, which means that $\phi = 1$.

SOLUTION METHOD

The method to solve the transformed model is, among others, given by Aris and Amundson [33]. It is assumed that Equations 13.10 and 13.11 have a solution of the following form:

$$\psi_a(\zeta,\tau) = \frac{\psi_c(\zeta,\tau) - \psi_{c0}}{\psi_c(0,\tau) - \psi_{c0}}\, \psi_{a0}(\tau) \qquad (13.16)$$

This expression indicates that the concentration of component A depends on the inlet concentration $\psi_{a0}(\tau)$, but also depends in a complicated way on the coke concentration. Combination of Equations 13.11 and 13.16 results in:

$$\frac{\partial \psi_c(\zeta,\tau)}{\partial \tau} = g(\psi_c)\, \frac{\psi_c(\zeta,\tau) - \psi_{c0}}{\psi_c(0,\tau) - \psi_{c0}} \cdot \psi_{a0}(\tau) \qquad (13.17)$$

For the boundary condition at $\zeta = 0$, this equation reduces to:

$$\frac{\partial \psi_c(0,\tau)}{\partial \tau} = g[\psi_c(0,\tau)] \cdot \psi_{a0}(\tau) \qquad (13.18)$$

which can also be derived directly from Equation 13.11. As Equation 13.18, contains only functions of τ, this equation can be integrated to τ:

$$\int_0^{\tau} \psi_{a0}(\tau')\mathrm{d}\tau' = \int_{\psi_{c0}}^{\psi_c(0,\tau)} \frac{\mathrm{d}w}{g(w)} \qquad (13.19)$$

Equation 13.16 can simply be differentiated to ζ:

$$\frac{\partial \psi_a(\zeta,\tau)}{\partial \zeta} = \frac{\psi_{a0}(\tau)}{\psi_c(0,\tau) - \psi_{c0}} \cdot \frac{\partial \psi_c(\zeta,\tau)}{\partial \zeta} \qquad (13.20)$$

from which follows the derivative of the coke concentration to the distance coordinate ζ:

$$\frac{\partial \psi_c(\zeta,\tau)}{\partial \zeta} = \frac{\psi_c(0,\tau) - \psi_{c0}}{\psi_{a0}(\tau)} \cdot \frac{\partial \psi_a(\zeta,\tau)}{\partial \zeta} \tag{13.21}$$

Hence, from Equations 13.10 and 13.11:

$$\frac{\partial \psi_a(\zeta,\tau)}{\partial \zeta} = -\frac{\partial \psi_c(\zeta,\tau)}{\partial \tau} \tag{13.22}$$

Combination of Equations 13.17, 13.21 and 13.22 gives:

$$\frac{\partial \psi_c(\zeta,\tau)}{\partial \zeta} = -g[\psi_c(\zeta,\tau)]\,[\psi_c(\zeta,\tau) - \psi_{c0}] \tag{13.23}$$

Integration of Equation 13.23 results in:

$$-\zeta = \int_{\psi_c(0,\tau)}^{\psi_c(\zeta,\tau)} \frac{dw}{(w - \psi_{c0})g(w)} \tag{13.24}$$

which gives a functional implicit relationship between $\psi_c(\zeta,\tau)$, ζ and $\psi_c(0,\tau)$. For a given $g(\psi_c)$ the solution of Equations 13.16, 13.19 and 13.24 now exists with as results $\psi_a(\zeta,\tau)$ and $\psi_c(\zeta,\tau)$.

EXPONENTIAL ACTIVITY FUNCTION

When taking an exponential activity function (Equation 13.12), assuming that the feed concentration is constant, which means that $\psi_{a0}(\tau) = 1$, and starting with a clean catalyst ($\psi_c(\zeta, 0) = \psi_{c0} = 0$), Equation 13.19 becomes:

$$\int_0^{\tau} d\tau' = \int_0^{\psi_c(0,\tau)} \frac{dw}{k_0 e^{-\gamma w}} \tag{13.25}$$

or

$$k_0\tau = \frac{1}{\gamma} e^{\gamma w} \Big/_{0}^{\psi_c(0,\tau)} = \frac{1}{\gamma}\,[e^{\gamma \psi_c(0,\tau)} - 1] \tag{13.26}$$

The expression for $\psi_c(0,\tau)$ follows from Equation 13.26:

$$\psi_c(0,\tau) = \frac{1}{\gamma} \ln (\gamma k_0 \tau + 1) \tag{13.27}$$

Equation 13.24 now becomes:

$$-k_0 \zeta = \int_{\psi_c(0,\tau)}^{\psi_c(\zeta,\tau)} \frac{dw}{we^{-\gamma w}} \tag{13.28}$$

This integral cannot be evaluated directly. For small coke concentrations, the following approximation can be made:

$$e^{\gamma w} \approx 1 + \gamma w \tag{13.29}$$

thus:

$$-k_0 \zeta = \int_{\psi_c(0,\tau)}^{\psi_c(\zeta,\tau)} \frac{1+\gamma w}{w} dw = \int_{\psi_c(0,\tau)}^{\psi_c(\zeta,\tau)} \frac{dw}{w} + \int_{\psi_c(0,\tau)}^{\psi_c(\zeta,\tau)} \gamma dw$$

$$= \ln \frac{\psi_c(\zeta,\tau)}{\psi_c(0,\tau)} + \gamma [\psi_c(\zeta,\tau) - \psi_c(0,\tau)] \tag{13.30}$$

As it is assumed that the feed concentration is constant ($\psi_a(0,\tau) = 1$) and the catalyst does not contain any coke ($\psi_c(\zeta,0) = 0$), Equation 13.16 can be written as:

$$\psi_a(\zeta,\tau) = \frac{\psi_c(\zeta,\tau)}{\psi_c(0,\tau)} \tag{13.31}$$

or

$$\ln \psi_a(\zeta,\tau) = \ln \frac{\psi_c(\zeta,\tau)}{\psi_c(0,\tau)} \tag{13.32}$$

Combination of Equations 13.30 and 13.32 results in:

$$\psi_a(\zeta,\tau) = \exp\{-k_0\zeta - \gamma[\psi_c(\zeta,\tau) - \psi_c(0,\tau)]\} \tag{13.33}$$

The solution is now given by Equations 13.27, 13.30 and 13.33. First, $\psi_c(0,\tau)$ can be calculated from Equation 13.27 for different values of τ. Then $\psi_c(\zeta,\tau)$ can be solved by iteration from Equation 13.30, for different values of ζ and τ. Finally $\psi_a(\zeta,\tau)$ can be determined from Equation 13.33. When it is assumed that there is no coke deposition on the catalyst, Equation 13.33 reduces to:

$$\psi_a(\zeta,\tau) = \exp(-k_0\zeta) \tag{13.34}$$

When Equations 13.10 and 13.12 are combined for $\phi = 1$ and $\psi_c = 0$, the result is:

$$\frac{\partial\psi_a}{\partial\zeta} = -k_0\psi_a \tag{13.35}$$

It can be seen that Equation 13.34 is the solution of Equation 13.35.

HYPERBOLIC ACTIVITY FUNCTION

Again, a clean catalyst and constant feed concentration are assumed, Equation 13.19 becomes:

$$\int_0^{\tau} d\tau' = \int_0^{\psi_c(0,\tau)} \frac{dw}{\dfrac{k_0}{1 + k_1 w}} \tag{13.36}$$

thus:

$$k_0\tau = \left(w + \frac{1}{2}k_1 w^2\right)\Bigg/_{0}^{\psi_c(0,\tau)} = \psi_c(0,\tau) + \frac{1}{2}k_1\psi_c^2(0,\tau) \tag{13.37}$$

Equation 13.37 can also be written as:

$$k_1\psi_c^2(0,\tau) + 2\psi_c(0,\tau) - 2k_0\tau = 0 \tag{13.38}$$

from which:

$$\psi_c(0,\tau) = \frac{-1 + \sqrt{1 + 2k_0k_1\tau}}{k_1} \tag{13.39}$$

Equation 13.24 can be written as:

$$-k_0\zeta = \int_{\psi_c(0,\tau)}^{\psi_c(\zeta,\tau)} \frac{dw}{\dfrac{w}{1 + k_1w}} = \int_{\psi_c(0,\tau)}^{\psi_c(\zeta,\tau)} \frac{dw}{w} + k_1 \int_{\psi_c(0,\tau)}^{\psi_c(\zeta,\tau)} dw$$

$$= ln\,\frac{\psi_c(\zeta,\tau)}{\psi_c(0,\tau)} + k_1[\psi_c(\zeta,\tau) - \psi_c(0,\tau)] \tag{13.40}$$

Also now ψ_a can be calculated with the aid of Equation 13.32:

$$\psi_a(\zeta,\tau) = \exp\{-k_0\zeta - k_1[\psi_c(\zeta,\tau) - \psi_c(0,\tau)]\} \tag{13.41}$$

When we compare Equations 13.30 and 13.33 with Equations 13.40 and 13.41, we see that γ has been replaced by k_1. The hyperbolic and exponential activity functions are also identical for low values of coke concentrations:

$$e^{-\gamma\psi_c} \simeq \frac{1}{1 + \gamma\psi_c} \tag{13.42}$$

hence the result is obvious.

However Equations 13.39, 13.40 and 13.41 are valid generally and not restricted to low coke concentrations.

EXAMPLE OF HYPERBOLIC DEACTIVATION

For a reaction in which a hydrocarbon is cracked, the following data are given:

$k'_0 = 2.5\ \text{sec}^{-1}$;
$\bar{v} = 0.5$ m/sec;
$\ell = 0.4$ m;
$\epsilon = 0.40$;
$\bar{c}_{ai} = 1.5\ \text{kg/m}^3$;
$k'_1 = 200$ kg of catalyst/kg of coke;
$\alpha = 10^{-5}$ kg of coke/kg of A; and
$\rho_c = 1180\ \text{kg/m}^3$.

By applying Equation 13.9 we find:

$$k_0 = \frac{k'_0 \ell}{v} = \frac{2.5 \times 0.4}{0.5} = 2$$

$$k_1 = \frac{k'_1}{\rho_c} \bar{c}_{ai} \frac{\epsilon}{1 - \epsilon} = \frac{200}{1800} \cdot 1.5 \cdot \frac{0.4}{0.6} = 0.17$$

Initially, at time t = 0, when no coke is present:

$$\bar{v} \frac{dc_a}{dz} = -k'_0 c_a \tag{13.43}$$

or

$$\frac{c_{a,out}}{c_{a,in}} = e^{-k_0} \tag{13.44}$$

thus

$$\frac{c_{a,out}}{c_{a,in}} = e^{-2} = 0.14 \tag{13.45}$$

The conversion C is:

$$C = 1 - \frac{c_{a,out}}{c_{a,in}} = 1 - 0.14 = 0.86 \tag{13.46}$$

As time increases, the conversion will decrease because the catalyst area is more and more occupied by coke. The coke concentration at z = 0 (at the entrance of the reactor) can be calculated with the aid of Equation 13.39:

$$\psi_c(0,\tau) = \frac{-1 + \sqrt{1 + 2k_0k_1\tau}}{k_1} = \frac{-1 + \sqrt{1 + 0.68\tau}}{0.17} \qquad (13.47)$$

With the aid of Equation 13.40, $\psi_c(\zeta,\tau)$ is calculated for different values of ζ and τ. With the aid of the transformations of Equation 13.9, the calculations can easily be transformed to the original coordinates. Finally, ψ_a is calculated with the aid of Equation 13.41 for different values of ζ and τ. The concentration profile of the reactant as a function of time is shown in Figure 13.2, in the middle of the reactor ($z/\ell = 0.5$) and at the outlet ($z/\ell = 1$).

It can be seen that as time increases, the concentration of A increases, which means that less A is converted. Figure 13.3 shows the reactant concentration as a function of distance in the reactor. At $t = 0$ the exponential profile is clear, according to:

$$c_a = \bar{c}_{a,in}e^{-2z/\ell} \qquad (13.48)$$

For larger values of time the curve is smoothed; for increasing catalyst deactivation less component is converted, especially in the beginning. Figures 13.4 and 13.5 show the coke concentration as function of location and time respectively. Finally in Figure 13.6 the decreasing conversion is shown for the first 200 hr.

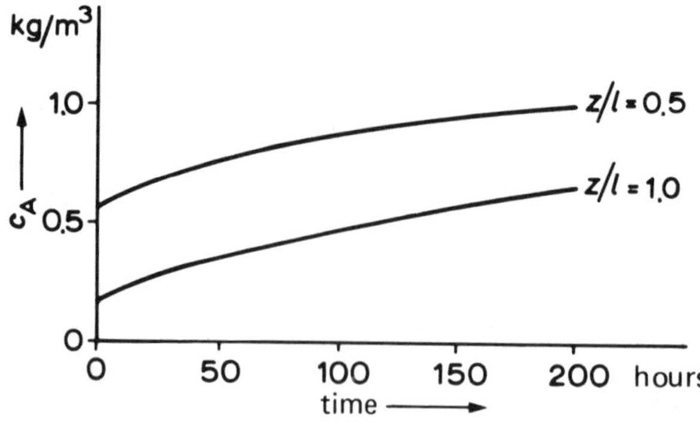

Figure 13.2 Reactant concentration as a function of time in the case of the hyperbolic activity function.

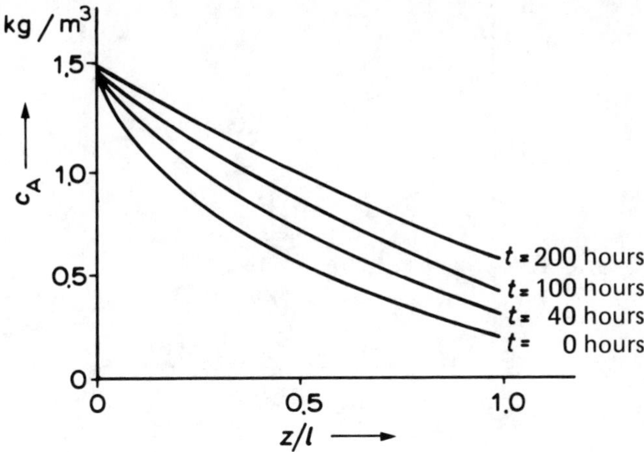

Figure 13.3 Reactant concentration as a function of location with a hyperbolic activity function.

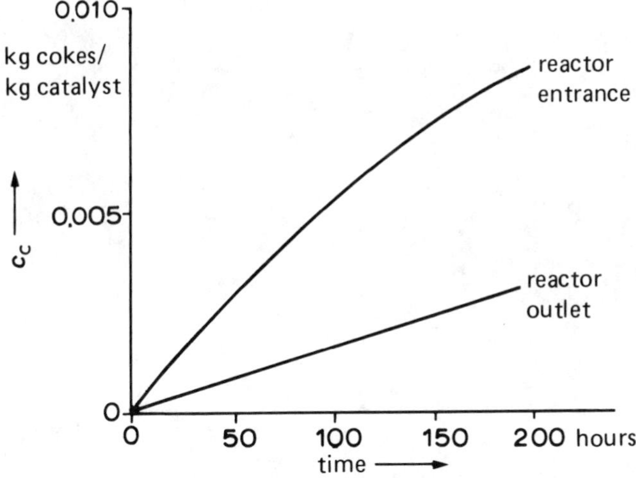

Figure 13.4 Coke concentration as a function of location with hyperbolic activity function.

CONCLUSIONS

It can be seen that the model for catalyst deactivation in an isothermal tubular reactor can be solved analytically in the case of an hyperbolic activity

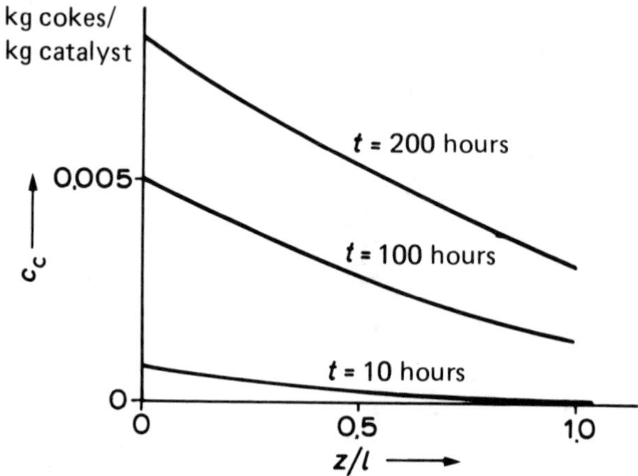

Figure 13.5 Coke concentration as a function of time at the reactor entrance and exit in the case of a hyperbolic activity function.

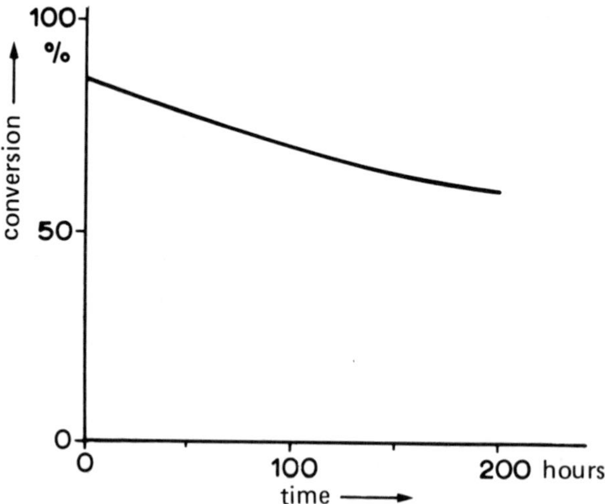

Figure 13.6 Conversion as a function of time for a hyperbolic activity function.

function. When the reaction mechanism is no longer of first order in the reacting components and when conditions are no longer isothermal, the partial nonlinear differential equations cannot be handled analytically anymore. A suitable numerical solution method must then be selected to perform computer calculations (see Chapter 15).

CHAPTER 14

DISCRETIZATION AND CONTINUIZATION

In the previous chapters, attention was paid to the description of the dynamic behavior of distributed systems by means of partial differential equations. The equations could be solved "by using pencil and paper," if the geometrics are simple.

In many practical problems, however, the geometrics are complicated. A river, for example, often has a position-dependent cross-sectional area, resulting in position-dependent coefficients. A lake has an arbitrary shape, resulting in complicated geometric boundary conditions. Industrial heat exchangers have a baffle system at the shell side, resulting in a complicated mixing pattern somewhere between plug flow and ideally mixed sections. This often leads to a subdivision of the system into sections, which are each assumed to be ideally mixed. In this way, partial differential equations are approximated by sets of simultaneous ordinary differential equations.

EXAMPLE

The following partial differential equation holds for an isothermal tubular flow reactor without dispersion with a first-order reaction:

$$\frac{\partial c}{\partial t} + v\frac{\partial c}{\partial z} + k_0 c = 0 \tag{14.1}$$

When the system is divided into sections, $\partial c/\partial z$ can be replaced by:

$$\frac{\partial c}{\partial z} \simeq \frac{c_i - c_{i-1}}{\Delta z} \tag{14.2}$$

a "backward difference," thus, Equation 14.1 becomes, for section i:

$$\frac{dc_i}{dt} + \frac{v}{\Delta z}(c_i - c_{i-1}) + k_0 c_i = 0 \tag{14.3}$$

which is a first-order linear differential equation. Different terms are used in the literature to indicate the division into sections, such as quantization, discretization, finite sections approach or lumping.

An important question is: in how many sections must the system be divided? This will be discussed in a later section of this chapter, including by determining the accuracy of the approximation. If the geometrics are relatively simple, the set of simultaneous ordinary difference equations which originates from the discretization can be solved with pencil and paper. This can give insight into the accuracy of the discretization. Introduction of the operator s into Equation 14.3 results in:

$$sc_i + \frac{v}{\Delta z}(c_i - c_{i-1}) + k_0 c_i = 0 \tag{14.4}$$

or

$$\left[s + \frac{v}{\Delta z} + k_0\right] c_i - \frac{v}{\Delta z} c_{i-1} = 0 \tag{14.5}$$

For linear difference equations it is known that the general solution has the form $c_i = C \cdot \beta^i$. Further, $c_{i-1} = C\beta^{i-1} = C\beta^i/\beta = c_i/\beta$, thus, Equation 14.5 can be written as:

$$(a_0 - a_1\beta^{-1})c_i = 0 \tag{14.6}$$

The general solution is determined by the characteristic equation $a_0 - a_1\beta^{-1} = 0$, thus $\beta = a_1/a_0$, and the general solution becomes:

$$c_i = C\beta^i = C\left(\frac{a_1}{a_0}\right)^i \tag{14.7}$$

The unknown constant C is determined from the boundary condition for the first section (i = 0), for which $c_i = C = c_0$ thus:

$$c_i = c_0\left(\frac{v/\Delta z}{s + \frac{v}{\Delta z} + k_0}\right)^i \tag{14.8}$$

For i = n, Equation 14.8 can be written as:

$$\frac{c_n}{c_0} = \left[\frac{\frac{v}{\Delta z}\Big/\left(\frac{v}{\Delta z} + k_0\right)}{\frac{1}{\frac{v}{\Delta z} + k_0} s + 1}\right]^n = \frac{K^n}{(\tau s + 1)^n} \tag{14.9}$$

If n is large, the numerator can be approximated by:

$$\left[\frac{v}{\Delta z}\Big/\left(\frac{v}{\Delta z} + k_0\right)\right]^n = \left(\frac{1}{1 + \frac{k_0 \Delta z}{v}}\right)^n \simeq e^{-k_0 n \Delta z/v} \cong e^{-k_0 z/v} \tag{14.10}$$

Similarly, the following approximation can be made:

$$\left(\frac{1}{1 + \tau s}\right)^n \simeq e^{-n\tau s} \tag{14.11}$$

with

$$n\tau = \frac{n}{\frac{v}{\Delta z} + k_0} = \frac{n\Delta z}{v + k_0 \Delta z} = \frac{z}{v + k_0 \Delta z} \tag{14.12}$$

If n is large and Δz consequently small, the term $k\Delta z$ in Equation 14.12 can be ignored compared to v. Equation 14.9 can then finally be written as:

$$\frac{c_n}{c_0} \simeq e^{-k_0 z/v} e^{-sz/v} \tag{14.13}$$

which is also the exact solution of Equation 14.1. When n is smaller, Δz is larger, and errors will occur in Equations 14.10 and 14.12. Before introducing more examples, we shall discuss solution methods for difference equations.

DIFFERENCES

In the previous example, the term $\partial c/\partial z$ was replaced by $(c_i - c_{i-1})/\Delta z$. This is called a backward difference. The approximation can be judged by developing c_{i-1} around the point c_i into a Taylor series:

$$c_{i-1} = c_i + \left(\frac{\partial c}{\partial z}\right)_i (z_{i-1} - z_i) + \frac{1}{2!}\left(\frac{\partial^2 c}{\partial z^2}\right)_i (z_{i-1} - z_i)^2 + \ldots . \qquad (14.14)$$

Hence, for the first derivative:

$$\left(\frac{\partial c}{\partial z}\right)_i = \frac{c_i - c_{i-1}}{\Delta z} + \frac{1}{2!}\left(\frac{\partial^2 c}{\partial z^2}\right)_i \Delta z + \theta(\Delta z)^2 + \ldots . \qquad (14.15)$$

The second and following terms in the right side indicate the error inherent in this approximation. However, there are also other possibilities to approximate the derivative. Development of c_{i+1} around c_i gives:

$$c_{i+1} = c_i + \left(\frac{\partial c}{\partial z}\right)_i (z_{i+1} - z_i) + \frac{1}{2!}\left(\frac{\partial^2 c}{\partial z^2}\right)_i (z_{i+1} - z_i)^2 + \ldots . \qquad (14.16)$$

from which the forward difference equation follows:

$$\left(\frac{\partial c}{\partial z}\right)_i = \frac{c_{i+1} - c_i}{\Delta z} + \frac{1}{2!}\left(\frac{\partial^2 c}{\partial z^2}\right)_i \Delta z + \theta(\Delta z)^2 \qquad (14.17)$$

This alternative, however, has as a disadvantage that the numerical solution can easily become unstable (see also Chapter 15). Another approximation which is sometimes applied is the central difference. This equation is obtained by summing Equations 14.15 and 14.17:

$$\left(\frac{\partial c}{\partial z}\right)_i = \frac{c_{i+1} - c_{i-1}}{2\Delta z} + \theta(\Delta z)^2 \qquad (14.18)$$

This alternative seems to be more accurate, but has the disadvantage that problems can arise in the approximation of the boundary conditions and instabilities can occur. Other difference schemes, also for higher-order derivatives, can be derived simply from the previous equations. For example, from Equations 14.14 and 14.16:

$$c_{i+1} + c_{i-1} = 2c_i + \left(\frac{\partial c}{\partial z}\right)_i (-\Delta z + \Delta z) + \left(\frac{\partial^2 c}{\partial z^2}\right)_i \Delta z + \theta(\Delta z)^2 \qquad (14.19)$$

hence the second derivative:

$$\left(\frac{\partial^2 c}{\partial z^2}\right)_i = \frac{c_{i+1} - 2c_i + c_{i-1}}{(\Delta z)^2} + \theta(\Delta z) \tag{14.20}$$

Other difference schemes are given in Table 14.1.

LINEAR DIFFERENCE EQUATIONS

A linear difference equation of order n has the following form:

$$a_0 y_i + a_1 y_{i+1} + a_2 y_{i+2} + \dots + a_n y_{i+n} = \phi(i) \tag{14.21}$$

The parameters a_0 to a_n can be constants or functions of i. The same applies to ϕ. The solution of this equation is analogous to the solution of linear differential equations: the solution consists of general and particular parts. To make the subject not too complicated, we shall restrict ourselves to a second-order linear difference equation with constants coefficients:

$$y_{i+2} - Ay_{i+1} + By_i = \phi(i) \tag{14.22}$$

A further restriction will be made to the general solution, which can be found by setting the right side of Equation 14.22 to zero. Assume that the solution has the following form:

Table 14.1 Difference Schemes for the First Derivative

Difference	Equation	Error Term	Determined for a Value of z Between
Forward (1st order)	$\frac{c_{i+1} - c_i}{\Delta z}$	$\frac{1}{2}\left(\frac{\partial^2 c}{\partial z^2}\right)_i \Delta z$	$i\Delta z \leqslant z \leqslant (i+1)\Delta z$
Backward (1st order)	$\frac{c_i - c_{i-1}}{\Delta z}$	$\frac{1}{2}\left(\frac{\partial^2 c}{\partial z^2}\right)_i \Delta z$	$(i-1)\Delta z \leqslant z \leqslant i\Delta z$
Central (2nd order)	$\frac{c_{i+1} - c_{i-1}}{2\Delta z}$	$\frac{1}{6}\left(\frac{\partial^3 c}{\partial z^3}\right)_i (\Delta z)^2$	$(i-1)\Delta z \leqslant z \leqslant (i+1)\Delta z$
Forward (2nd order)	$\frac{-c_{i+2} + 4c_{i+1} - 3c_i}{2\Delta z}$	$\frac{1}{3}\left(\frac{\partial^3 c}{\partial z^3}\right)_i (\Delta z)^2$	$i\Delta z \leqslant z \leqslant i\Delta z$

$$y_i = C\beta^i \tag{14.23}$$

in which C is a constant. Substitution of Equation 14.23 into Equation 14.22 gives:

$$C\beta^{i+2} - AC\beta^{i+1} + BC\beta^i = 0 \tag{14.24}$$

or

$$\beta^2 - A\beta + B = 0 \tag{14.25}$$

Equation 14.25 is called the characteristic equation. From this equation β follows:

$$\beta_{1,2} = \frac{A \pm \sqrt{A^2 - 4B}}{2} \tag{14.26}$$

As there are two roots, the general solution is:

$$y_i = C_1\beta_1^i + C_2\beta_2^i \tag{14.27}$$

in which C_1 and C_2 are constants which can be determined from the boundary conditions of the problem. The roots β_1 and β_2 can be real, different or equal, or complex. In the case of real and different roots, Equation 14.27 is the final general solution. When the roots are equal, the general solution is:

$$y_i = (C_1 + C_2 i)\beta^i \tag{14.28}$$

When the roots are complex, we may write:

$$\beta_1 = \alpha + j\gamma = re^{j\phi} \text{ and } \beta_2 = \alpha - j\gamma = re^{-j\phi} \tag{14.29}$$

where α = the real part of the complex number
γ = the imaginary part of the complex number
r = the modulus of the complex number
ϕ = the phase angle.

Now Equation 14.27 can be written as:

$$y_i = r^i(C_1 \cos i\phi + C_2 \sin i\phi) \tag{14.30}$$

EXAMPLE

Consider a stage of an extraction section as shown in Figure 14.1. The flows in the extract phase and raffinate phase are denoted S and F, respectively, the mole fractions of the dissolved substance are y and x, respectively. It is assumed that the flow rates do not change during the extraction. The partial mass balance for section i is:

$$Sy_{i-1} + Fx_{i+1} = Fx_i + Sy_i \tag{14.31}$$

Furthermore, it is assumed that for the component which is transferred, the following equilibrium relationship holds:

$$y_{i+1} = mx_{i+1} \tag{14.32}$$

where m is the distribution coefficient. Substitution of Equation 14.32 into 14.31 gives:

$$mSy_{i-1} + Fy_{i+1} = Fy_i + mSy_i \tag{14.33}$$

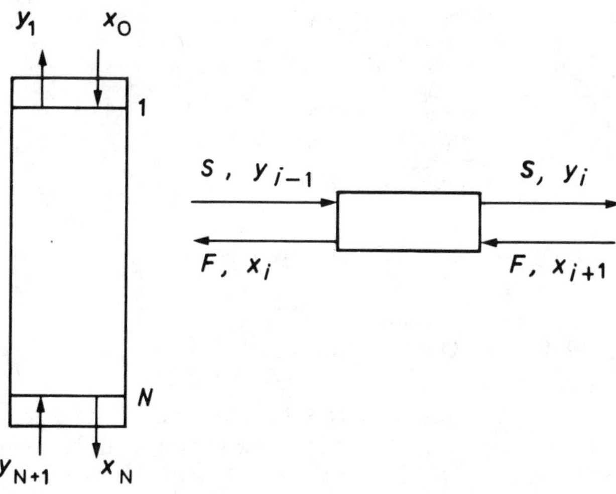

Figure 14.1 Extraction process.

thus:

$$y_{i+1} - y_i \left(\frac{mS + F}{F}\right) + y_{i-1} \frac{mS}{F} = 0 \tag{14.34}$$

which can be written as:

$$y_{i+1} - (\alpha_0 + 1)y_i + \alpha_0 y_{i-1} = 0 \tag{14.35}$$

When a solution of Equation 14.23 is to be found, the characteristic equation is:

$$\beta^2 - (\alpha_0 + 1)\beta + \alpha_0 = 0 \tag{14.36}$$

which results in $\beta_1 = 1$ and $\beta_2 = \alpha_0$, thus:

$$y_i = C_1 + C_2\alpha_0^i \tag{14.37}$$

The boundary conditions can be determined for i = 0 and i = N + 1 (inlet flows). For i = 0, y_0 is equal to $y_0 = C_1 + C_2$ and for $i = N + 1$, x_{N+1} is given, which can be converted into a value for $y_{N+1} = C_1 + C_2\alpha_0^{N+1}$, thus:

$$y_{N+1} - y_0 = C_2(\alpha_0^{N+1} - 1) \tag{14.38}$$

from which:

$$\frac{y_i - y_0}{y_{N+1} - y_0} = \frac{\alpha_0^i - 1}{\alpha_0^{N+1} - 1} \tag{14.39}$$

where y_0 is the concentration belonging to x_0 according to $y_0 = mx_0$.

DISCRETIZATION ERROR AND STABILITY

In the first section of this chapter some attention has already been paid to the discretization error. In Table 14.1 also the error term is given; this is the remainder term of the Taylor series which is ignored in the approximation of the partial derivative. The calculation of the discretization error will be illustrated by means of two examples.

Example

Consider a cascade of ideally mixed tanks, in which heating coils are installed. The heating medium is condensing steam. For tank i the following static energy balance holds:

$$\phi c_f \bar{\theta}_{i-1} - \phi c_f \bar{\theta}_i + hA(\bar{\theta}_s - \bar{\theta}_i) = 0 \tag{14.40}$$

where ϕ = fluid flow (kg/sec)
c_f = specific heat of the fluid (J/kg-°K)
$\bar{\theta}$ = static temperature (K)
h = heat transfer coefficient (W/m^2-°K)
A = heat transfer area (m^2)

In Equation 14.40 the first term is the energy flow from tank i − 1, the second term the energy flow from tank i and the third term the heat flow from the condensing steam to the fluid. This equation can be rewritten in the following form:

$$-a_- \bar{\theta}_{i-1} + a\bar{\theta}_i = \bar{\theta}_s \tag{14.41}$$

with $a_- = \phi c_f/hA$ and $a = 1 + \phi c_f/hA$. The general solution of Equation 14.41 is determined by the characteristic equation $-a_-\beta^{-1} + a = 0$, thus:

$$\beta = \frac{a_-}{a} = \frac{1}{1 + \Psi} \tag{14.42}$$

with $\Psi = hA/\phi c_f$. The general solution is: $\bar{\theta}_i = C\beta^i$. The particular solution is very simple here because the steam temperature is independent of i. Therefore, the particular solution is found to be $\bar{\theta}_i = \bar{\theta}_s$, resulting in the total solution:

$$\bar{\theta}_i = C\beta^i + \bar{\theta}_s \tag{14.43}$$

The unknown constant can be determined from the initial condition: $\bar{\theta}_0 = C + \bar{\theta}_s$, thus $C = \bar{\theta}_0 - \bar{\theta}_s$. The final solution can now be written as:

$$\bar{\theta}_i = \bar{\theta}_s - (\bar{\theta}_s - \bar{\theta}_0)\frac{i}{(1 + \Psi)^i} \tag{14.44}$$

This result can be compared to the static behavior of a heat exchanger, in which liquid flows through the pipes. In Chapter 10, the following result was found:

$$\bar{\theta}_i = \bar{\theta}_s - (\bar{\theta}_s - \bar{\theta}_0)e^{-z/vT_f} \tag{14.45}$$

with $T_f = M_f^* c_f/hA^*$. M_f^* was defined as the liquid mass per unit length and A^* as the tube area per unit length. In Equation 14.45, z/v is the delay time of the liquid from the entrance to point i, for which we may write:

$$\frac{z}{v} = \frac{M_f^*}{\phi}\left(\frac{i}{n}\right)\ell \tag{14.46}$$

Substitution of Equation 14.46 and the expression for T_f into Equation 14.45 gives, together with the expression for Ψ:

$$\bar{\theta}_i = \bar{\theta}_s - (\bar{\theta}_s - \bar{\theta}_0)e^{-\Psi i} \tag{14.47}$$

where $A = A^*\ell/n$, the area per section.

Comparison of Equations 14.44 and 14.47 gives an indication of the approximation:

$$e^{-\Psi i} \approx (1 + \Psi)^{-i} \tag{14.48}$$

The exponential function can be developed into a series as:

$$e^{-\Psi i} = 1 - \Psi i + \frac{(\Psi i)^2}{2!} + \ldots. \tag{14.49}$$

and the right side of Equation 14.48 can be written as:

$$(1 + \Psi)^{-i} = 1 - \Psi i + \frac{\Psi^2 i(i-1)}{2!} + \ldots. \tag{14.50}$$

The deviation which originates from the approximation becomes visible in the third term on the right sides of Equations 14.49 and 14.50. The difference is equal to $\Psi^2 i/2 = i/2(hA/\phi c_f)^2$. This deviation gives an impression of the magnitude of the discretization error.

In the previous example we started from a static model. Also in the dynamic case we can get an impression of the discretization error. As a second example, the discretized dynamic behavior of a heat exchanger will be discussed. The distributed dynamic behavior was already discussed in Chapter 10. The liquid temperature inside the tubes was given by the partial differential equation:

$$\frac{\partial\theta}{\partial t} + v\frac{\partial\theta}{\partial z} = T_f^{-1}(\theta_s - \theta) \tag{14.51}$$

where θ = liquid temperature (°K)
v = liquid velocity (m/sec)
θ_s = steam temperature (°K)

T_f was defined previously. Approximation by a discrete model gives the following set of ordinary differential equations:

$$\frac{d\theta_i}{dt} + v\frac{\theta_i - \theta_{i-1}}{\Delta z} \approx \bar{T}_f^{-1}(\theta_s - \theta_i) \tag{14.52}$$

where $i = 1, 2 \ldots n$

The accuracy of Equation 14.52 can be judged by continuization of the local variable $\theta_{i-1}(t)$ to a distributed variable $\theta_{i-1}(z,t)$ over the interval $(i-1)\Delta z \leqslant z \leqslant i\Delta z$. We can then introduce a Taylor series:

$$\theta_{i-1}(z,t) = \theta_{i-1}(i\Delta z,t) + \frac{z - i\Delta z}{1!}\left(\frac{\partial\theta_{i-1}}{\partial z}\right)_{i\Delta z} + \frac{(z - i\Delta z)^2}{2!}\left(\frac{\partial^2\theta_{i-1}}{\partial z^2}\right)_{i\Delta z} + \ldots. \tag{14.53}$$

Now it is required that θ_{i-1} and θ_i are connected smoothly. This means:

$$\left.\begin{aligned} \theta_{i-1}(i\Delta z,t) &= \theta_i(i\Delta z,t) \\ \left(\frac{\partial\theta_{i-1}}{\partial z}\right)_{i\Delta z} &= \left(\frac{\partial\theta_i}{\partial z}\right)_{i\Delta z} \\ \left(\frac{\partial^2\theta_{i-1}}{\partial z^2}\right)_{i\Delta z} &= \left(\frac{\partial^2\theta_i}{\partial z^2}\right)_{i\Delta z} \end{aligned}\right\} \tag{14.54}$$

etc. When this all is substituted into Equation 14.53 for the point $z = (i-1)\Delta z$, the result is:

$$\theta_{i-1}[(i-1)\Delta z,t] = \theta_i(i\Delta z,t) - \frac{\Delta z}{1!}\left(\frac{\partial\theta_i}{\partial z}\right)_{i\Delta z} + \frac{(\Delta z)^2}{2!}\left(\frac{\partial^2\theta_i}{\partial z^2}\right)_{i\Delta z} \tag{14.55}$$

Substitution of Equation 14.55 into Equation 14.51 leads to:

$$\left(\frac{\partial\theta_i}{\partial t}\right)_{i\Delta z} + v\left(\frac{\partial\theta_i}{\partial z}\right)_{i\Delta z} - \frac{v\Delta z}{2}\left(\frac{\partial^2\theta_i}{\partial z^2}\right)_{i\Delta z} \simeq \bar{T}_f^{-1}(\theta_s - \theta_i) \tag{14.56}$$

When we compare Equation 14.56 with Equation 14.51 we see that $\theta_i(i\Delta z)$ corresponds to $\theta(i\Delta z)$, but that Equation 14.56 contains an additional series of higher-order derivatives. If Δz is small, the term with the second derivative is the most important one. A second derivative to the geometric coordinate indicates a dispersion effect. Apparently, discretization is accompanied by an extension of the mathematical model, which could be called discretization dispersion. Hence plug flow cannot correctly be represented anymore. This picture is in fact well known from reactor engineering: the behavior of a cascade of ideally mixed tanks only approaches the behavior of plug flow if the number of tanks goes to infinity. In the case of a small number, ideal mixing plays the role of dispersion.

In the previous two examples, we analyzed the discretization error, but we did not go into detail about the stability of the discretized solution. Stability is related to the propagation of discretization errors. It can be advised, when analyzing system behavior, to investigate the discretization error first and determine the stability next, before making computer calculations. The determination of the stability of the discretized solution will be illustrated by means of an example.

Consider the differential equation:

$$\frac{dy}{dz} = -K_0 y \tag{14.57}$$

When the differential quotient is approximated by a forward difference, we may write:

$$\frac{y_{i+1} - y_i}{\Delta z} = -K_0 y_i \tag{14.58}$$

The accuracy of the approximation can be determined by developing y_{i+1} around y_i into a Taylor series:

$$y_{i+1} = y_i + \Delta z y_i' + \frac{(\Delta z)^2}{2!} y_i'' + \ldots \tag{14.59}$$

Substitution of Equation 14.59 into Equation 14.58 results in:

$$y_i' = -K_0 y_i - \frac{\Delta z}{2!} y_i'' + \epsilon_i \tag{14.60}$$

where the prime denotes a derivative and ϵ_i the error. The left side and the first term in the right side of Equation 14.60 represent the original equation (see Equation 14.57). The total error is therefore equal to the remainder of Equation 14.60. This error is of the order Δz. The stability of Equation 14.58 can be determined by repeated substitution.

$$y_i = y_0(1 - K_0 \Delta z)^i \tag{14.61}$$

where y_0 is the initial condition for $i = 0$. The solution is stable if $-1 \leqslant 1 - K_0 \Delta z \leqslant 1$, hence $0 \leqslant K_0 \Delta z \leqslant 2$.

By way of comparison, a central difference will now be introduced instead of a forward difference:

$$\frac{y_{i+1} - y_{i-1}}{2\Delta z} = -K_0 y_i \tag{14.62}$$

from which the following difference equation results:

$$y_{i+1} + 2K_0 \Delta z y_i - y_{i-1} = 0 \tag{14.63}$$

The roots of the characteristic equation are:

$$\beta_{1,2} = -K_0 \Delta z \pm \sqrt{K_0^2 (\Delta z)^2 + 1} \tag{14.64}$$

When $K_0 \Delta z$ is much smaller than one, which gives the best approximation, one root is slightly smaller than 1 and the other slightly smaller than -1, which means that the solution $Y_i = C_1 \beta_1^i + C_2 \beta_2^i$ will be unstable. It can be seen that not every discretization scheme gives a stable solution.

DIFFERENTIAL DIFFERENCE EQUATIONS

Another type of difference equation, which was only introduced but not analyzed yet, is the differential difference equation. When analyzing the static behavior of a cascade of ideally mixed tanks, a set of difference equations results. However, when a step disturbance is introduced in the inlet concentration of the first tank, there is a composition change in the flows through all reactors. Therefore, the difference equation contains a derivative with time. In this case we talk about a differential difference equation. The analytical solution of this type of equation can strongly be simplified by Laplace transformation, solving the resulting equation and transforming back. This will be illustrated by the following example.

Example

Consider a cascade of N identical ideally mixed tanks, each with a volume V_i. Suppose the tanks are fed by a flow of ϕ m^3/sec. The inlet concentration to the first tank changes stepwise, with magnitude c_0. The partial mass balance for stage i is:

$$\phi c_{i-1} - \phi c_i = V_i \frac{dc_i}{dt} \tag{14.65}$$

When starting from an equilibrium situation, Laplace transformation gives:

$$\phi c_{i-1} - \phi c_i = V_i s c_i \tag{14.66}$$

The solution of this difference equation is:

$$c_i = A \cdot \left(\frac{\phi}{\phi + V_i s}\right)^i \tag{14.67}$$

in which A is a constant. As the inlet concentration changes with stepwise:

$$c_0(s) = \frac{c_0(t)}{s} \tag{14.68}$$

thus:

$$c_i = \frac{c_0}{s}\left(\frac{\phi}{\phi + V_i s}\right)^i \tag{14.69}$$

from which for i = N

$$c_N = \frac{c_0}{s}\left(\frac{\phi}{\phi + V_i s}\right)^N \qquad (14.70)$$

From a Laplace transformation table the following transformation can be obtained:

$$£\left[\frac{t^{N-1}e^{-at}}{(N-1)!}\right] = \left(\frac{1}{s+a}\right)^N \qquad (14.71)$$

Now it is known that:

$$£^{-1}\left[\frac{1}{s}\,f(s)\right] = \int_0^t f(t)dt \qquad (14.72)$$

thus:

$$£^{-1}\left[\frac{1}{s(s+a)^N}\right] = \int_0^t \frac{t^{N-1}e^{-at}}{(N-1)!}\,dt$$

$$= \frac{1}{(N-1)!}\left(\frac{-t^{N-1}}{a}\,e^{-at} + \frac{N-1}{a}\int t^{N-2}e^{-at}dt\right)_0^t$$

$$= \frac{-t^{N-1}e^{-at}}{a(N-1)!} + \frac{1}{a(N-2)!}\int t^{N-2}e^{-at}dt = \ldots.$$

$$= \frac{-t^{N-1}e^{-at}}{a(N-1)!} - \frac{t^{N-2}e^{-at}}{a^2(N-2)!} - \ldots - \frac{e^{-at}}{a^N} + \frac{1}{a^N} \qquad (14.73)$$

which can be obtained from repeated partial integration for which the well known rule holds:

$$\int uv' = uv - \int vu' \qquad (14.74)$$

The prime denotes a derivative. Equation 14.70 can be written with the aid of Equation 14.73 as:

$$c_N = c_0\left[1 - e^{-\phi t/V_i} - \frac{\phi t}{V_i}\,e^{-\phi t/V_i} - \ldots. - \frac{(\phi t/V_i)^{N-1}}{(N-1)!}\,e^{-\phi t/V_i}\right]$$

thus

$$c_N = c_0 - c_0 e^{-\phi t/V_i} \left[1 + \frac{\phi t}{V_i} + \frac{1}{2!}\left(\frac{\phi t}{V_i}\right)^2 + \ldots . \frac{1}{(N-1)!}\left(\frac{\phi t}{V_i}\right)^{N-1} \right] \quad (14.75)$$

When N = 1, Equation 14.75 reduces to:

$$\begin{aligned} c_1 &= c_0 - c_0 e^{-\phi t/V_i} \\ &= c_0 (1 - e^{-t/\tau}) \end{aligned} \quad (14.76)$$

in which τ is the residence time in the tank. For N = 3 Equation 14.75 becomes:

$$c_3 = c_0 - c_0 e^{-\phi t/V_i} \left[1 + \frac{\phi t}{V_i} + \frac{1}{2}\left(\frac{\phi t}{V_i}\right)^2 + \frac{1}{6}\left(\frac{\phi t}{V_i}\right)^3 \right]$$

which can be written with the total volume $V_t = 3V_i$, as:

$$c_3 = c_0 - c_0 e^{-3\phi t/V_t} \left[1 + \frac{3\phi t}{V_t} + \frac{9}{2}\left(\frac{\phi t}{V_t}\right)^2 + \frac{27}{6}\left(\frac{\phi t}{V_t}\right)^3 \right] \quad (14.77)$$

It is difficult to evaluate Equation 14.75 for the case where N approaches infinity, but substitution of $V_t = NV_i$ into Equation 14.70 gives:

$$c_N = \frac{c_0}{s}\left(1 + \frac{V_t s}{N\phi}\right)^{-N} \quad (14.78)$$

which can be developed according to the binomial theorem to:

$$c_N = \frac{c_0}{s}\left[1 - \frac{V_t s}{\phi} + \frac{N+1}{2N}\left(\frac{V_t s}{\phi}\right)^2 - \frac{(N+1)(N+2)}{3!N^2}\left(\frac{V_t s}{\phi}\right)^3 + \ldots \right] \quad (14.79)$$

When N is large, the expression between brackets can be approximated by:

$$c_N = \frac{c_0}{s}\left[1 - \frac{V_t s}{\phi} + \frac{1}{2!}\left(\frac{V_t s}{\phi}\right)^2 - \frac{1}{3!}\left(\frac{V_t s}{\phi}\right)^3 + \ldots\right]$$

$$\simeq \frac{c_0}{s}\, e^{-V_t s/\phi} \qquad (14.80)$$

Back transformation gives:

$$c_N = 0 \quad \text{for} \quad 0 < t < V_t/\phi$$

$$c_N = c_0 \quad \text{for} \quad t > V_t/\phi$$

In Figure 14.2, c_N/c_0 is given as a function of $\phi t/V_t$ for different values of N, as calculated with the aid of Equation 14.75. It can be seen that a large number of small ideally mixed tanks shows almost the same response as a tube through which liquid flows and which has the same volume as all the tanks.

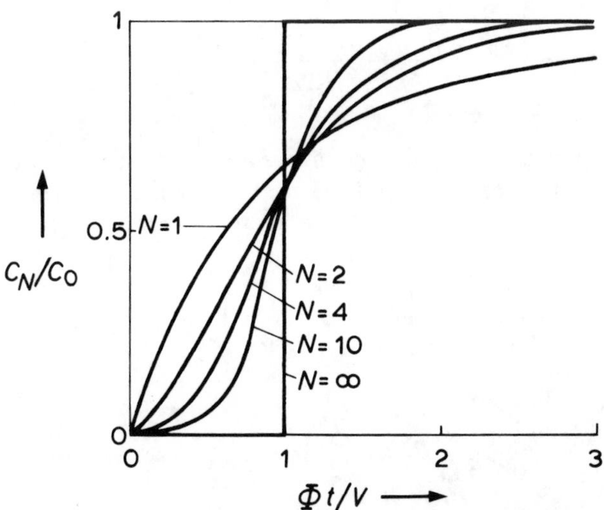

Figure 14.2 Response of N tanks.

POLLUTION IN A RIVER

Until now we have analyzed equations in which the highest partial derivative was first-order. Equations with higher-order derivatives, however, are handled in a similar way. As an example we shall consider pollution in a river. The model with initial and boundary conditions was discussed extensively in Chapter 11; where the following partial differential equation was found.

$$\frac{\partial c}{\partial t} + v\frac{\partial c}{\partial z} - D_f\frac{\partial^2 c}{\partial z^2} + kc = 0 \tag{14.81}$$

When a backward difference is applied for $\partial c/\partial z$ and a central difference for $\partial^2 c/\partial z^2$, these terms are replaced by:

$$\left.\begin{aligned} \frac{\partial c}{\partial z} &\cong \frac{c_i - c_{i-1}}{\Delta z} \\ \frac{\partial^2 c}{\partial z^2} &\cong \frac{c_{i+1} - 2c_i + c_{i-1}}{(\Delta z)^2} \end{aligned}\right\} \tag{14.82}$$

Combination of Equations 14.81 and 14.82 gives:

$$\frac{dc_i}{dt} + v\frac{c_i - c_{i-1}}{\Delta z} - D\frac{c_{i+1} - 2c_i + c_{i-1}}{(\Delta z)^2} + kc_i = 0 \tag{14.83}$$

It can be expected that for the continuous case, the dispersion coefficient is equal to the physical D; therefore, the dispersion coefficient in Equation 14.81 has a subscript f. Because approximations have been made in Equation 14.82, it can be expected that the dispersion coefficient in Equation 14.82 will not be equal to D_f.

When c_{i-1} and c_{i+1} are developed into a Taylor series around point c_i, Equation 14.83 can be written as:

$$\left(\frac{\partial c_i}{\partial t}\right)_{i\Delta z} + v\left(\frac{\partial c_i}{\partial z}\right)_{i\Delta z} - \left(D + \frac{v\Delta z}{2}\right)\left(\frac{\partial^2 c_i}{\partial z^2}\right)_{i\Delta z} + kc_i(i\Delta z,t) = 0 \tag{14.84}$$

with an error of the order $(\Delta z)^2$. As could be expected, the dispersion coefficient is increased with the discretization dispersion $v\Delta z/2$. However,

Δz can be selected in such a way that the discretization dispersion is just equal to the physical dispersion:

$$\frac{\Delta z}{\ell} = \frac{2D_f}{v\ell} = \frac{2}{Pe} \tag{14.85}$$

where $Pe = v\ell/D_f$, a type of Péclet number
ℓ = the length of the considered river section.

When Equations 14.81 and 14.84 are compared, it can be seen that:

$$D_f \simeq D + \frac{v\Delta z}{2} \quad \text{or} \quad D = D_f - \frac{v\Delta z}{2} \tag{14.86}$$

When $\Delta z/\ell = 2/Pe$ as proposed in Equation 14.85, D_f will be equal to $v\Delta z/2$, resulting in $D = 0$. Equation 14.83 then reduces to:

$$\frac{dc_i}{dt} + v\frac{c_i - c_{i-1}}{\Delta z} + kc_i = 0 \tag{14.87}$$

This selection of Δz results in a very simple model. However, when D is large, the approximation is too inaccurate, because third- and higher-order terms in the Taylor series have too much influence. Then it is better to select Δz not too large, and replace D in Equation 14.83 by:

$$D' = D_f - \frac{v\Delta z}{2} \tag{14.88}$$

Solution of the Discretized Equation

The equation for the river (Equation 14.83) will now be solved. Laplace transformation of this equation gives:

$$sc_i + \frac{v}{\Delta z}(c_i - c_{i-1}) - \frac{D'}{(\Delta z)^2}(c_{i+1} - 2c_i + c_{i+1}) + kc_i = c_i(z,0^+) \tag{14.89}$$

in which D is replaced by D′ according to Equation 14.88. When the initial condition is set equal to zero, the result is:

$$a_- c_{i-1} - ac_i + a_+ c_{i+1} = 0 \tag{14.90}$$

in which

$$\left.\begin{aligned} a_- &= \frac{v}{\Delta z} + \frac{D'}{(\Delta z)^2} = \frac{v}{2\Delta z} + \frac{D_f}{(\Delta z)^2} \\ a &= s + k + \frac{2D'}{(\Delta z)^2} + \frac{v}{\Delta z} = s + k + \frac{2D_f}{(\Delta z)^2} \\ a_+ &= \frac{D'}{(\Delta z)^2} = -\frac{v}{2\Delta z} + \frac{D_f}{(\Delta z)^2} \end{aligned}\right\} \tag{14.91}$$

General Solution

The general solution of the difference equation has the following form:

$$c_i = A_1(J_1)^i + A_2(J_2)^i \tag{14.92}$$

in which $J_{1,2}$ follows from the characteristic equation:

$$a_- J^{-1} - a + a_+ J = 0 \tag{14.93}$$

thus

$$J_{1,2} = \frac{a \pm \sqrt{a^2 - 4a_- a_+}}{2a_+} \tag{14.94}$$

Substitution of Equation 14.91 into Equation 14.94 results in:

$$J_{1,2} = \frac{s + k + \dfrac{2D_f}{(\Delta z)^2} \pm \sqrt{(s+k)^2 + \dfrac{4D_f}{(\Delta z)^2}(s+k) + \dfrac{v^2}{(\Delta z)^2}}}{2\dfrac{D_f}{(\Delta z)^2} - \dfrac{v}{\Delta z}} \tag{14.95}$$

Boundary Conditions

The entrance section (i = 1) determines the inlet boundary condition (Figure 14.3):

$$\Delta z \frac{dc_1}{dt} = vc_0 + D' \frac{c_2 - c_1}{\Delta z} - vc_1 - kc_1 \Delta z \tag{14.96}$$

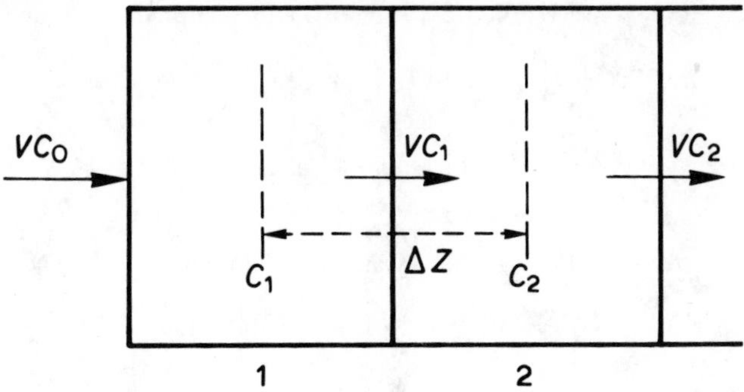

Figure 14.3 Determination of the boundary condition at the entrance.

Laplace transformation with an initial condition equal to zero gives:

$$a_0 c_0 - a^* c_1 + a_+ c_2 = 0 \tag{14.97}$$

in which

$$\left.\begin{aligned} a_0 &= \frac{v}{\Delta z} \\ a^* &= s + k + \frac{v}{2\Delta z} + \frac{D_f}{(\Delta z)^2} \end{aligned}\right\} \tag{14.98}$$

At the end (i = n) we have:

$$\Delta z \frac{dc_n}{dt} = vc_{n-1} - vc_n - kc_n\Delta z - D' \frac{c_n - c_{n-1}}{\Delta z} \tag{14.99}$$

Laplace transformation of this equation with zero initial condition results in:

$$a_- c_{n-1} - a^* c_n = 0 \tag{14.100}$$

Substitution of the Boundary Conditions

Substitution of Equation 14.92 into Equations 14.97 and 14.100 results in:

$$(a_+ J_1 - a^*)J_1 A_1 + (a_+ J_2 - a^*)J_2 A_2 = -a_0 c_0 \tag{14.101}$$

$$(a_- - a^* J_1)J_1^{n-1} A_1 + (a_- - a^* J_2)J_2^{n-1} A_2 = 0 \tag{14.102}$$

Solution of A_1 and A_2 gives with $J_1 J_2 = a_-/a_+$:

$$A_1 = J_1^{-1} J_2^{n-2}(a_- - a^* J_2) a_0 c_0 / \text{denominator} \tag{14.103}$$

$$A_2 = -J_1^{n-2} J_2^{-1}(a_- - a^* J_1) a_0 c_0 / \text{denominator} \tag{14.104}$$

in which the denominator is given by

$$\text{denominator} = J_1^{n-3}(a_- - a^* J_1)^2 - J_2^{n-3}(a_- - a^* J_2)^2 \tag{14.105}$$

Substitution again into Equation 14.92 finally results in:

$$\frac{c_i}{a_0 c_0} = \left(\frac{a_-}{a_+}\right)^{i-1} \frac{J_1^{n-i}(a^* - a_+ J_2) + J_2^{n-i}(a_+ J_1 - a^*)}{J_1^{n-1}(a^* - a_+ J_2)^2 - J_2^{n-1}(a_+ J_1 - a^*)^2} \tag{14.106}$$

in which the relationship between J_1, J_2 and s is determined by Equation 14.95. Numerical solution gives the possibility to analyze the accuracy of the discretization by comparison with the continuous case.

CONTINUIZATION

The transition of a discrete equation into a continuous one can be denoted with the term continuization (Friedly [34] talks about "inverse quantization"). Continuization is attractive for multistage processes, e.g., cascades of reactors or distillation in tray columns. Continuization sometimes leads to a better understanding of and insight to the process. This will be illustrated by means of an example: the analysis of the behavior of a binary distillation column as given by Rosenbrock (see Gould [35]).

Model

A tray is characterized by the following partial mass balance (Figure 14.4):

$$\frac{d(M_i x_i)}{dt} = V_{i-1} f(x_{i-1}) - V_i f(x_i) + L_{i+1} x_{i+1} - L_i x_i \tag{14.107}$$

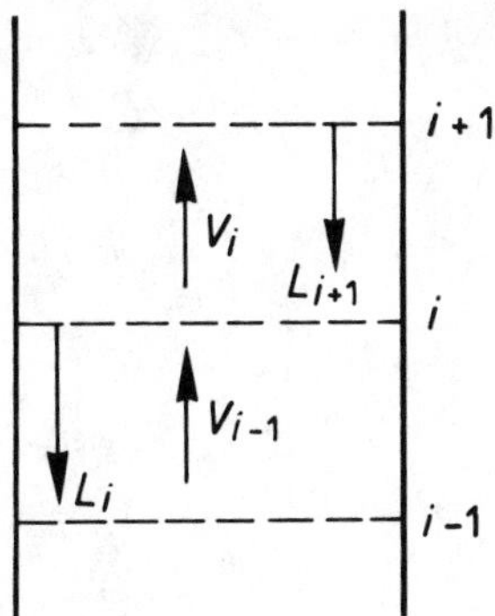

Figure 14.4 Schematic representation of a part of a distillation column.

where M_i = mass of the liquid on tray i (vapor mass has been ignored) (mol)
x_i = concentration of the key component in the liquid (mol/mol)
V_i = vapor flow which leaves tray i (mol/sec)
L_i = liquid flow leaving tray i (mol/sec)
$f(x_i)$ = vapor concentration, belonging to the liquid concentration (if theoretical trays are assumed, f is the equilibrium relationship)

The discrete variables x_i, L_i, V_i and M_i can now be expressed in terms of distributed variables x(z,t), L(z,t), V(z,t) and M(z,t), where z is a position coordinate along the column with the tray distance as unit of length. The partial flow from tray i + 1 can now be expressed in the variables on tray i with the aid of a Taylor series:

$$L_{i+1}x_{i+1} = L(i+1,t)x(i+1,t) = L(i,t)x(i,t) + \frac{\partial}{\partial z}[L(i,t)x(i,t)] + \frac{1}{2}\frac{\partial^2}{\partial z^2}[L(i,t)x(i,t)] \qquad (14.108)$$

For the partial flow from tray i − 1 one has:

$$V_{i-1}f(x_{i-1}) = V(i-1,t)f[x(i-1,t)] = V(i,t)f[x(i,t)] - \frac{\partial}{\partial z}\{V(i,t)f[x(i,t)] + \frac{1}{2}\frac{\partial^2}{\partial z^2}[V(i,t)f\ldots. \qquad (14.109)$$

Substitution of Equations 14.108 and 14.109 into Equation 14.107 gives, by approximation:

$$\frac{\partial(Mx)}{\partial t} \simeq \frac{\partial}{\partial z}[Lx - Vf(x)] + \frac{1}{2}\frac{\partial^2}{\partial z^2}[Lx + Vf(x)] \qquad (14.110)$$

in which all the variables are functions of z and t.

Boundary Conditions

At the top of the distillation column, vapor is condensed in a condenser. The condensate flows to an accumulator (Figure 14.5). The following partial mass balance holds:

$$\frac{d(M_{n+1}x_{n+1})}{dt} = V_n f(x_n) - L_{n+1}x_{n+1} - Dx_{n+1} \qquad (14.111)$$

After continuization of the variables, this becomes:

$$\frac{\partial(M_{n+1}x)}{\partial t} \simeq Vf(x) - Lx - D(t)x - \frac{\partial}{\partial z}[Vf(x)] + \frac{1}{2}\frac{\partial^2}{\partial z^2}[Vf(x)] \qquad (14.112)$$

in which z = n + 1. D(t) is the top product flow. At the bottom of the distillation column, part of the liquid is evaporated in the reboiler (Figure 14.6). The following partial mass balance holds:

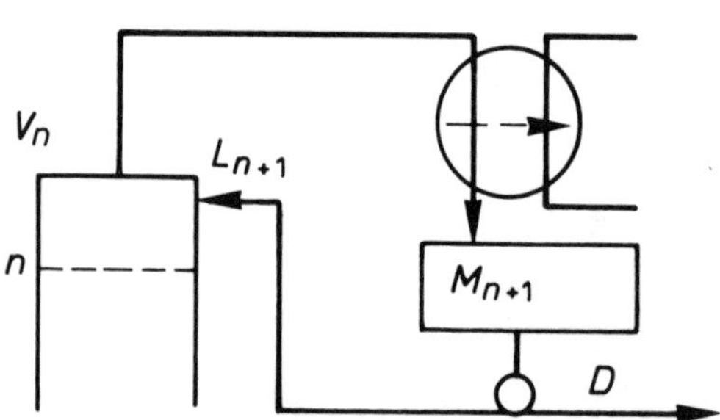

Figure 14.5 Top of the distillation column.

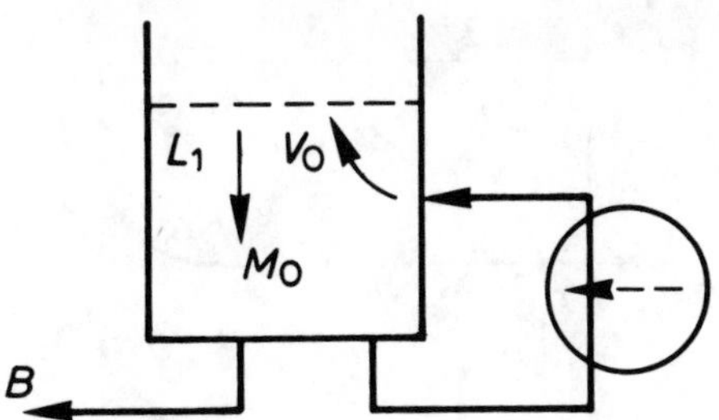

Figure 14.6 Bottom of the distillation column.

$$\frac{d(M_0 x_0)}{dt} = L_1 x_1 - V_0 f(x_0) - B x_0 \tag{14.113}$$

After continuization of the variables this equation can be written as:

$$\frac{\partial(M_0 x)}{\partial t} \simeq Lx - Vf(x) - B(t)x + \frac{\partial}{\partial z}(Lx) + \frac{1}{2}\frac{\partial^2}{\partial z^2}(Lx) \tag{14.114}$$

where z = 0, B(t) is the bottom product flow. At the feed tray the feed flow causes jumps in V and L. Thus, the equations for both column sections have to be formulated separately, where the partial mass balances around the feed tray are used as intermediate boundary conditions (Figure 14.7). For the vapor flow:

$$V_f'' f(x_f'') = V_f' f(x_f') + F_v f(x_f) \tag{14.115}$$

where V_f' = vapor flow from tray f
V_f'' = vapor flow to tray f + 1
F_v = the vapor flow in the feed
x_f = concentration in the liquid feed

For the liquid flow:

$$L_{f+1}' x_{f+1}' = L_{f+1}'' x_{f+1}'' + F_\ell x_f \tag{14.116}$$

where L_{f+1}'' = liquid flow from tray f + 1
L_{f+1}' = liquid flow to tray f
x_{f+1}'' = liquid concentration to tray f
x_{f+1}' = liquid concentration on tray f + 1
F_ℓ = liquid flow in the feed

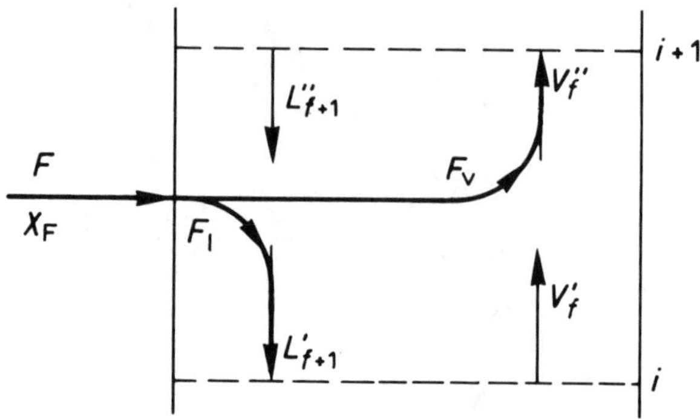

Figure 14.7 Situation around the feed tray.

The variables with two primes belong to the upper part of the column (the rectifying section); the variables with one prime belong to the lower part (the stripping section).

Combination of Equations

If, by approximation, the flows are equal throughout a column section, and if the mass inventories are constant, Equations 14.110, 14.111 and 14.114 to 14.116 can be written, for the top:

$$M_{N+1}\frac{\partial x''}{\partial t} \simeq V''f(x'') - L''x'' - Dx'' - V''\frac{\partial f(x'')}{\partial z} + \frac{1}{2}V''\frac{\partial^2 f(x'')}{\partial z^2} \qquad (14.117)$$

for the rectifying section; $f < z < n + 1$:

$$M''\frac{\partial x''}{\partial t} \simeq L''\frac{\partial x''}{\partial z} - V''\frac{\partial f(x'')}{\partial z} + \frac{1}{2}L''\frac{\partial^2 x''}{\partial z^2} + \frac{1}{2}V''\frac{\partial^2 f(x'')}{\partial z^2} \qquad (14.118)$$

for the feed tray; $z = f$:

$$V''f(x'') \simeq V'f(x') + F_v f(x_f) \qquad (14.119)$$

and

$$L'x' \simeq L''x'' + F_\ell x_f \qquad (14.120)$$

For the stripping section; $0 < z < f$:

$$M' \frac{\partial x'}{\partial t} \simeq L' \frac{\partial x'}{\partial z} - V' \frac{\partial f(x')}{\partial z} + \frac{1}{2} L' \frac{\partial^2 x'}{\partial z^2} + \frac{1}{2} V' \frac{\partial^2 f(x')}{\partial z^2} \tag{14.121}$$

For the bottom; $z = 0$:

$$M_0 \frac{\partial x'}{\partial t} \simeq L'x' - V'f(x') - Bx' + L' \frac{\partial x'}{\partial z} + \frac{1}{2} L' \frac{\partial^2 x'}{\partial z^2} \tag{14.122}$$

Furthermore, the following equations hold:

$$V'' = V' + F_v \tag{14.123}$$

$$L' = L'' + F_\ell \tag{14.124}$$

Equations 14.118 and 14.121 are two simultaneous partial differential equations with four boundary conditions: Equations 14.117, 14.119, 14.120 and 14.122. It will not be tried here to find a solution, but only to use them for qualitative analysis.

Interpretation

For small variations, the partial differential equations may be linearized. Equations 14.118 and 14.121 then become:

$$\frac{M''}{V''} \frac{\partial(\delta x'')}{\partial t} \simeq \frac{\partial}{\partial z} \left\{ \left[\frac{L''}{V''} - m''(z) \right] \delta x'' \right\} + \frac{1}{2} \frac{\partial^2}{\partial z^2} \left\{ \left[\frac{L''}{V''} + m''(z) \right] \delta x'' \right\} \tag{14.125}$$

$$\frac{M'}{V'} \frac{\partial(\delta x')}{\partial t} \simeq \frac{\partial}{\partial z} \left\{ \left[\frac{L'}{V'} - m'(z) \right] \delta x' \right\} + \frac{1}{2} \frac{\partial^2}{\partial z^2} \left\{ \left[\frac{L'}{V'} + m'(z) \right] \delta x' \right\} \tag{14.126}$$

with

$$m(z) = \frac{df(z)}{dz} \tag{14.127}$$

(the slope of the vapor-liquid equilibrium line).

Both equations are quite similar to the equation for the river with dispersion. By analogy, the "transport velocities" are proportional to

$$m''(z) - \frac{L''}{V''} \text{ and } m'(z) - \frac{L'}{V'} \text{ (respectively)} \qquad (14.128)$$

and the dispersion coefficients are proportional to

$$\frac{L''}{V''} + m''(z) \text{ and } \frac{L'}{V'} + m'(z) \text{ (respectively)} \qquad (14.129)$$

The slope of the vapor-liquid equilibrium line usually decreases monotonically when x goes from 0 to 1, hence also if z goes from 0 to n + 1. In the rectifying section L''/V'' is smaller than one. In the lower part of this section the "transport velocity" (see Equation 14.128) will usually be positive, but higher in this section it becomes zero and still higher up even negative. This means that disturbances cannot disappear by transportation, but only by dispersion.

Similar reasoning can be set up for the stripping section. There $L' > V'$ which means that near the feed tray the transport is downward. Further down, however, $m'(z)$ increases, resulting in zero transport velocity. Apparently, the continuized description gives a certain insight in the behavior of the distillation column, which cannot be obtained so easily in another way. It is surprising that the dynamic behavior of a distillation process can be approximated by that of tubular flow reactors with axial dispersion.

CHAPTER 15

NUMERICAL SOLUTION OF PARTIAL DIFFERENTIAL EQUATIONS

In previous chapters, partial differential equations were reduced to a form which could be solved with pencil and paper. Also, the difference equations which resulted from discretization were of such a structure that a solution could easily be determined. In many cases, analytical solutions do not exist, e.g., when coefficients are functions of the geometric independent variable. Thanks to the digital computer, however, this no longer need be a problem. Different methods are available in standard routines for the solution of ordinary differential equations, for example a third- or fourth-order Runge-Kutta method or a predictor-corrector method with variable step length.

Thus, if the model can be transformed to a set of ordinary differential equations, using a computer is no longer a problem. This, however, does not mean that analytical methods are obsolete; they remain important as a check on computer results, as a means to obtain insight and as a way to reduce the number of computer runs.

DIFFERENTIAL DIFFERENCE EQUATIONS

This type of equation, which came into the picture when discussing multistage processes, has the form:

$$\frac{dy_i}{dt} = f(y_{i-1}\ ,\ y_i\ ,\ y_{i+1})\ ,\ i = 1 \ldots . n \tag{15.1}$$

The values y_i are given for the initial value of time. An interval h in t is assumed. The third-order Runge-Kutta method can be evaluated for every i, as:

$$\left.\begin{aligned} k_{i,1} &= hf(y_{i-1}\ ,\ y_i\ ,\ y_{i+1}) \\ k_{i,2} &= hf\left(y_{i-1} + \frac{1}{2}\,k_{i-1,1}\ ,\ y_i + \frac{1}{2}\,k_{i,1}\ ,\ y_{i+1} + \frac{1}{2}\,k_{i+1,1}\right) \\ k_{i,3} &= hf(y_{i-1} + 2k_{i-1,2} - k_{i-1,1}\ ,\ y_i + 2k_{i,2} - k_{i,1}\ , \\ &\qquad y_{i+1} + 2k_{i+1,2} - k_{i+1,1}) \end{aligned}\right\} \qquad (15.2)$$

The value of y_i at $t = t + h$ is given by:

$$y_i(h) = y_i + \frac{1}{6}\,(k_{i,1} + 4k_{i,2} + k_{i,3}) \qquad (15.3)$$

For a fourth-order Runge-Kutta method, the same expressions hold for $k_{i,1}$ and $k_{i,2}$. For $k_{i,3}$ and $k_{i,4}$ the following equations hold:

$$\left.\begin{aligned} k_{i,3} &= hf\left(y_{i-1} + \frac{1}{2}\,k_{i-1,2}\ ,\ y_i + \frac{1}{2}\,k_{i,2}\ ,\ y_{i-1} + \frac{1}{2}\,k_{i+1,2}\right) \\ k_{i,4} &= hf(y_{i-1} + k_{i-1,3}\ ,\ y_i + k_{i,3}\ ,\ y_{i+1} + k_{i+1,3}) \end{aligned}\right\} \qquad (15.4)$$

for $t = t + h$, y_i becomes:

$$y_i(h) = y_i + \frac{1}{6}\,(k_{i,1} + 2k_{i,2} + 2k_{i,3} + k_{i,4}) \qquad (15.5)$$

When the interval h is very small, $k_{i,2}$, $k_{i,3}$ and $k_{i,4}$ will not differ very much from $k_{i,1}$. Equations 15.3 and 15.5 can then be approximated by:

$$y_i(h) = y_i + hf(y_{i-1}\ ,\ y_i\ ,\ y_{i+1}) \qquad (15.6)$$

In fact, Equation 15.6 represents a Taylor series in which second- and higher-order terms are ignored. However, it usually pays to incorporate Equations 15.2 or 15.4, as this allows a much higher value of h.

Example

In an absorption column with six theoretical trays, a component is absorbed from the gasflow by the liquid flow. Initially, the process is in a static situation. Suddenly, the concentration of the component to be absorbed increases 10%. The amount of liquid M on a tray is equal to

36 mol, the gasflow G is 2 mol/sec and the liquid flow L is 0.2 mol/sec. The liquid-gas relationship can be approximated by the equation y = 0.2x. How does the composition in the gas phase change as a function of time?

In Figure 15.1, tray i of the absorption column is shown. The partial mass balance for this tray is:

$$M \frac{dx_i}{dt} = Lx_{i+1} - Lx_i + Gy_{i-1} - Gy_i \tag{15.7}$$

where x = the liquid fraction of the component
y = the gas fraction of the component

When using the equilibrium relationship y = mx, Equation 15.7 can be written as:

$$\frac{M}{L}\frac{dy_i}{dt} = y_{i+1} - y_i + \alpha y_{i-1} - \alpha y_i \tag{15.8}$$

with

$$\alpha = \frac{mG}{L} \tag{15.9}$$

The derivative dy_i/dt could, according to Equation 15.6, be replaced by:

$$\frac{dy_i}{dt} \simeq \frac{y_i(t+h) - y_i(t)}{h} \tag{15.10}$$

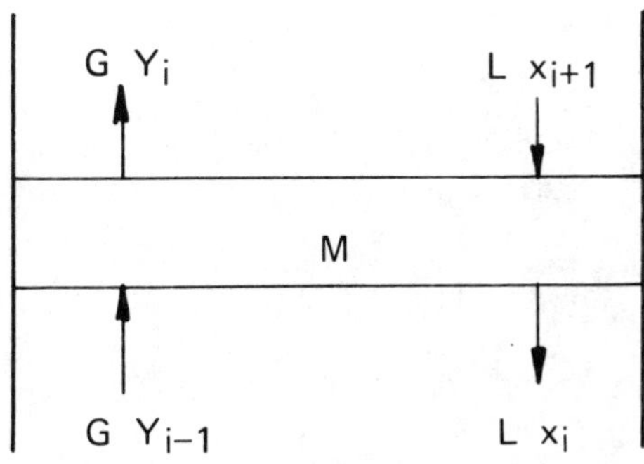

Figure 15.1 Tray i of the absorption column.

where h is the step length in time. Combination of Equations 15.8 and 15.10 results in:

$$y_i(t+h) = \frac{\alpha Lh}{M} y_{i-1}(t) + \left[1 - \frac{Lh}{M}(1+\alpha)\right] y_i(t) + \frac{Lh}{M} y_{i+1}(t) \qquad (15.11)$$

It can be shown that a stable solution exists if the coefficient of $y_i(t)$ is positive; however, the approximation can still be inaccurate.

The condition for stability of Equation 15.11 can be determined as follows. Write for convenience the equation for errors, which is analogous to Equation 15.11, as:

$$\delta y_i(t+h) = \lambda_1 \delta y_{i-1}(t) + (1 - \lambda_1 - \lambda_2)\delta y_i(t) + \lambda_2 \delta_{i+1}(t) \qquad (15.12)$$

The error in the i direction can be correlated. Say that this correlation can be expressed by a correlation coefficient ν. The values of ν may vary between −1 and +1. A value of ν equal to −1 means that the error of two successive steps in the i direction are equal but have a different sign. Hence if $\delta y_i(t)$ and $\delta y_{i-1}(t)$ are correlated we may write:

$$\delta y_{i-1}(t) = \nu \delta y_i(t) \qquad (15.13)$$

In a similar way:

$$\delta y_{i+1}(t) = \nu \delta y_i(t) \qquad (15.14)$$

Thus, Equation 15.12 may be written as:

$$\begin{aligned} \delta y_i(t+h) &= \lambda_1 \nu \delta y_i(t) + (1 - \lambda_1 - \lambda_2)\delta y_i(t) + \lambda_2 \nu \delta y_i(t) \\ &= [1 - (1-\nu)(\lambda_1 + \lambda_2)]\delta y_i(t) \end{aligned} \qquad (15.15)$$

Equation 15.15 will converge if the coefficient of $\delta y_i(t)$ in absolute magnitude is smaller than one, or

$$-1 \leqslant 1 - (1-\nu)(\lambda_2 + \lambda_2) \leqslant 1 \qquad (15.16)$$

from which

$$0 \leqslant (1-\nu)(\lambda_1 + \lambda_2) \leqslant 2 \qquad (15.17)$$

The most rigorous requirement from this condition is imposed for $\nu = -1$, thus:

$$\lambda_1 + \lambda_2 \leqslant 1 \qquad (15.18)$$

Substitution of the values for λ_1 and λ_2 results in:

$$1 - \frac{Lh}{M}(1 + \alpha) \geqslant 0 \qquad (15.19)$$

thus the step length in time:

$$h \leqslant \frac{M}{L(1 + \alpha)} \qquad (15.20)$$

Substitution of data gives a value of α equal to 2; thus, for a step length of 30 sec, Equation 15.11 becomes:

$$y_i(t + h) = \frac{1}{3}\, y_{i-1}(t) + \frac{1}{2}\, y_i(t) + \frac{1}{6}\, y_{i+1}(t) \qquad (15.21)$$

Initially, the column is in equilibrium:

$$t = 0 \;;\quad y_i = \bar{y}_i \qquad (15.22)$$

In the bottom of the column, a step is introduced at $t = 0$:

$$i = 0 \;;\quad y_0 = \bar{y}_0 + 0.10 \qquad (15.23)$$

Finally, in the top of the column clean liquid is added:

$$i = 7 \;;\quad y_7 = 0 \qquad (15.24)$$

In Figure 15.2, the concentration variation is given as function of time. δy_i is the variation in y_i compared to the static value $\bar{y}_i$, thus $\delta y_i = y_i - \bar{y}_i$ and $\delta y_0 = y_0 - \bar{y}_0 = 0.10$, according to the condition in Equation 15.23.

The error in the calculation of y_i can be estimated from a Taylor series development of $y_i(t + h)$ around the point $y_i(t)$. The result is:

$$y_i(t + h) = y_i(t) + h\,\frac{dy_i}{dt} + \frac{h^2}{2!}\,\frac{d^2y_i}{dt^2} + \ldots \qquad (15.25)$$

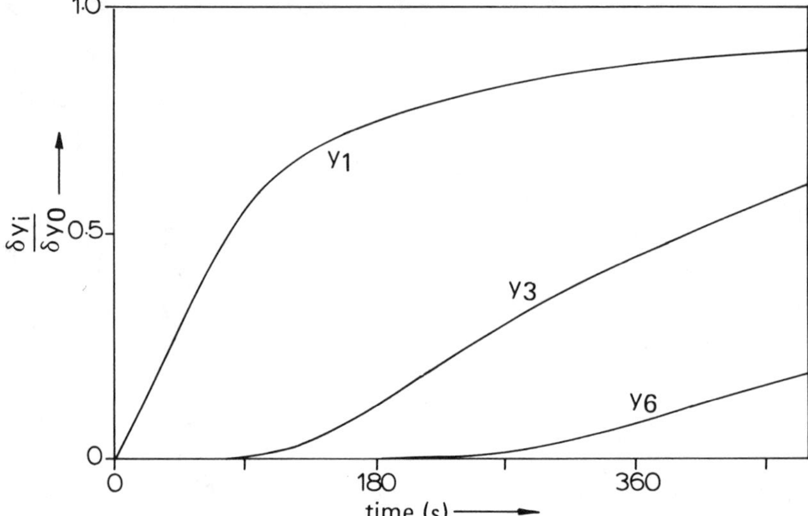

Figure 15.2 Concentration profile in the column as a function of time.

Table 15.1 Calculation of the Error in y_1

t	y_1	Δy_1	$\Delta^2 y_1$	error
0	0	0.190		
30	0.190	0.205	0.015	0.008
60	0.395	0.160	-0.045	-0.023
90	0.555	0.095	-0.065	-0.033
120	0.650	0.060	-0.035	-0.018
150	0.710	0.040	-0.020	-0.010
180	0.750	0.034	-0.006	-0.003
210	0.784	0.027	-0.007	-0.004
240	0.811	0.019	-0.008	-0.004
270	0.830	0.019	0	0
300	0.849	0.013	-0.006	-0.003
330	0.862	0.012	-0.001	-0.001
360	0.874			

From Equations 15.10 and 15.25 it is clear that the second-order term is a measure of error. In Table 15.1, the values of y_1 are given as function of time. The first derivative can be approximated by taking the difference between two successive values of y_1, denoted with Δy_1 and dividing these values by the step length in time h. The second derivative of y_1 in a certain point can be approximated by taking the difference between two successive values of Δy_1, denoted with $\Delta^2 y_1$, and dividing these values by h^2. Then the error measure is:

$$\frac{h^2}{2!}\frac{d^2 y_1}{dt^2} \simeq \frac{h^2}{2}\frac{\Delta^2 y_1}{h^2} \simeq \frac{\Delta^2 y_1}{2} \tag{15.26}$$

These values are also given in Table 15.1. After 11 time steps, the accumulated error has a value equal to 0.09, which is about 10% of the absolute value. It is clear that the approximation is rather inaccurate. The accuracy can best be improved by applying a higher-order numerical method.

PARTIAL DIFFERENTIAL EQUATIONS

After having analyzed the numerical solution of ordinary differential equations, we shall now turn to partial differential equations. However, a complete treatment of numerical solution methods would be too extensive. Therefore, only an example will be given: the parabolic differential equation for heat conduction in a rod (see Chapter 9):

$$\frac{\partial}{\partial z}\left(k_t \frac{\partial T}{\partial z}\right) = \rho c_p \frac{\partial T}{\partial t} \tag{15.27}$$

where T = temperature
k_t = thermal conductivity
ρ = density
c_p = specific heat

For given boundary conditions this equation has been solved in Chapter 9 with the aid of Laplace transform. However, when the boundary conditions are not constant, but depend, for example, on time, analytical solution usually is no longer possible and the equation has to be solved numerically. If the thermal conductivity is constant, Equation 15.27 can be written as:

$$\frac{\partial T}{\partial t} = \beta \frac{\partial^2 T}{\partial z^2} \tag{15.28}$$

with

$$\beta = k_t / c_p \rho \tag{15.29}$$

DISCRETIZATION

In Chapter 14, the following approximation was derived for the second-order derivative:

$$\frac{\partial^2 T}{\partial z^2} = \frac{T_{i+1} - 2T_i + T_{i-1}}{(\Delta z)^2} + \theta(\Delta z)^2 \tag{15.30}$$

If the term $\theta(\Delta z)^2$ is ignored, substitution of Equation 15.30 into 15.28 gives:

$$\frac{\partial T_i}{\partial t} = \frac{\beta}{(\Delta z)^2} [T_{i+1} - 2T_i + T_{i-1}] \tag{15.31}$$

When a central difference is applied with respect to time, $\partial T_i/\partial t$ can be replaced by:

$$\frac{\partial T_i}{\partial t} \simeq \frac{T_{i,j+1} - T_{i,j-1}}{2\Delta t} \tag{15.32}$$

thus Equation 15.31 can be written as:

$$\frac{T_{i,j+1} - T_{i,j-1}}{2\Delta t} = \frac{\beta}{(\Delta z)^2} [T_{i+1,j} - 2T_{i,j} + T_{i-1,j}] \tag{15.33}$$

or

$$T_{i,j+1} = T_{i,j-1} + \frac{2\beta\Delta t}{(\Delta z)^2} [T_{i+1,j} - 2T_{i,j} + T_{i-1,j}] \tag{15.34}$$

where the index i refers to the length coordinate and the index j to the time coordinate. In this way a grid has been put over the (z,t) area (Figure 15.3). On the intersections of the z and t coordinates the temperatures are calculated. For the calculation of a new point, the information of the four previous points is used. A stability analysis shows, however, that Equation 15.34 does not give a stable solution for the calculation of the temperature. One can generally say, that when a point is calculated from a number of other points, and one of the coefficients of these points has a negative sign, the solution can be unstable. Therefore, we have to select another discretization scheme. The application of a forward difference in the time gives:

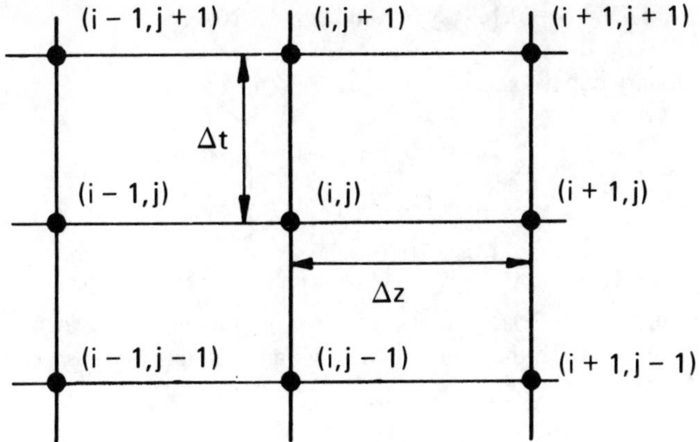

Figure 15.3 Grid for the calculation of temperatures.

$$\frac{\partial T_i}{\partial t} \simeq \frac{T_{i,j+1} - T_{i,j}}{\Delta t} \tag{15.35}$$

Substitution into Equation 15.31 gives:

$$T_{i,j+1} = \lambda T_{i+1,j} + (1 - 2\lambda)T_{i,j} + \lambda T_{i-1,j} \tag{15.36}$$

in which

$$\lambda = \frac{\beta \Delta t}{(\Delta z)^2} \tag{15.37}$$

The value of λ now determines the stability and convergence of the solution; the solution is stable when all coefficients in Equation 15.36 are positive, thus λ has to be smaller than 0.5 [36].

Example

Consider a rod with a length of 0.2 m for which $\beta = 2 \times 10^{-4}$ m²/sec. When the length is divided into four sections, $\Delta z = 0.05$ m, and when time steps are selected equal to 2 sec, $\lambda = 0.16$, which is far below the limit value of 0.5.

Initially, the rod has a temperature of 20°C; suddenly the temperature at both ends of the rod is increased to 60°C. The temperature distribution in the rod can now be calculated with the aid of Equation 15.36:

$$T_{i,j+1} = 0.16T_{i+1,j} + 0.68T_{i,j} + 0.16T_{i-1,j} \tag{15.38}$$

with boundary conditions:

$$\left.\begin{array}{l} T_{0,j} = 60 \quad \text{for } j > 0 \\ T_{4,j} = 60 \quad \text{for } j > 0 \\ T_{i,0} = 20 \quad \text{for all } i \end{array}\right\} \tag{15.39}$$

Some results of the calculations are given in Figure 15.4. When the value of λ is greater than ½, the solution will become unstable rather soon.

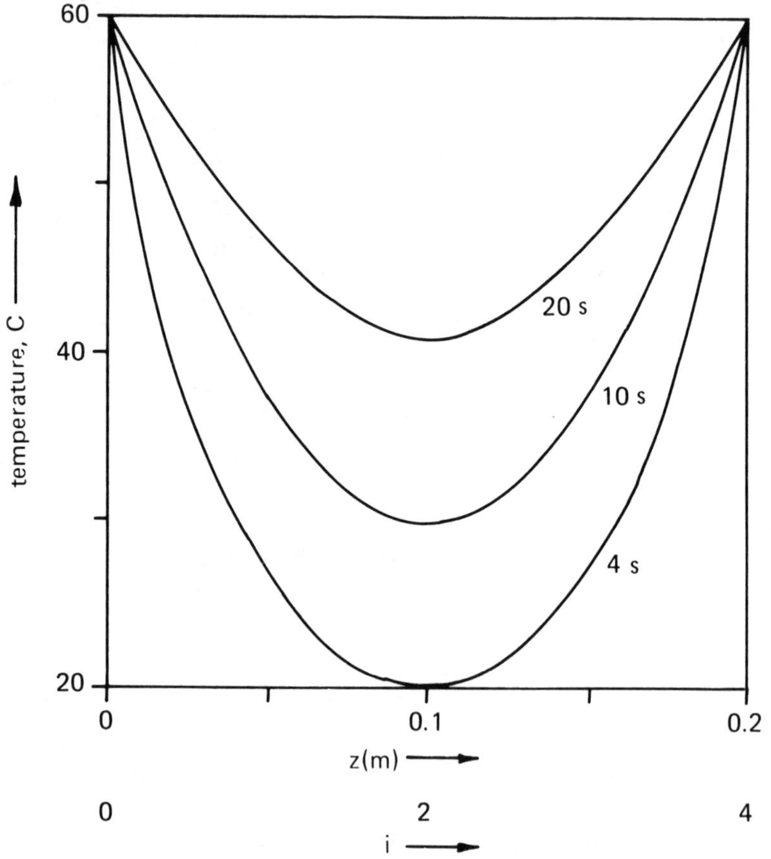

Figure 15.4 Temperature profiles in a rod.

To show the effects of instability, we shall now choose Δt equal to 10 sec, which increases λ to 0.8 (well above 0.5). The calculation proceeds according to:

$$T_{i,j+1} = 0.8T_{i+1,j} - 0.6T_{i,j} + 0.8T_{i-1,j} \tag{15.40}$$

The results for j = 0 to 3 are given in Table 15.2. After 30 sec, the calculations give a temperature in the middle of the rod that is already higher than both temperatures at the end, which is physically impossible. Continued calculations even result in negative temperatures. It is once and again clear that much attention has to be paid to the stability problem.

CRANK-NICHOLSON METHOD

The previous discretization method resulted in discretization errors of the order $\theta(\Delta t) + \theta(\Delta z)^2$. The Crank-Nicholson method is a more accurate discretization method that reduces the error in the approximation of the first derivative with time from $\theta(\Delta t)$ to $\theta(\Delta t)^2$. The grid over the z,t area is now divided as shown in Figure 15.5. The time interval is divided once again. Application of the central difference in the time, which has an accuracy of the order $\theta(\Delta t)^2$ gives for a point halfway:

$$\left.\frac{\partial T}{\partial t}\right/_M \simeq \frac{T_{i,j+1} - T_{i,j}}{\Delta t} \tag{15.41}$$

The second-order derivative with respect to position is calculated according to:

Table 15.2 Calculation of the Temperature Profiles in a Rod for λ = 0.8

Time (sec)	j	Length (m); i: 0; 0	5; 1	10; 2	15; 3	20; 4
0	0	20	20	20	20	20
10	1	60	20	20	20	60
20	2	60	52	20	52	60
30	3	60	32.8	71.2	32.8	60

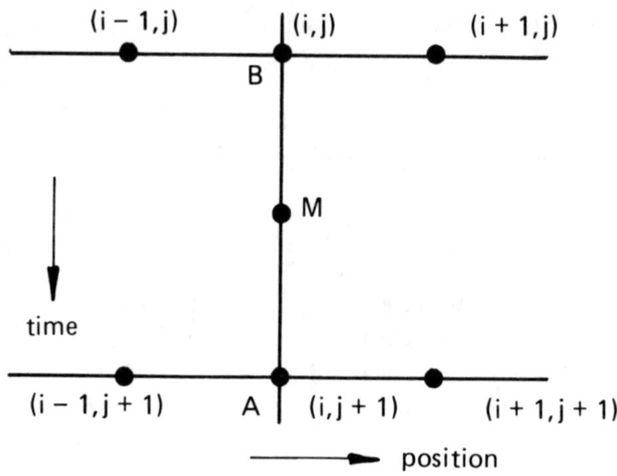

Figure 15.5 Grid for the application of the Crank-Nicholson method.

$$\left.\frac{\partial^2 T}{\partial z^2}\right/_M = \frac{1}{2}\left[\left.\frac{\partial^2 T}{\partial z^2}\right/_A + \left.\frac{\partial^2 T}{\partial z^2}\right/_B\right]$$

$$= \frac{1}{2(\Delta z)^2}\left[T_{i+1,j} - 2T_{i,j} + T_{i-1,j} + T_{i+1,j+1} - 2T_{i,j+1} + T_{i-1,j+1}\right] \tag{15.42}$$

which can easily be obtained when using Equation 15.30. Substitution of Equations 15.41 and 15.42 into Equation 15.28 results in:

$$\frac{T_{i,j+1} - T_{i,j}}{\Delta t} = \frac{\beta}{2(\Delta z)^2}\left[T_{i+1,j} - 2T_{i,j} + T_{i-1,j} + T_{i+1,j+1} - 2T_{i,j+1} + T_{i-1,j+1}\right] \tag{15.43}$$

Introduction of λ according to Equation 15.37 now gives:

$$(1 + \lambda)T_{i,j+1} = (1 - \lambda)T_{i,j} + \frac{1}{2}\lambda(T_{i+1,j} + T_{i-1,j}) + \frac{1}{2}\lambda(T_{i+1,j+1} + T_{i-1,j+1}) \tag{15.44}$$

As the discretization error in Equation 15.41 now is of the order $\theta(\Delta t)^2$ and in Equation 15.42 of the order $\theta(\Delta z)^2$, the total discretization error in Equation 15.44 is now of the order $\theta(\Delta z)^2 + \theta(\Delta t)^2$, and the error has been reduced compared to Equation 15.36. In Equation 15.44, the last two terms are unknown. Therefore a first approximation is assumed for $T_{i,j+1}$ for all values of i. The right side of Equation 15.44 is then calculated, resulting in a better approximation for $T_{i,j+1}$. The calculation is repeated until a certain relative error is obtained in the calculation of the temperature distribution. It is possible to show that Equation 15.44 is stable for all values of λ. This method is therefore often applied in the solution of initial value problems. For the application of other methods reference is made to the literature [37,38].

In all cases, special care must be devoted to obtain difference equations which give a stable solution.

CHAPTER 16

CONTROL QUALITY

The purpose of this chapter is to give insight into the control quality that can be obtained by simple control loops. This will be done for a simple dynamic model, which is a good approximation for many process control loops. This model is characterized by two dimensionless parameters: the ratios between three time constants. The control quality is judged on the basis of the speed of control and the static control accuracy. Here the selection of controller behavior comes into the picture: proportional control action is often insufficient, which implies extension with other control actions. Finally some formulas will be discussed that characterize the response of a control loop to disturbances.

SIMPLE DYNAMIC MODEL

The complete control loop contains process and instrumentation, consisting of the measuring unit, transmission, controller, and transmission correcting unit. As we have seen in previous chapters, the dynamics of all of these units may be important for control quality.

If the control loop may be linearized, the sequence of the lags in the control loop is unimportant. Hence they can be represented by a common block in the information stream diagram (Figure 16.1). Further, there are the control actions that must be selected with respect to the given control loop lags. The control actions correspond to an idealization of the actual controller behavior, as the latter also contains some lags. Because these lags cannot independently be adjusted, they will be included in the control loop lags (Figure 16.1). As a result of this arrangement, u is not completely identical with the output signal of the actual controller.

On the other hand, y_m corresponds reasonably well with the measured

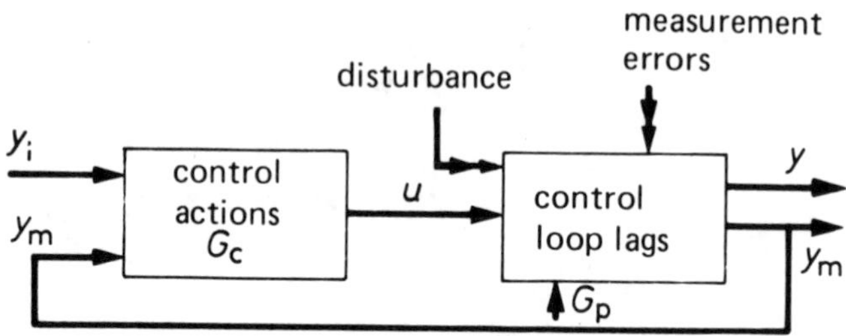

Figure 16.1 Information flow diagram of a linear control loop.

value of the controlled variable (as it appears at the controller input) and y_i with the set value. Disturbances and measurement errors enter somewhere in the loop dynamics.

In many cases, the control loop lags are first-order with a large time constant (which may be related to the total residence time in the process). However, this is not sufficient for description of control quality. At least one secondary lag must be taken into account to get realistic results. There are two kinds of elementary lags: first-order and dead time (transportation lag, pure time delay). To generate a spectrum of possibilities, they will both be included in the model. In this way we arrive at two first-order elements and a dead time, which yield the following control loop dynamics in operational notation:

$$\frac{\delta y_m}{\delta u} = G_p = \frac{K_p e^{-s\tau}}{(1 + s\tau_1)(1 + s\tau_2)} \tag{16.1}$$

where K_p = the static process gain
τ = dead time
τ_1, τ_2 = time constants

The values of the parameters can be determined experimentally from the step response. For this purpose, the controller is set on manual, and a small step change is introduced into the signal to the control valve. The measured value of the controlled condition is recorded. If the result is an S-shaped curve, a good agreement with Equation 16.1 may be expected.

It can also happen, however, that the step response deviates from the S curve. We already have seen an example of this when studying the dynamic behavior of chemical reactors: nonminimum phase behavior, where the response starts going into a wrong direction (Figure 16.2). One could work

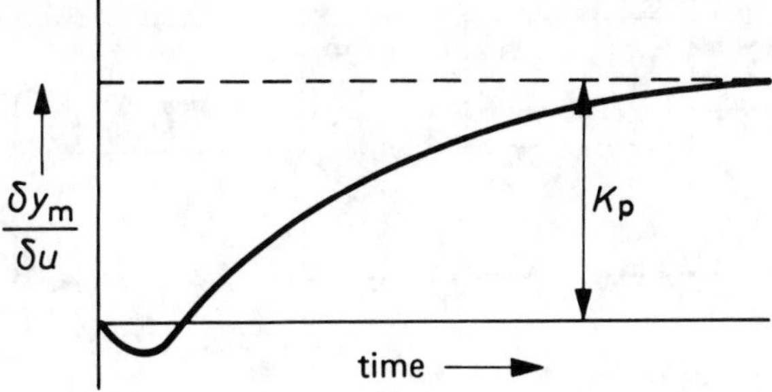

Figure 16.2 Example of nonminimum-phase behavior.

here with a modified model where the dead time is replaced by a nonminimum phase time constant:

$$G_p = K_p \frac{(1 - s\tau_3)}{(1 + s\tau_1)(1 + s\tau_2)} \tag{16.2}$$

Here at the limit of stability, the proportional controller gain is given by the characteristic equation:

$$s^2\tau_1\tau_2 + s(\tau_1 + \tau_2 - K_cK_p\tau_3) + 1 + K_cK_p = 0 \tag{16.3}$$

from which follows:

$$K_cK_p = \frac{\tau_1 + \tau_2}{\tau_3}$$

It is clear that τ_3 limits the adjustable closed-loop gain.

REDUCTION OF DISTURBANCES

A disturbance v influences the actual value of the controlled variable y via transfer function G_v; the correcting variable acts on y via transfer function G_u. This is shown in Figure 16.3. For small variations we have:

$$\delta y = G_v\delta v + G_u\delta u \tag{16.4}$$

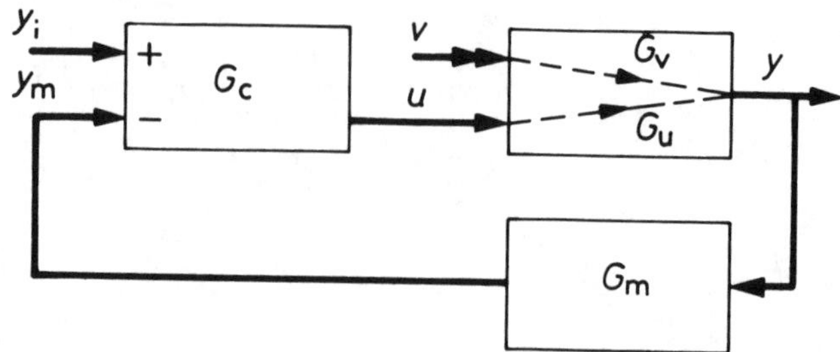

Figure 16.3 Information flow diagram showing the influence of disturbances.

For the measuring system:

$$\delta y_m = G_m \delta y \tag{16.5}$$

and for the controller, if the set value is constant:

$$\delta u = -G_c \delta y_m \tag{16.6}$$

Elimination of δy_m and δu from Equations 16.4 to 16.6 gives:

$$(\delta y) \text{ with control} = \frac{G_v \delta v}{1 + G_c G_p} \tag{16.7}$$

where

$$G_p = G_u G_m \tag{16.8}$$

Without control, $(G_c = 0)\delta y$ would be:

$$(\delta y) \text{ without control} = G_v \delta v \tag{16.9}$$

The ratio of the variations with and without control now becomes:

$$\frac{(\delta y) \text{ with control}}{(\delta y) \text{ without control}} = \frac{1}{1 + G_c G_p} \tag{16.10}$$

This deviation ratio [39] is independent of G_v and therefore holds for all kinds of disturbances. The stability can be judged from the characteristic equation:

$$1 + G_c(s)G_p(s) = 0 \tag{16.11}$$

PROPORTIONAL CONTROL ACTION

Proportional control action has already extensively been discussed in the previous chapters. P action is given by:

$$G_c = K_c \tag{16.12}$$

For the model according to Equation 16.1, the open-loop transfer function now becomes:

$$G_cG_p = \frac{Ke^{-s\tau}}{(1 + s\tau_1)(1 + s\tau_2)} \tag{16.13}$$

where

$$K = K_cK_p \tag{16.14}$$

The limit of stability can be found by dividing Equation 16.11 by $s^2 + \omega_u^2$. This, however, is thwarted by the presence of the exponential function. The following method leads to the desired result: separate G_p into a dead time and a second-order system. Suppose that y_m shows a sinusoidal oscillation:

$$\delta y_m = A \sin \omega_u t \tag{16.15}$$

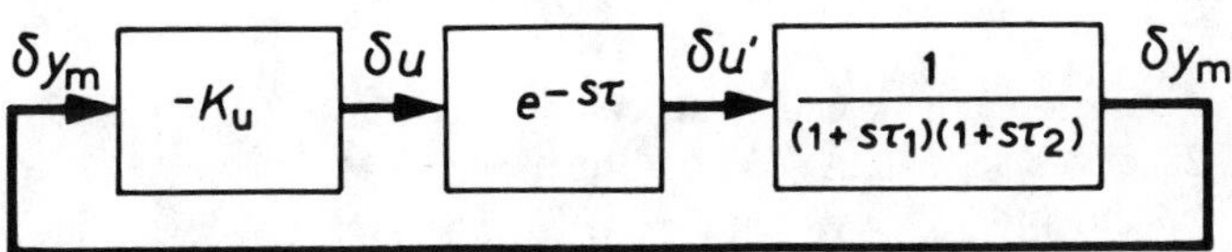

Figure 16.4 Control loop with proportional control action.

Then the variation in δu is:

$$\delta u = -AK_u \sin\omega_u t \tag{16.16}$$

After the dead time:

$$\begin{aligned} \delta u' &= -AK_u \sin\omega_u(t-\tau) \\ &= -AK_u \sin\omega_u t \cos\omega_u\tau + AK_u \cos\omega_u t \sin\omega_u\tau \end{aligned} \tag{16.17}$$

The second-order system is represented by the differential equation:

$$\tau_1\tau_2 \frac{d^2(\delta y_m)}{dt^2} + (\tau_1 + \tau_2)\frac{d(\delta y_m)}{dt} + \delta y_m = \delta u' \tag{16.18}$$

From Equations 16.15 and 16.18 it follows that:

$$-\omega_u^2\tau_1\tau_2 A \sin\omega_u t + \omega_u(\tau_1+\tau_2)A\cos\omega_u t + A\sin\omega_u t = \delta u' \tag{16.19}$$

Equations 16.17 and 16.19 must be valid for all values of time t. This is the case when the sine terms agree:

$$-\omega_u^2\tau_1\tau_2 + 1 = -K_u \cos\omega_u\tau \tag{16.20}$$

and the cosine terms:

$$\omega_u(\tau_1+\tau_2) = K_u \sin\omega_u\tau \tag{16.21}$$

Elimination of K_u results in:

$$\omega_u(\tau_1+\tau_2) = (-1 + \omega_u^2\tau_1\tau_2)\tan\omega_u\tau \tag{16.22}$$

For given values of τ_1, τ_2 and τ, ω_u can be determined with the aid of Equation 16.22 and K_u with the aid of Equation 16.21. Figure 16.5 shows the result as a function of τ_1/τ and τ_2/τ. The controller gain at the limit of stability $K_{c,u}$ can be determined by dividing K_u by the process gain K_p.

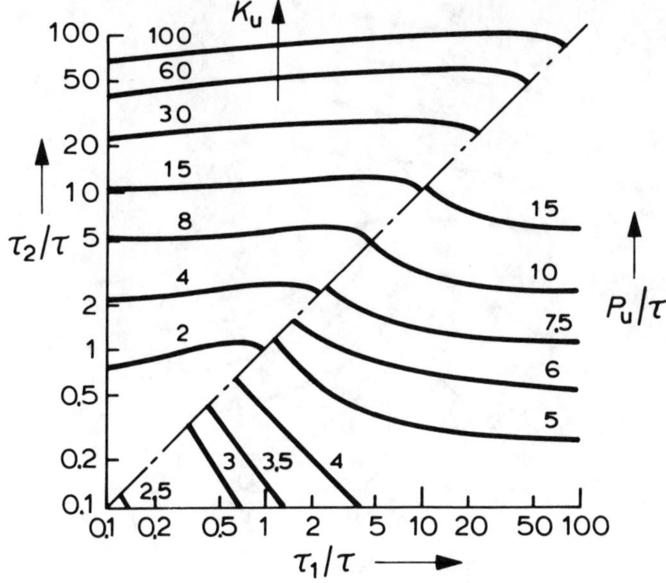

Figure 16.5 Limit of stability for proportional control.

The symmetry with respect to τ_1 and τ_2 has been used by plotting P_u/τ in the lower part of the figure and K_u in the upper part. For loops with a first-order character ($\tau_1 \gg \tau$ and $\tau \gg \tau_2$), K_u is very large, hence P action can give here considerable disturbance reduction. This is also clear from the results for damped oscillatory behavior ($\zeta = 0.25$) which are shown in Figure 16.6. For other values of τ_1 and τ_2, the closed loop gain is insufficient, and P action has to be supplemented with another control action.

INTEGRAL CONTROL ACTION

Sustained deviations can be eliminated by integral (I) action. This is characterized by the relationship between deviation and output signal of the controller:

$$\delta u = \frac{K_c}{\tau_i} \int (\delta y_i - \delta y_m) dt \tag{16.23}$$

where τ_i = integration time

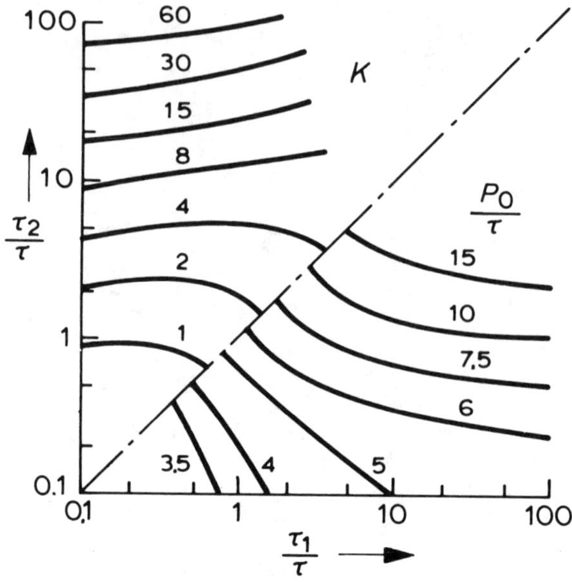

Figure 16.6 Proportional control for damped behavior, $\zeta = 0.25$.

Introducing operators and neglecting set value charges ($\delta y_i = 0$):

$$\frac{\delta u}{\delta y_m} = -G_c = -\frac{K_c}{s\tau_i} \tag{16.24}$$

A remaining deviation, however small it may be, causes an output signal which increases or decreases continuously. Consequently, the process finally is influenced so strongly that the deviation disappears (unless the value reaches the end of its stroke). Integral action is therefore a useful addition to proportional action; and the combination is used rather frequently in practice. It has the following transfer function:

$$G_c = K_c \left(1 + \frac{1}{s\tau_i}\right) \tag{16.25}$$

The response to a stepwise control deviation is shown in Figure 16.7, with a simple method to determine τ_i.

Integral action operates gradually, hence it introduces an additional control lag, and therefore decreases the speed of control. Thus, it is not desirable

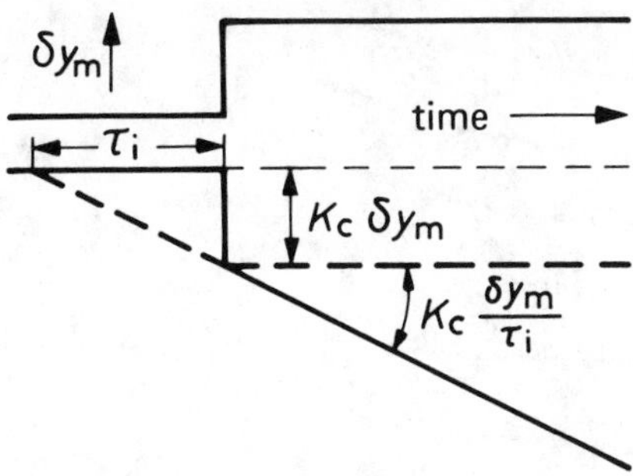

Figure 16.7 Response of PI controller to constant deviation.

to make τ_i very small. On the other hand, we do not want to make τ_i too large, because the rate of disturbance reduction is then slow. A good solution is to relate the value of τ_i to the oscillation period P_0 at $\zeta = 0.25$. Then the speed of control is not influenced too strongly.

Some adjustment procedures, e.g., those of Ziegler and Nichols [40] relate τ_i to P_u, the oscillation period at the limit of stability. This, however, gives incorrect results for control loops with a second-order character ($\tau_1 \gg \tau$, $\tau_2 \gg \tau$).

Figure 16.8 shows P_0/τ and K for $\zeta = 0.25$ and $\tau_i = P_0$. Compared to proportional action, the speed of control is somewhat decreased, but there is an elimination of sustained deviations. The speed of control can be judged on the basis of the oscillation period P_0 for $\zeta = 0.25$. A large value of P_0 means slow control, a small value means fast control. Figure 16.9 shows the result for integral action only. For loops with a dead time character (τ_1 and $\tau_2 \ll \tau$) the speed of control is only about twice as slow as for PI action, but in other cases PI action is much better.

DERIVATIVE CONTROL ACTION

Besides P and I action, derivative action is sometimes used. This action is characterized by the relationship:

$$\delta u = -K_c \tau_d \frac{d\delta y_m}{dt} \qquad (16.26)$$

where τ_d = derivative time

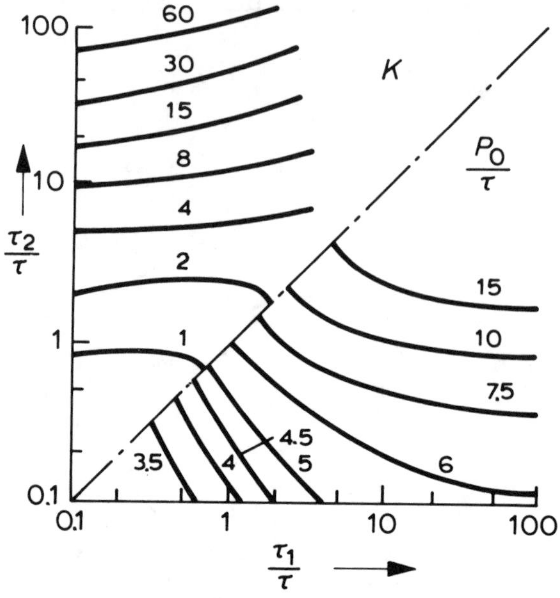

Figure 16.8 PI control for damped behavior, $\zeta = 0.25$.

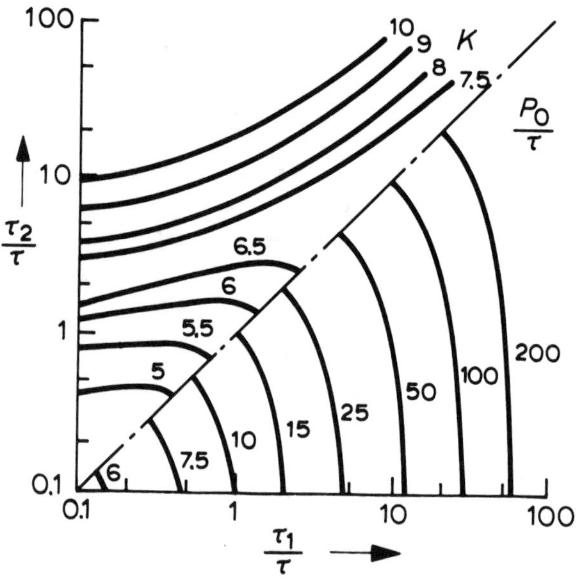

Figure 16.9 Integral control for damped behavior, $\zeta = 0.25$.

D action only reacts to variations and not to sustained deviations. Therefore, it must be combined with P and possibly I action. The combination of PD action is characterized by:

$$\delta u = -K_c\left(\delta y_m + \tau_d \frac{d\delta y_m}{dt}\right) \qquad (16.27)$$

or, with operators:

$$\frac{\delta u}{\delta y_m} = -K_c(1 + s\tau_d) \qquad (16.28)$$

D action can be used, for example, to compensate a first-order lag. This is clear from the combination of both transfer functions:

$$\frac{1}{1 + s\tau_2} \cdot K_c(1 + s\tau_d) \qquad (16.29)$$

This expression becomes a constant when τ_d is made equal to τ_2. In this way, derivative action can reduce a loop with a second-order character (τ_1 and $\tau_2 \gg \tau$) to one with a first-order character ($\tau_1 \gg \tau$, $\tau_2 \simeq 0$), resulting in a considerable increase in the speed of control and a larger value of K_c. The latter provides an adjustment criterion for τ_d: adjust τ_d in such a way that K_c is maximized under the condition that $\zeta = 0.25$.

Figure 16.10 shows the result for PD action. Compared to Figure 16.6, there is also some improvement for loops with a first-order character.

Ideal D action cannot be realized in practice. Actual behavior is modified by a first-order lag:

$$\frac{\delta u}{\delta y_m} = -K_c \frac{1 + s(1 + \beta)\tau_d}{1 + s\beta\tau_d} \qquad (16.30)$$

where the modification factor $\beta \ll 1$. In practice, the effect of D action is often disappointing. The causes are not quite clear, possibly the enlarged sensitivity to measurement noise and small-signal nonlinearities (such as mechanical friction) play a role.

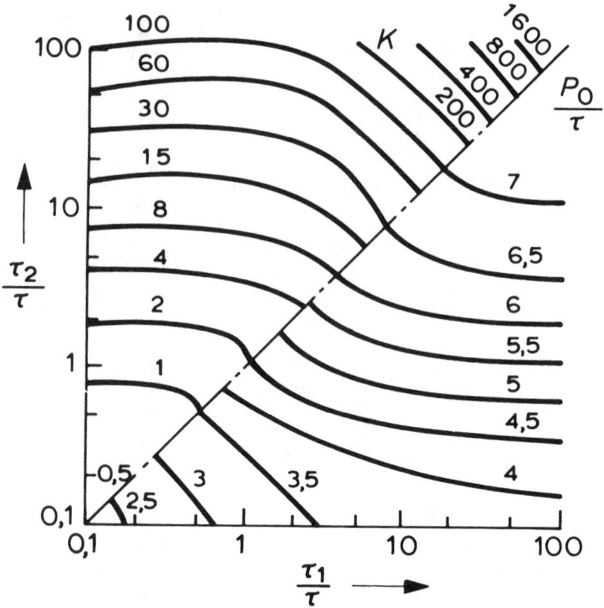

Figure 16.10 PD controller for damped behavior, $\zeta = 0.25$.

PID CONTROL

To eliminate remaining deviations, PD control action is usually extended to PID control action. The transfer function may take several forms: the so-called three-term type is often considered desirable.

$$\frac{\delta u}{\delta y_m} = -K_c\left(1 + s\tau_d + \frac{1}{s\tau_i}\right) \tag{16.31}$$

Many actual controllers are of the product type:

$$\frac{\delta u}{\delta y_m} = -K_c'\left[(1 + s\tau_d')\left(1 + \frac{1}{s\tau_i'}\right)\right] \tag{16.32}$$

This formula can also be written in three-term form:

$$\frac{\delta u}{\delta y_m} = -K_c'\left(1 + \frac{\tau_d'}{\tau_i'}\right)\left[1 + s\,\frac{\tau_i'\tau_d'}{\tau_i' + \tau_d'} + \frac{1}{s(\tau_i' + \tau_d')}\right] \tag{16.33}$$

However, derivative action can only be adjusted over a limited range. This is clear from Equations 16.31 and 16.33, from which the following relation for τ_d/τ_i follows:

$$\frac{\tau_d}{\tau_i} = \frac{\tau_i'\tau_d'}{(\tau_i' + \tau_d')^2} = \frac{\tau_d'/\tau_i'}{(1 + \tau_d'/\tau_i')^2} \tag{16.34}$$

Equation 16.34 has a maximum value equal to 0.25 for $\tau_d'/\tau_i' = 1$, while there are no restrictions in Equation 16.31.

Figure 16.11 shows the speed of control and static gain for PID control action. Evidently, the result is better than for one or two control actions. Figure 16.12 gives an impression of the applicability of the different types of control actions. On the boundaries, P_0 makes a jump with a (arbitrary)

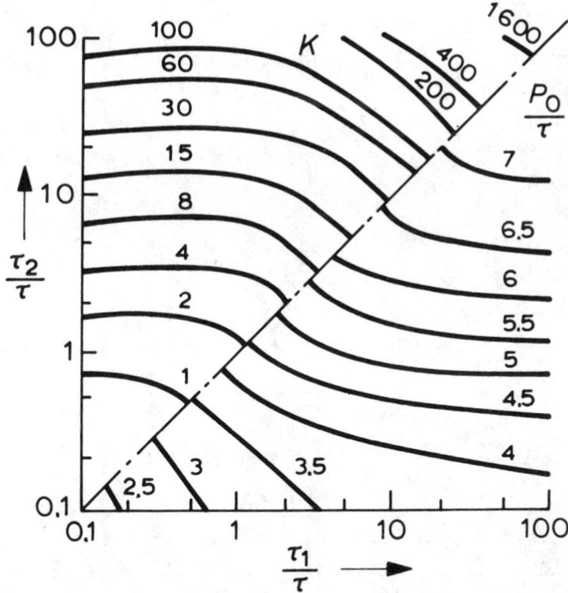

Figure 16.11 PID controller for damped behavior, $\zeta = 0.25$.

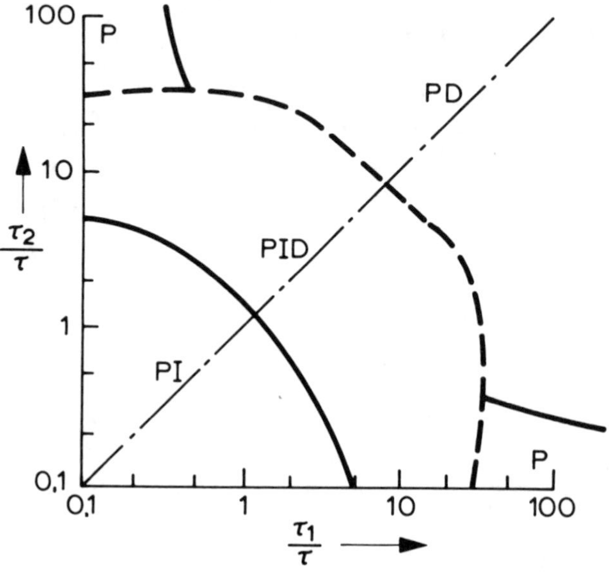

Figure 16.12 Application of various controller actions.

factor of 1.5. On the dotted boundary between P and PD on the one side and PID on the other side, K without I action has just reached 30, in other words, a lower value of K requires I action. Figure 16.12 has to be used with some care, because in an actual situation, the nature of the disturbances, the objective for control, etc., play a role.

ADJUSTMENT OF CONTROL ACTIONS

In the preceding section we used a certain "recipe" for adjustment of the controller actions:

$$\zeta = 0.25 \; ; \quad \tau_i = P_0 \; ; \quad \tau_d \text{ such that } K = \max \qquad (16.35)$$

In the literature one can find many different recipes, nearly as many as there are authors. There is no recipe that is optimal under all conditions: fast disturbances require an adjustment different from slow ones, a large control deviation can be disastrous immediately or only after some time, non-linear effects make an optimal adjustment for high throughput, suboptimal for small throughput, etc. Fortunately, the adjustment seems to be not very critical, and errors of say 50% are usually allowable.

In practice, one usually tries first to find the proportional gain at the limit of stability. This gain is reduced so that a reasonable damping is obtained. The period of the damped oscillation can be measured to adjust τ_i. After this has happened the proportional gain has to be reduced somewhat to get again a reasonable damping. If the ratio of the periods for damped and undamped (at the limit of stability) oscillation seems to be large, D action may be advantageous (second-order process, compare Figures 16.5 and 16.6). Then the controller is first adjusted without I action, for example, by increasing τ_d step by step and finding the limit of stability every time, until the maximal value of K_c is reached. Then K_c has to be reduced again and τ_i adjusted in a way as just described.

RESPONSES TO STEP DISTURBANCES

We shall examine the response of the controlled condition to step changes for three cases (Figure 16.13):

1. a step in the set value;
2. a step at the outlet of the process; and
3. a step at the inlet of the process;

Figure 16.13a shows the general character of the response to a step change in the set value. The measured value approaches the set value via a damped oscillation. For a desired degree of damping, the response is approximately characterized by the oscillation period. This is the sense of using the latter as a measure for the speed of control.

Figure 13b shows the response to a step change at the outlet of the process. The result is the mirror image of Figure 16.13a, hence the interpretation is quite similar. Figure 16.13c shows the response to a step change at the process inlet, in other words, in the correcting variable. This looks much more gradual than the previous response because the step is delayed and smoothed by the process. The maximum deviation of the measured value can be estimated by means of the following rule of thumb:

$$\frac{\epsilon_{max}}{\text{step size in correcting condition}} \simeq \frac{1}{K_v K_c} \qquad (16.36)$$

where K_v = valve gain
K_c = proportional controller gain

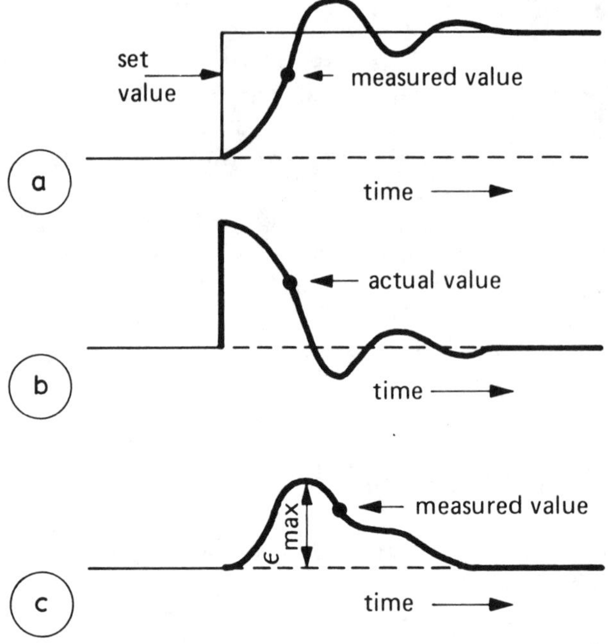

Figure 16.13 Responses to step disturbances.

Equation 16.36 is valid for P, PD, PI and PID control actions. Without I action the remaining deviation is also approximately equal to Equation 16.36.

D action has an indirect influence in Equation 16.36, because this action allows a larger actual value of K_c. In Figure 16.13, a distinction is made between the actual value and the measured value of the controlled condition. This difference may be very important when the measurement lag is relatively large. Then the measured value may give a favorable picture, while the real value shows large deviations ("masking effect").

CHAPTER 17

SELECTION OF THE CONTROL SCHEME

In its widest sense, process control can be considered as the art and science of realizing desired process operation, automatically and manually. It encompasses a wide variety of circumstances, objectives and approaches, which can be arranged in a number of categories, for instance, in the following way:

1. Scheduling. Finding and implementing the best sequence of process runs for the required variety of feedstocks and products. In this sequence, periods have to be included for catalyst regeneration, cleaning, general maintenance, etc. Special care must be taken to satisfy supply, storage and delivery constraints.
2. Optimizing Control. Determining and implementing the "best" values for process conditions for a given process run. "Best" is usually interpreted as optimal with respect to a function of product values, feedstock and utility costs, plant deterioration, and the like. The best values can change with time.
3. Regulatory Control. Keeping process conditions near to "best" values, for instance as determined by optimizing control (see 2). This requires a sufficient reduction of the influences of process disturbances. An important goal of regulatory control is product quality control, because strict quality specifications usually must be met.
4. Sequence Control. Startup, switchover and shutdown of continuous processes are essentially dynamic operations that should be realized in a certain sequence and within certain constraints. The same applies to all operation phases in batch processes.
5. Off-Normal Handling. Off-normal handling provides several barriers against undesirable process conditions which can result in equipment damage, severe pollution of the environment and danger to human beings. In practice, automatic and manual actions (by the operators) alternate, to compensate for shortcomings in each and to attain high reliability. The first barrier is provided by automatic regulatory and/or sequence control, which keeps process conditions near desired values, hence away from undesired ones. However, automatic control can be ineffective due to control valve saturation, equipment or instrument malfunctioning. Therefore, the operators check the operation of the automatic control system. The second barrier is initiated by the alarm system, which warns the operators that off-normal conditions are about to occur, or have already been attained.

This gives the operators the opportunity to take action. The third barrier becomes effective when dangerous conditions are imminent. Then an automatic protection action takes place, which at least locally, interrupts normal process operation and causes a shutdown or a transition to standby. Here, too, the operators have an opportunity (the last one) to intervene, if the protection system does not function properly (see Chapter 19).

In practice, these five categories of process control deserve attention, particularly with respect to their distribution over automation equipment and process operators (the "man/machine allocation") and their integration into an overall control system. In this chapter, however, we shall mainly restrict ourselves to automatic regulatory control, which corresponds to automatic control in the narrow sense of the word.

Objectives for control will be discussed, resulting in a number of process variables to be controlled. When discussing the selection of correcting variables, more controlled variable will be discovered. When controlled and correcting variables have been found, the basic control scheme can be constructed. Finally, to improve control quality, the basic control scheme can be extended with supplementary control actions.

DEFINITION OF OBJECTIVES

Priority Sequence

Objectives for control are related to desired product properties, desired process conditions, costs, values and limitations in process apparatus. Therefore, they have to be formulated in consultation with material experts, process engineers and apparatus experts. The corresponding classification, however, does not line up with the proper priority sequence. The latter can be found by examining the consequences if a certain objective cannot be realized.

The following sequence results:

1. Consequences for Safety. Development of explosive mixtures in process equipment, leakage of dangerous gases etc., constitute danger for human beings in and near the plant. In some cases, an automatic protection system is applied, to take the process out of operation or to bring the process in a safe state. One is, for example, obliged by law to install safety valves on pressure vessels to prevent explosion of process apparatus.
2. Immediate Adverse Effects on Process Apparatus. Pumping of compressors, pressure buildup due to sudden steam formation, etc., can cause immediate damage to process equipment.
3. Consequences for the Environment. Opening of safety valves as a result of too high pressures in process apparatus, too high or low pH of wastewater, etc., can cause inadmissible pollution of the environment. The priority is strongly dependent on legislation and rules imposed by plant management. The priority indicated here corresponds to practice under modern anti-pollution legislation with effective inspection.

4. Consequences for Product Quality. Examples are: too much residue in gas oil as a result of bad separation in the distillation of crude oil, purity below 99% in the preparation of technical solvents, too much cross-linking as a result of high conversion in the preparation of synthetic rubber.
5. Consequences for the Product Quantity. The product quantity must approximately agree with the results of the plant scheduling activities.
6. Consequences for the Costs of Process Operation and Slow Deterioration of Process Apparatus. Although these categories are different, the priority sequence is about the same: an increased deactivation of the catalyst or aging of process apparatus can be desired if this leads to a sufficient increase in process efficiency.

However, in the definition of objectives, it is easier to follow a more systematic approach, where one finds the different types of constraints for process operation. Thereafter, the priority sequence can be determined.

Material Qualities

We talk here about material qualities instead of product qualities because requirements are given for intermediate products, auxiliary compounds and waste flows. The desired properties of material flows can sometimes be translated into limit values for physical properties (density, viscosity, boiling point, etc.) or in maximal concentrations of impurities. In other cases, measurements have to be taken, which are more or less similar to how the material is used in practice (e.g., a motor for the determination of the octane number of fuel).

Traditionally, controller set values were adjusted manually on the bases of laboratory analysis. This can still be found in chemical plants but in addition use is made of "offline" analysis instruments, which can be operated by the operator. This is often called semiautomatic analysis, as the sample must be handled by the operator. In some cases use can be made of "online" quality instruments which can be arranged in an automatic control loop if they are sufficiently reliable. In other cases, there is a reasonable correlation between the desired material properties and other parameters which can be measured more easily (e.g., conversion and product contamination). Then, automatic control can be applied, often with the aid of an online computer.

In still other cases it is not possible to formulate the desired material properties independent of the process operation. Then one can say that certain process situations are undesirable (for example, too high temperatures in polymerization reactions), or there may be an empirical recipe for process operation (as is the case for many batch reactors). Sometimes one can already indicate some controlled conditions here.

In some industries, especially in the food and oil industry, final products are obtained by blending of auxiliary products. A too-low quality of a certain product can then be corrected by changing the blending ratio. The product quality can then, under extreme cases, be arranged under priority 6 (see above). This means that there is no sharp condition for the material property in question; a change in quality will have only economic consequences. A typical example is the preparation of cattle feed, where a shortage of a certain type of protein in one intermediate product can be compensated by increasing the concentration of another intermediate product.

Constraints in Process Operation

The feasible region for process operation is surrounded by constraints. Passing these constraints will not always have visible external consequences, but will interfere with normal process operation. Table 17.1 gives some examples.

Constraints in Process Apparatus

Constraints in process apparatus (Table 17.2) fall into two classes: those that should not be passed, and those that cannot be passed. The latter correspond to the ranges of correcting units; they do not lead to variables to be controlled.

Operation Costs

If there are still some degrees of freedom left, one can try to improve process operation by decreasing operation costs. An example is the optimal division of a feed over parallel units (furnaces, heat exchangers, etc.). Another example is the balance between product yield and heating costs in distillation

Table 17.1 Some Constraints in Process Operation

Maximum level in accumulators, separators, etc.
Minimum level in accumulators, separators, etc.
Minimum pressure in vessels, columns, etc. (for example avoiding vacuum)
Avoiding undesired two-phase flow patterns (e.g., in furnace pipes)
Reversal of flow direction (e.g., of fluidized matter in downcomers and risers)
Maximum cooling water temperature (to avoid scaling)
Maximum process temperatures (e.g., to avoid coke formation)
Maximum loading of distillation trays, packed beds, etc. (to avoid flooding)
Minimum loading of distillation trays, packed beds, etc. (to avoid "raining")
Maximum temperature difference in evaporators (to avoid film boiling)
Maximum or minimum oxygen concentrations (to avoid explosions)

processes (provided operation is maintained within the feasible operation region of the trays).

As is clear from the foregoing, formulation of the objectives already results in a number of controlled and correcting variables. In the following section a more systematic method will be developed for the selection of the correcting variables; in other words, the locations of control valves (or correction units for pump speed).

SELECTION OF CORRECTING VARIABLES

In the literature, the following method is often proposed for the determination of the number of degrees of freedom for control [41]. Count the number of variables, subtract the number of equations, and the difference will correspond to the number of degrees of freedom. Some of these, such as cooling water temperature and feed composition, are independent of process operation; the remainder can be used for locating the control valves.

This method, however, has some pitfalls:

1. For a complicated process, there are many variables and equations. One can easily make mistakes, resulting in an incorrect solution of the problem. The literature presents some striking examples [42].
2. The number of degrees of freedom is not an absolute number; it depends on the selection of the controlled variables. It is, for example, possible to install

Table 17.2 Some Constraints in Process Apparatus

Constraints Which Cannot Be Passed
- Control valve entirely opened
- Control valve entirely closed

Constraints Which May Not Be Passed:

With immediate consequences
- Maximum pressure in vessels, columns, etc.
- Maximum speed of rotation in rotating apparatus
- Minimum speed of rotation in rotating apparatus (critical speed of rotation for compressor)
- Maximum power of electric motors
- Minimum flow through compressors (to avoid pumping)

With more gradual consequences:
- Maximum furnace temperature (in relation to mechanical deterioration of tubes)
- Low or high pH (in relation to corrosion)
- Maximum process temperature (in relation to coke formation, etc.)

several control valves in a process flow line. Figure 17.1 shows a simple example, where the pressure control loop attenuates upstream disturbances ("environmental control"), and the flowrate is adjusted by the temperature controller.

3. The selection of variables and equations requires a good understanding of the process. This understanding, however, can also be used for a direct search for the degrees of freedom for control. This will be done in the next sections.

Incompressible Medium

If the process unit is entirely filled with an incompressible medium, the number of control valves is *at least* equal to the number of flows that can independently be selected. Any additional control valves have to be used for pressure or pressure difference control.

Figure 17.1 gave a simple example, where the number of flows that can be independently selected, is equal to one. Figure 17.2 shows a metering pump with bypass. The metering pump delivers an approximately constant flow ϕ_T. Now $\phi_T = \phi_1 + \phi_2$, thus if ϕ_1 is controlled, ϕ_2 is fixed. Evidently, the number of flows that can be selected independently is equal to one. The extra control valve in the bypass line can then be used only for pressure control. Usually this control is self-acting (e.g., realized by a reducing valve).

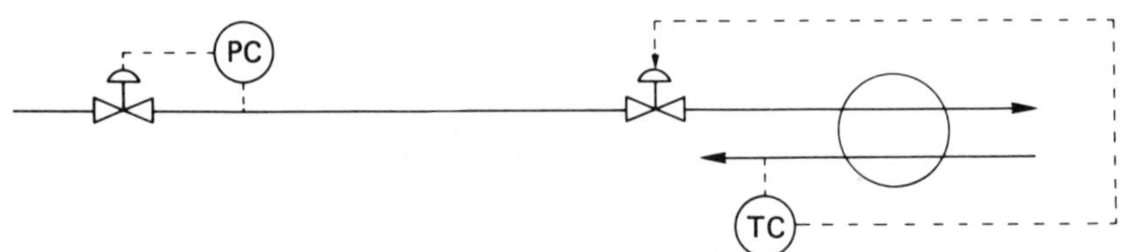

Figure 17.1 Several control valves in a pipe.

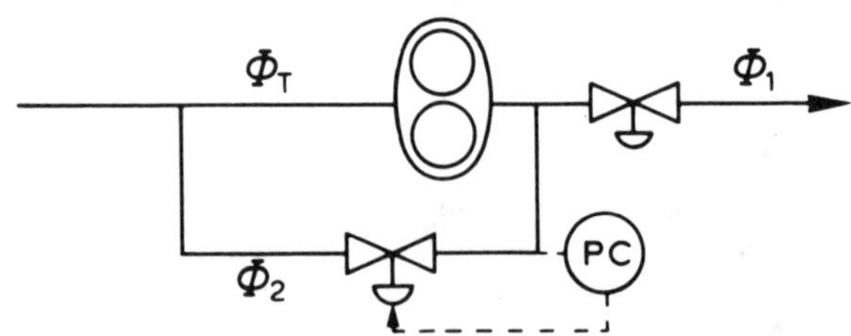

Figure 17.2 Metering pump with bypass.

A more complicated example is the in-line blending of intermediate products to a final product (Figure 17.3). Every intermediate product pipeline is equipped with flow control, the set value of which is proportional to a central set value. In this way the ratio between the auxiliary product flows are kept constant (FrC: flow ratio control). Therefore, the total flow is fixed, and a control valve in the main pipeline cannot be used anymore for flow control, nor for temperature control, for example, because the heat content and therefore the temperature of the main flow is fixed by the ratio controls. Pressure or pressure difference control, however, is possible. Figure 17.4 shows a hot liquid flow which transfers its heat content to a series of heat exchangers. The pressure drops across the n heat exchangers are controlled by adjusting the valves in the bypasses. The number of flows that can be independently selected is equal to n + 1, one for each heat exchanger and

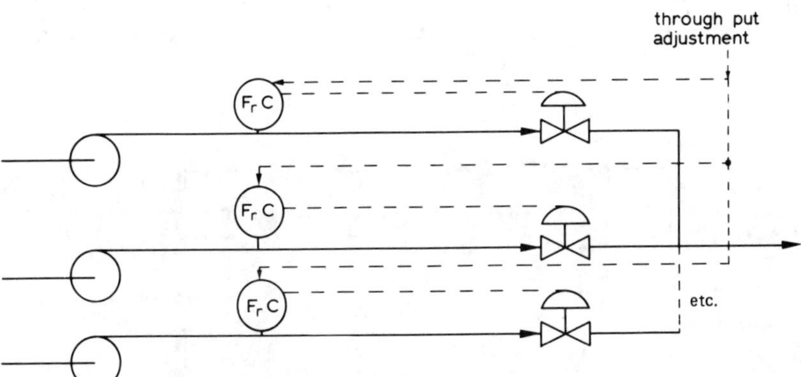

Figure 17.3 Inline blending.

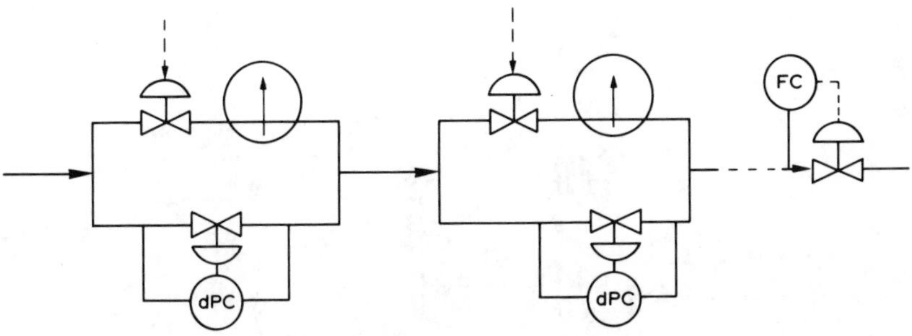

Figure 17.4 Series of heat exchangers.

one for the total flow. A special case is a pipeline with pumping stations (Figure 17.5). The purpose of control is to avoid too low suction pressures (vacuum) and too high discharge pressures, which can be realized with the aid of low value selectors. Each selector passes the minimum value to its speed controller, which adjusts the speed of rotation of the compressor. Within the allowable pressure range, the division of the load over the pumps can be adjusted from a central location.

Compressible Medium

If the volumes are relatively small, there are essentially no differences between incompressible and compressible media. However, if there is a large process volume, then one of the incoming or outgoing flows has to be used for the control of the mass content. As pressure is a practical measure of mass content, and as pressure control is usually desired to avoid opening of safety valves, the result is very much the same as that for incompressible media. Figure 17.6 shows a gas dryer. In this case, the pressure in the unit

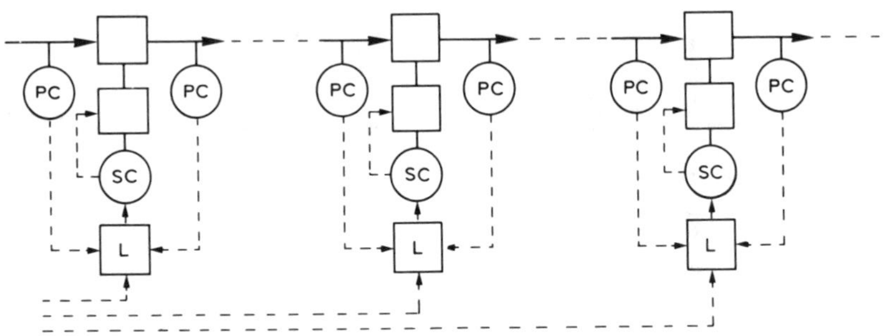

Figure 17.5 Pipeline with pumping stations.

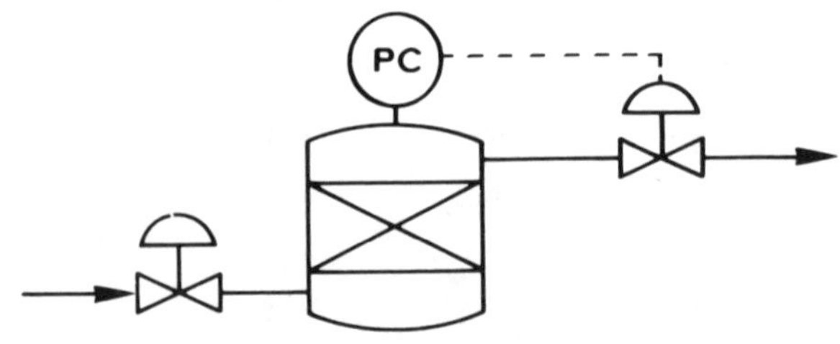

Figure 17.6 Gas dryer.

is controlled by adjusting the gas discharge flow. Another example is combustion pressure control in a furnace (Figure 17.7). A louver control valve in the stack can be used for control of the underpressure in the combustion chamber.

Free Liquid Level

This situation is rather similar to the previous one; now the mass content is represented by the height of the level. Figure 17.8 shows a vessel in which a number of components are being mixed to form a product. The inlet flows are independent; the outlet flow is adjusted by the level controller. In this case, the level control is not critical; its main task is to avoid overflowing or emptying the vessel.

In the case of a series of process units one often finds a series of level controllers, working cocurrent to the direction of flow (Figure 17.9). Here the throughput is determined by the raw material flow, while the production

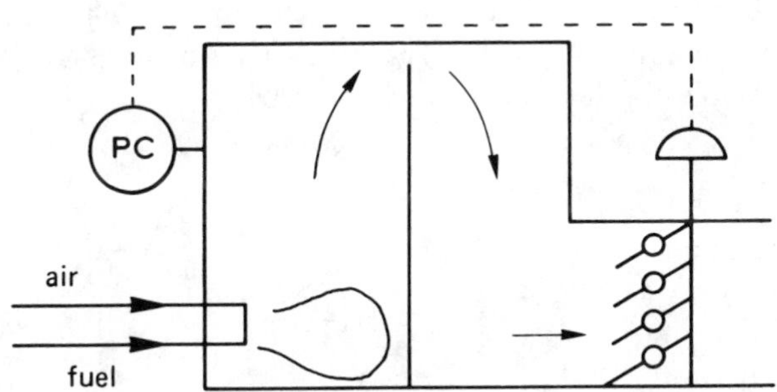

Figure 17.7 Combustion pressure control.

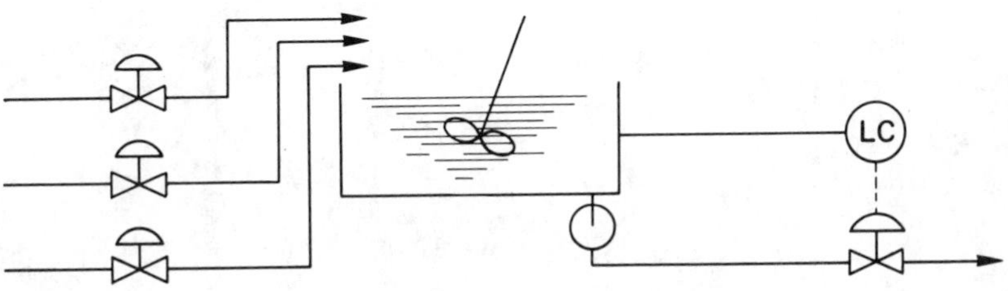

Figure 17.8 Mixing vessel.

is a dependent quantity. Buckley [43] reasoned that it is advantageous to apply level control countercurrent to the direction of flow (Figure 17.10). Production can then be adjusted directly in accordance with the demand. This idea, however, does not hold when the raw material stock is limited, or when more products are prepared from one raw material.

Taking care of feed- and product stocks can also be considered as a control problem. Then we are in the field of operations research, where techniques of inventory control are similar to feedback control of process operation. In other cases, level control is obtained by inherent regulation. Examples are the irrigation systems of the ancient civilizations (Egypt, Mesopotamia, etc.), where overflow weirs were used to control the water level. A more modern example is the liquid flow in a tray column where, just as in Figure 17.9 the inlet flow is independent and all overflows from the trays depend on inherent regulation of the tray inventory.

Two Liquid Phases

Two liquid phases are typical of extraction processes. Figure 17.11 shows a schematic drawing of an extraction column. It is assumed that the density of the feed is lower than the density of the solvent, and that feed and raffinate form the continuous phase. The level between the phases should be maintained below the feed inlet. In the vessel the pressure is a controlled

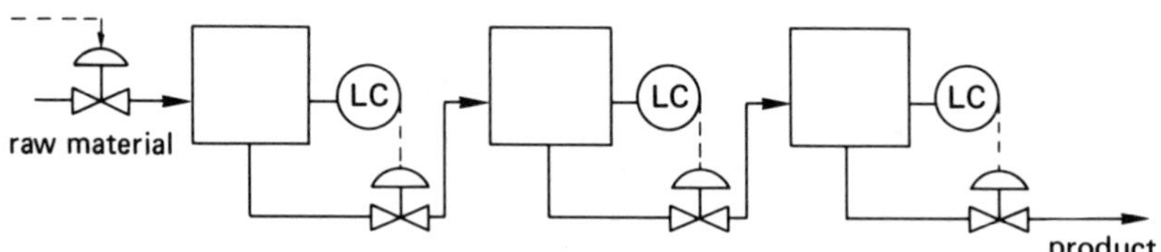

Figure 17.9 Level control, cocurrent to the direction of flow.

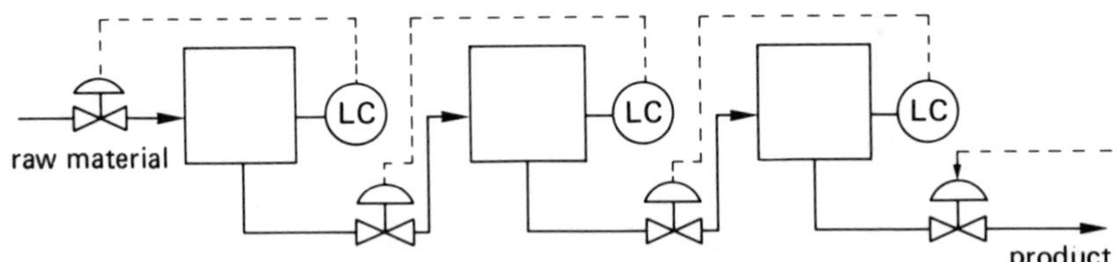

Figure 17.10 Level control, countercurrent to the direction of flow.

condition (just as in the section on incompressible media). Furthermore, the division of the volume over both phases, or, in other words, the interphase level, must be controlled. The remaining control valves determine the feed and solvent flows. It is not clear a priori which valve must be used for pressure control and which one for level control. One can say that it is logical to control the level with the extract flow and the pressure with the raffinate flow, but pressure responds rapidly to all flows, and the level belongs to the raffinate as well as to the extract phase. On the other hand, it makes sense to give the pressure control enough power of control, by selecting not too small a flow as the correcting condition.

One sometimes says that pressure control in extraction columns is impossible when the system is entirely filled with liquid. For continuous flows, however, there is a unique relationship between pressure drop and flow quantity, by which the pressure in the column is also determined (inherent regulation). It is possible that pressure pulses occur as a result of sudden flow changes. If this is not allowed, a gas cap can be installed for damping.

Liquid-Gas (Vapor)

Both phases can here be considered separately. Of course, there is mass transfer between the phases, but this depends on material properties, flow quantities, etc.; hence it is not suitable as correcting condition. The mass content of the gas (or the vapor) is characterized by the pressure and the mass content of the liquid by one or more levels. The supply or discharge

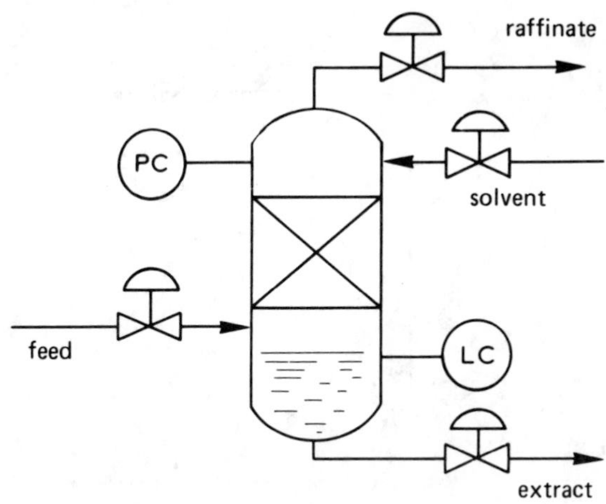

Figure 17.11 Extraction column.

flow of the corresponding phase can then be used for inventory control and the other stream determines the flowrate.

An example of a liquid-gas apparatus is an absorber (Figure 17.12). The vapor content is divided over the trays, the bottom and top section. If, however, the pressure drop over the trays is small compared to the absolute value of the pressure, it does not matter very much which pressure is controlled and which gas control valve is used.

Variations in the liquid flow, however, only propagate from the top to the bottom. The transfer function per tray is approximately first-order; hence, the transfer function from the liquid feed ($\phi_{\ell,n+1}$) to the liquid flow from the bottom tray ($\phi_{\ell,1}$) can be written:

$$\frac{\delta\phi_{\ell,1}}{\delta\phi_{\ell,n+1}} = \left(\frac{1}{1 + s\tau_\phi}\right)^n \tag{17.1}$$

with

$$\tau_\phi = \left(\frac{\partial M}{\partial L}\right)_V \tag{17.2}$$

the derivative of the mass content of the tray to the liquid flow at constant vapor flow.

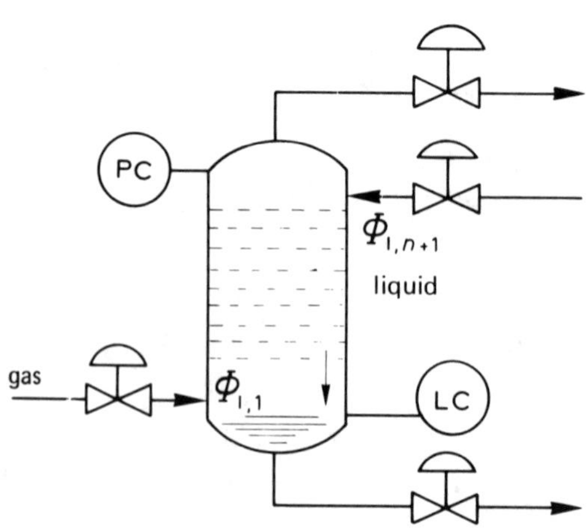

Figure 17.12 Absorber.

If the number of trays n is not too small, Equation 17.1 can be approximated by a dead time $n\tau_{\phi}$. Control of the bottom level by the liquid flow to the top of the column can therefore not be recommended.

An absorber is usually combined with a stripper, in which the absorbed components are separated from the absorption liquid (Figure 17.13). The absorption liquid circulates through both columns (see arrows in Figure 17.13). The total mass content of absorption liquid is constant, only the division over both columns can be controlled. This results in one free level (preferably the level associated with maximum holdup) and one controlled level.

Gas/Solid

The combination gas-solid occurs in fluidized flows and beds. Figure 17.14 shows a simplified flowchart for a catalytic cracking unit for the conversion of high-boiling oil fractions into more volatile components and fractions. The catalyst circulates in powdered form through the reactor and regenerator. The feed carries the hot regenerated catalyst from the downcomer under the regenerator. In the same way a catalyst flow is transported from the reactor to the regenerator by air, necessary for burning off the coke. Both the reactor and regenerator have a certain catalyst holdup, which can be measured with a pressure difference transmitter. One of these holdups has to be controlled, for which the reactor is the obvious choice. The second degree of freedom

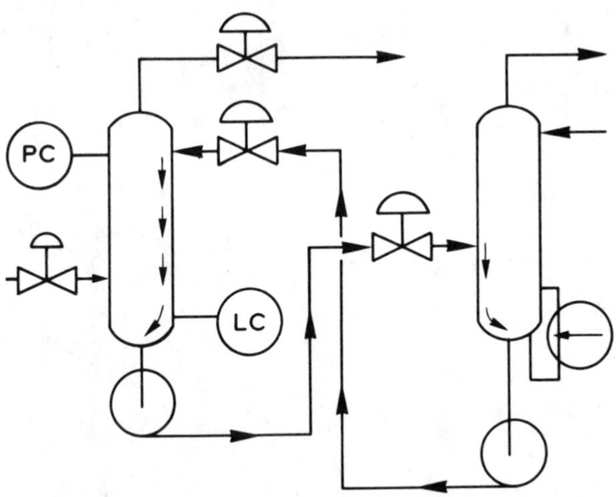

Figure 17.13 Absorber and stripper.

associated with the solid is the circulation rate, which can be used for reactor temperature control. It can be concluded that fluidized media can be handled in the same way as liquids.

Evaporation and Condensation

Until now, all thermal effects were left out of consideration. It happens however very often that a liquid is evaporated or a vapor is condensed. In terms of the two-phase model used above, evaporation is associated with an incoming liquid flow and an outgoing vapor flow. In the same way, condensation is associated with an outgoing liquid flow and incoming vapor flow.

An example is the water circuit of a steam boiler (Figure 17.15). The level in the drum is a measure for the water inventory and the pressure measures the steam inventory. The control valves are installed in the water and fuel supply. The steam discharge flow is an independent quantity. Because the steam inventory is hardly affected by the feed water flow, pressure control by means of the fuel (and air) is the only possibility. Then the control valve in the feed water flow remains for level control. Figure 17.15 also shows an injection of water after the first steam superheater. In this way the steam quality is controlled and steam production increased.

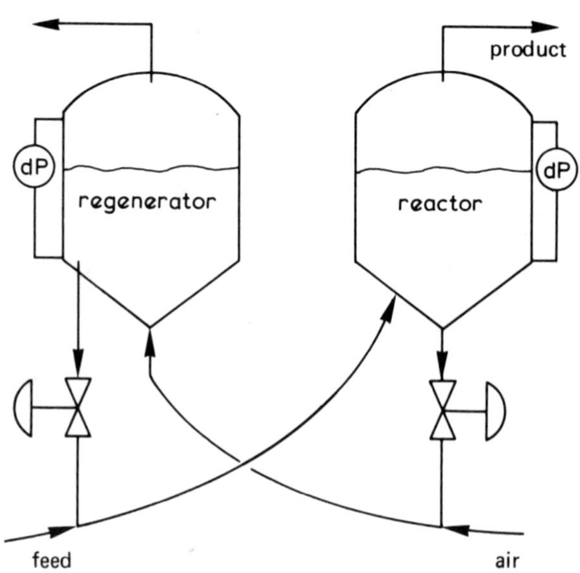

Figure 17.14 Catalytic cracking unit.

Sensible Heat

Sensible heat shows itself mainly as a temperature increase or decrease. Figure 17.16 shows a simple example: the mixing of a hot and cold flow. One of the control valves can be used for flow control; the other control valve is available for temperature control.

Figure 17.17 shows a number of ways to control the outlet temperature of a heat exchanger. Methods a and b can only be used if the flow of the corresponding stream may be varied. Method c has a very favorable dynamic behavior. If a flow is heated (cooled) by condensation (evaporation), several different control schemes are possible.

SELECTION OF CONTROLLED VARIABLES

In the previous sections, some controlled variables were found, while selecting control valves. It appeared that the number of control valves is not

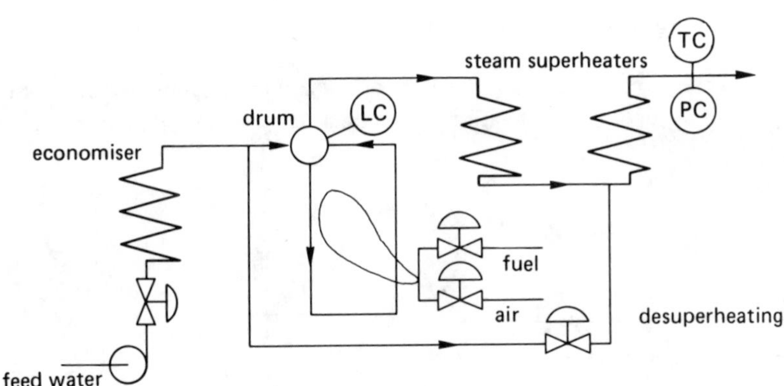

Figure 17.15 Steam boiler.

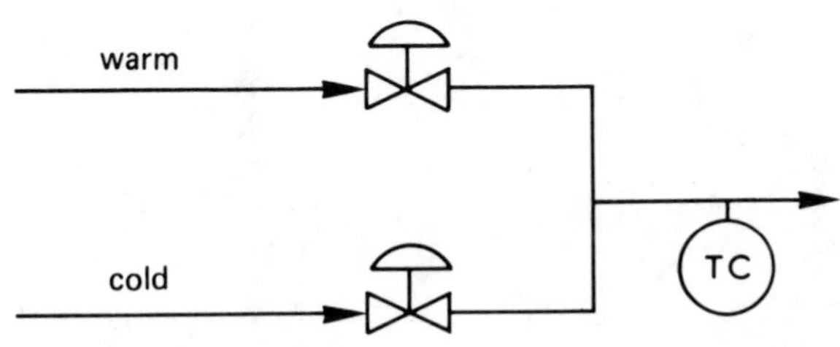

Figure 17.16 Mixing of a hot and cold flow.

constant, but depends on the type of controlled variables (one cannot talk about "*the* number of degrees of freedom for control"). The controlled variables will now be dealt with in a more systematic way.

Control of Mass Contents

In the previous sections, two types of control for inventories were discussed: pressure for gas and vapor and level for liquids and fluidized solid material. In principle, these controls need not be very accurate, provided too high and too low values are avoided. There might, however, be reasons to require a constant level, for example, to ensure a constant residence time in chemical reactors or a pressure near the safety valve setting. In tray columns where the pressure drop cannot be ignored, some attention should be paid to pressure measurement. A short distance between control valve and measurement point promotes the speed of control. If the pressure drop is relatively large, pressure controls can be installed near the top and the bottom, which, apart from the interaction between control loops, is similar to one pressure and one pressure difference control.

Controlled Variables Near Constraints

This category has already been discussed earlier. Examples are: pressure drop in tray columns in relation to maximal loading; outlet temperature of

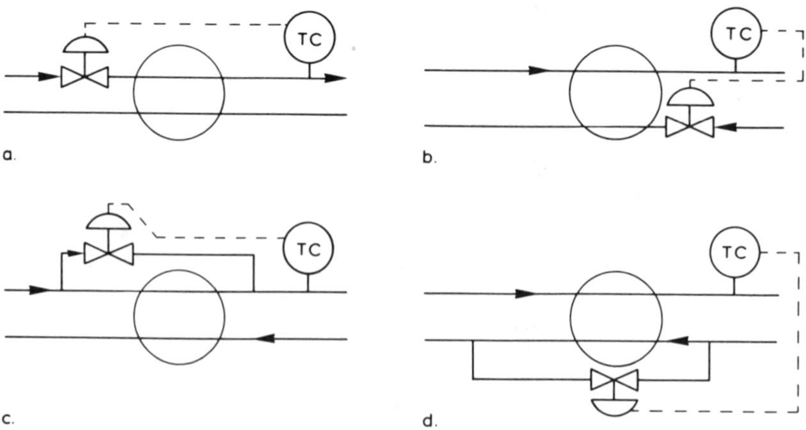

Figure 17.17 Control of the outlet temperature of a heat exchanger (sensible heat effect on both sides).

furnaces in relation to the life expectancy of the pipes; maximal concentration of a component in relation to product purity; and minimum flow through a compressor in relation to the pumping limit.

In the case of temperature and quality measurements, the selection of the measurement point is very important. One should avoid unnecessary delays, for example, between the outlet temperature of a heat exchanger and the measurement point. If one has to take a sample, a bypass is desirable to and from the quality measurement instrument (Figure 17.18), for example across a valve or across a pump.

A particular problem occurs in tray column operation, where a tray temperature is often controlled in combination with pressure, to keep the concentration on that tray more or less constant. This is, strictly speaking, only possible for binary mixtures, which rarely occur in practice. Sometimes the controlled variable must be determined by computation from a number of measured values. Implementation with modern computerized instrumentations is not difficult; the main problem is to set up a good model and to take care of off-normal conditions, such as failures of measuring units.

Flow Controls

Not all available degrees of freedom correspond to constraints. There are usually some which have to be adjusted to a desired value within the operational range. An example is the throughput of a process, which has to be adjusted in accordance with the expected demand of products. Only when capacity is smaller than demand, does the throughput also lie on a constraint.

Another example is the division of the load over parallel process units, for

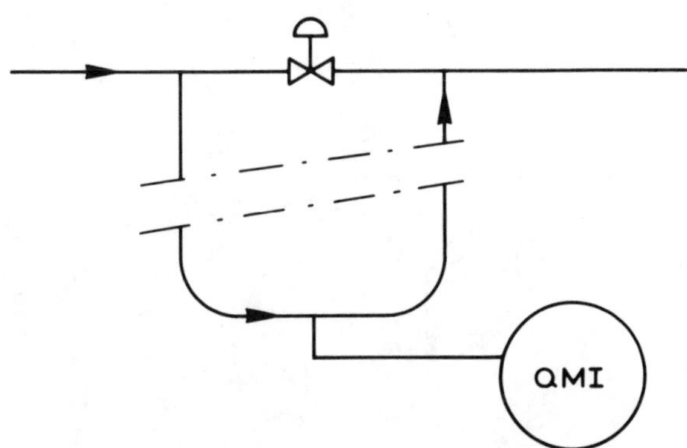

Figure 17.18 Bypass for quality measurement.

example, furnaces in ethylene plants, steam boilers or heat exchangers, which preheat the feed with the aid of product flows. The determination of the optimal division is a complicated problem, and becomes even more complicated when the dynamic behavior is taken into account. Here lies a possibility for the application of a digital online computer.

Environmental Controls

The controls discussed in the previous sections are usually not able to reduce the influence of all disturbances in a sufficient way. For example, if the pressure in a distillation column is controlled by condenser cooling, disturbances in the cooling water will not have much influence on the column. Disturbances in the heating system will affect evaporation in the reboiler, and will therefore also affect other column variables. To cope with this problem, "environmental" control can be used. Figure 17.19 shows two possibilities: one with a pressure control loop on the steam supply and one with a heat flow control loop on the hot oil supply. The latter works according to the equation:

$$Q = \phi\gamma(T_{in} - T_{out}) \tag{17.3}$$

where ϕ = flow (kg/sec)
γ = specific heat (J/kg-°K)

In distribution systems for cooling water, fuel, steam, etc. pressure controllers are often used to reduce or eliminate the effects of disturbances and interaction between individual users.

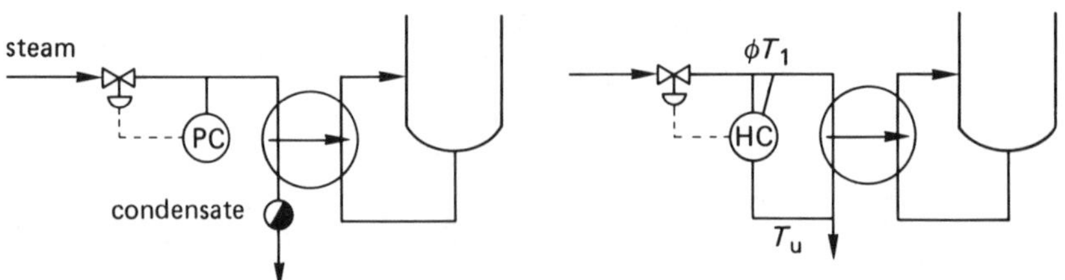

Figure 17.19 Environmental controls for an evaporator.

SELECTION OF THE BASIC CONTROL SCHEME

We have discussed the selection of correcting and controlled variables. The next step is to select the basic control scheme, which connects the controlled variables to appropriate correcting variables.

To find the basic control scheme, it is convenient to put all correcting and controlled variables in a table. A number of undesirable pairs can easily be eliminated by common sense. These pairs can be represented by an X in the table (see Table 17.3 for an example). From the remaining pairs, we must select a consistent set, which results in a favorable control scheme. Criteria are speed of control (corresponding to the period P_0 for $\zeta = 0.25$) and power of control. It is desirable to maximize the number of control loops with high control speed and power, or, which is often more suitable, to avoid loops with low control speed and power as much as is possible.

There are two limitations to this approach. First, there are usually insufficient time and data to calculate the oscillation period P_0 for each pair of correcting and controlled variables. Therefore, one has to live with a qualitative or semiquantitative estimate. Second, control loops can interact, which influences their control speeds and powers favorably or unfavorably (a simple example has already been analyzed). Hence, it is worthwhile to analyze actual control quality after the control system has been implemented, and introduce modifications if there is a significant discrepancy between design and reality.

In many cases, not all correcting variables are included in the basic control scheme. The remaining ones can be used for environmental control or process optimization.

EXTENDED CONTROL SCHEME

In this section, some examples will be given of extensions of the basic control scheme. The following items will be discussed: cascade or master-slave control, ratio control, feedforward control or anticipating action, application of selectors, and split range control.

Cascade Control

When we analyzed the dynamic behavior of an evaporator in Chapter 5, control of the heat flow to the evaporating medium within the tubes was shown in two ways: (1) with a valve in the steam supply; and (2) with a valve in the water discharge flow. In the first case, dynamic behavior is rather fast; in the second case the transfer function has a first-order character with a large time constant. One method to reduce the effect of a large time constant in the condensate level is the application of a slave control loop. Figure 17.20 shows two possibilities. In the first possibility, the steam flow is controlled by a flow controller (FC), which adjusts the control valve in the water discharge flow. The set point of this controller now becomes the correcting condition for the master controller (Figure 17.21). In the second possibility, the condensate level is controlled by a level controller (LC). Master-slave or cascade control can perform the following functions.

1. Increase the speed of control of the master control loop. This is particularly the case when the process contains two large time constants, one within the slave control loop and one within the master control loop (Figure 17.22). The transfer function from x to y is:

Table 17.3 Correcting and Controlled Conditions[a]

Correcting Conditions	Controlled Conditions			
	X_1	X_2	X_3	X_4
U_1	X			
U_2		X		
U_3		X		
U_4				X

[a]X denotes an undesirable combination.

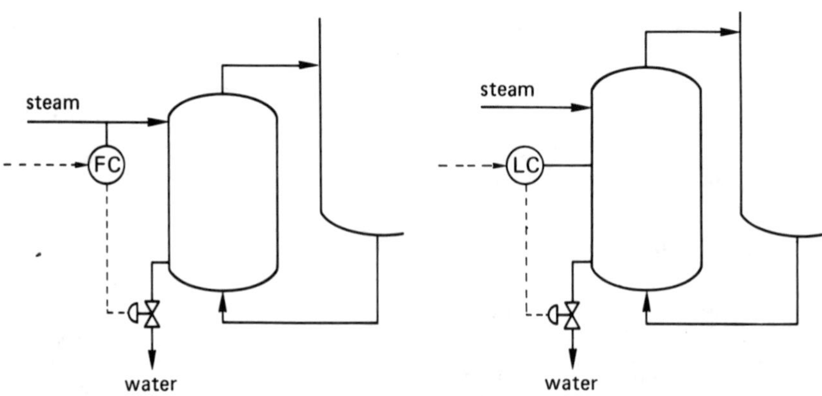

Figure 17.20 Slave control loops for an evaporator.

$$\frac{y}{x} = \frac{K}{1+K} \cdot \frac{1}{1+s\tau_1^*} \tag{17.4}$$

with

$$\tau_1^* = \frac{\tau_1}{1+K} \tag{17.5}$$

When τ_1 is equal to 600 sec and K is equal to 9, the behavior of the slave control loop corresponds to a first-order lag with time constant equal to 60 sec. The value of gain K is limited by secondary lags and nonlinearities in the slave loop, which restrict the stability region of the slave control loop.

2. Eliminate disturbances, which can be measured in the slave control loop. This only holds when the slave is faster than the master.
3. Limit automatically a process variable; in Figure 17.20, FC will maintain the steam flow at a constant value. LC cannot realize this and is therefore less suitable as a slave controller. When we want to avoid steam blowoff, we can extend the control scheme with a level measurement and low value selector, which passes the signal with the lowest value. This is shown in Figure 17.23.

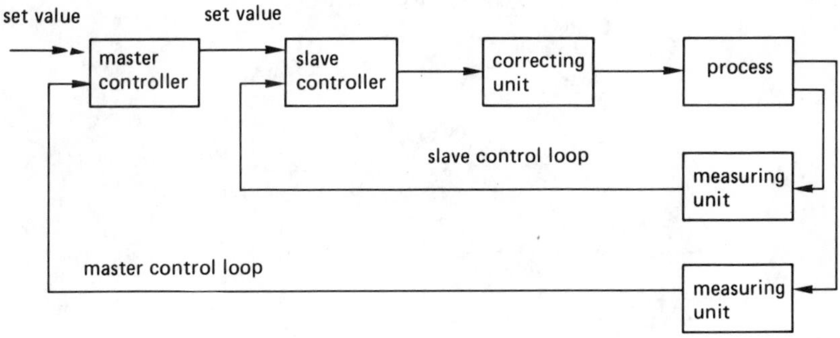

Figure 17.21 Information flow diagram of a cascade control.

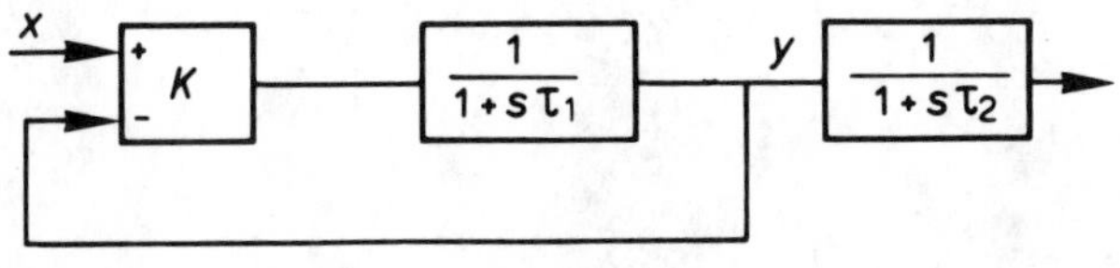

Figure 17.22 Simplified transfer functions in cascade control system.

Cascade Control, Ratio Control and Selectors

Complicated control systems, where cascade control, ratio control and selectors are applied, can be found in the control of tube furnaces. The dynamic behavior of this type of furnaces was analyzed in a previous chapter; control will now be discussed in more detail.

Usually, the outlet temperature T_{out} is automatically controlled. This is presumably also the highest temperature of the feed material in the furnace. For safety reasons, control is realized by adjusting the fuel supply, often via a slave control. Figure 17.24 shows a pressure control PC, controlling the burner pressure. The advantages of this cascade control are:

1. Disturbances in the fuel supply pressure are eliminated by the fast slave control loop.
2. When using liquid fuel, the burner pressure can be maintained above a minimal value P_{min} to guarantee good atomization. This can be achieved by a high value selector H: an instrument which passes the signal with the highest value.

Combustion air must be adjusted in value with the fuel flow. This can be done with a ratio controller, which compares the fuel and airflows and corrects the air supply. If the feed flow shows large fluctuations, temperature control may fail. Then it is possible to apply anticipating action or feed-forward control. Figure 17.25 shows an example.

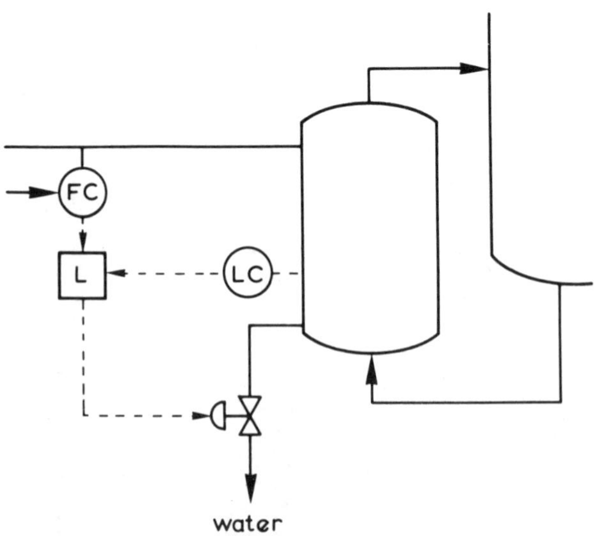

Figure 17.23 Extended control scheme for an evaporator.

A ratio control takes care of a constant ratio between fuel flow and feed flow. The temperature controller may correct the ratio set value. Feedforward control only works properly if the transfer functions along both paths are approximately identical. If, for example, the transfer function from feed flow to outlet temperature is faster than the one from fuel supply

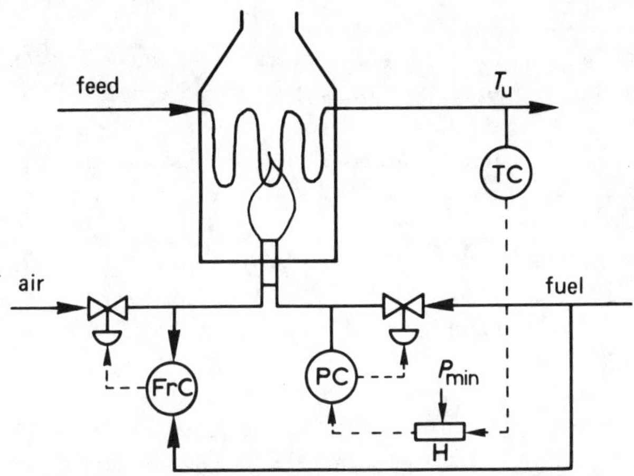

Figure 17.24 Control scheme for a furnace.

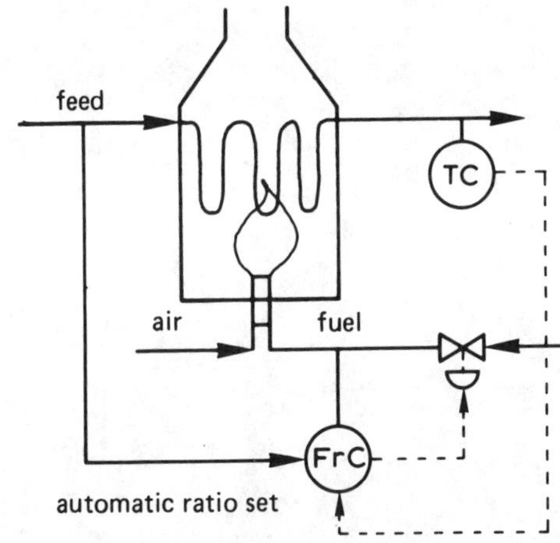

Figure 17.25 Feedforward control of a furnace.

to outlet temperature, anticipating action arrives too late, and the temperature controller still has to do all the work and act on the late anticipating action as well. In the reverse case, the anticipating action arrives too early and the result is not much better. In this respect, the ratio control system shown in Figure 17.25 looks favorable: the transfer functions from feed flow and from fuel flow to outlet temperature are rather similar as both have a direct influence over the total length of the furnace tubes. Differences originate mainly from an unequal distribution of heat flow and from the tube wall dynamics.

For many furnaces the feed is distributed in a number of parallel flows, passing through separate sections of tubes. Therefore there are several outlet temperatures which, preferably, all have to be controlled. Because the total feed flow however is an independent quantity, there are only n - 1 degrees of freedom available for feed distribution. One of the n outlet temperatures must therefore be controlled by means of the fuel flow. Moreover, all the outlet temperatures should be able to have an influence on the fuel supply, to avoid overheating under all circumstances. It is not easy to find a satisfactory control scheme for this process. A better approach is to utilize the inherent flexibility of the online computer, by formulating the control system in terms of a computer algorithm.

A general information flow diagram of feedforward control is given in Figure 17.26. A disturbance is measured and made to influence the controlled condition via the correcting condition. In this way, variations in the controlled condition due to a given disturbance are compensated entirely if the transfer functions along both paths compensate each other:

$$G_1 + G_2G_3 = 0 \tag{17.6}$$

The transfer function of the controller therefore must satisfy the relationship:

$$G_3 = -\frac{G_1}{G_2} \tag{17.7}$$

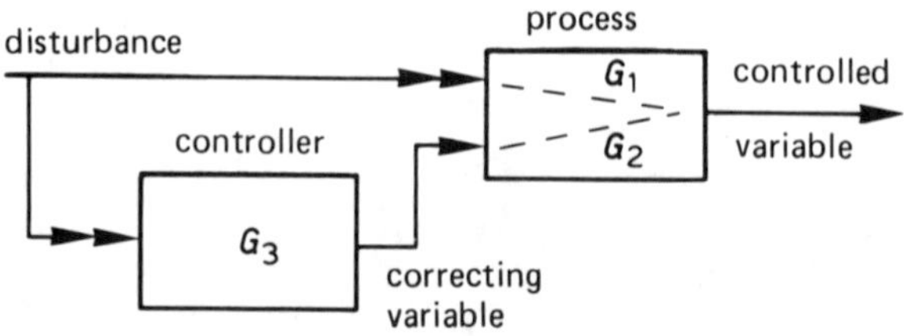

Figure 17.26 Information flow diagram of anticipating action or feedforward control.

Accurate compensation requires careful process modeling. Sometimes even this is physically impossible. For example, when G_2 contains dead time but G_1 does not, G_3 should contain a negative dead time for compensation. However, this lies more in the field of parapsychology than in science and technology.

Split Range Control

Figure 17.27 shows how the jacket temperature of a chemical reactor can be controlled by cooling water and steam. The controller actuates two control valves in split range: only after the water valve has been closed is the steam valve opened, and vice versa. For a gradual changeover, a small overlap is desirable (Figure 17.28). The distribution of the total range has

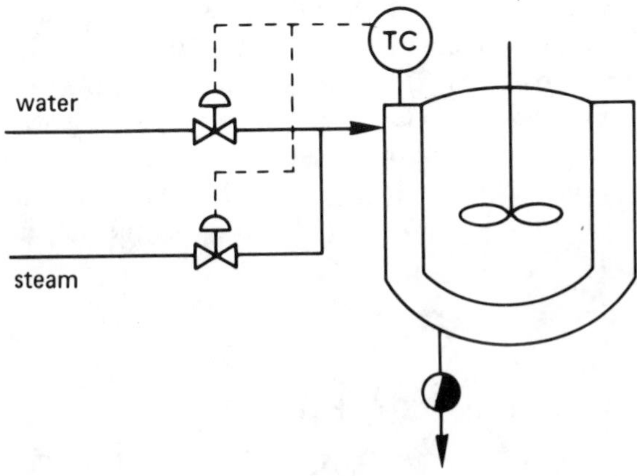

Figure 17.27 Shell temperature control.

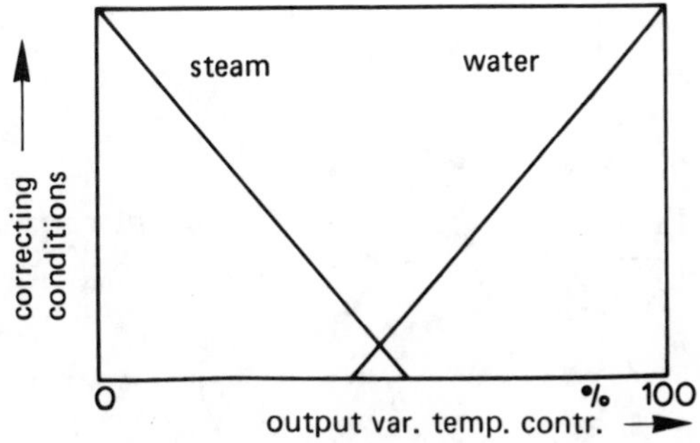

Figure 17.28 Overlapping ranges.

to be selected in such a way that the proportional gain of the controller can be tuned to the same value for water and steam.

CONTROL SCHEME FOR THE PRODUCTION OF AMMONIA

Process Description

Ammonia is an important raw material for the production of fertilizer and plastics. Economy of scale dictates large plants with high investment capital. The actual synthesis proceeds according to the equilibrium reaction:

$$3H_2 + N_2 \rightleftharpoons 2NH_3 + a\ \text{kJ/kmol} \tag{17.8}$$

An iron or iron-nickel catalyst is used; the pressure is between 200 and 300 bar; and the temperature is approximately 400–600°C.

For the preparation of the stoichiometric mixture of nitrogen and hydrogen, synthesis gas, there are many kinds of processes, but we shall restrict ourselves to the reforming process shown in Figure 17.29. Hydrogen is prepared by methane cracking. In the first cracking unit, steam and methane are introduced, and the following reactions occur in the presence of a nickel catalyst at 720°C:

$$\left.\begin{aligned} CH_4 + H_2O &\rightleftharpoons CO + 3H_2 - b\ \text{kJ/kmol} \\ CO + H_2O &\rightleftharpoons CO_2 + H_2 + c\ \text{kJ/kmol} \\ CH_4 + CO_2 &\rightleftharpoons 2CO + 2H_2 - d\ \text{kJ/kmol} \end{aligned}\right\} \tag{17.9}$$

The net effect of these reactions is endothermic, hence heat must be added. After the first cracker, air and steam are introduced in the process gas, and

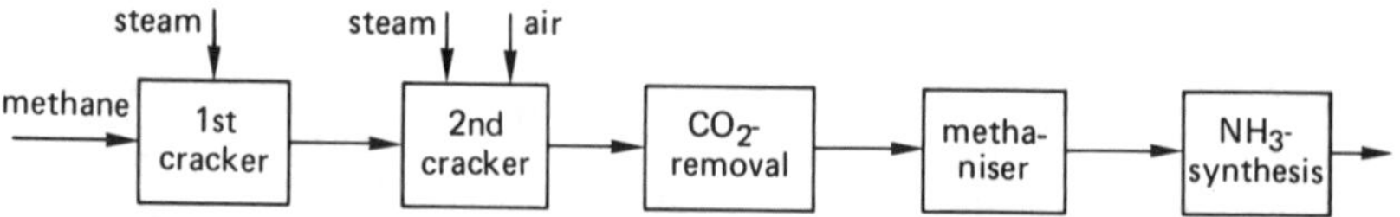

Figure 17.29 Block diagram of the process for the preparation of ammonia from methane.

the mixture flows to the second cracking unit. Here, at a temperature of about 950°C mainly the following reaction occurs:

$$CH_4 + O_2 \rightarrow CO_2 + H_2O + e \text{ kJ/kmol} \tag{17.10}$$

After the second cracker the process gas is fed to a waste heat boiler where steam of 100 bar is produced. The gas is cooled to about 450°C.

The steam is utilized for driving turbines and supplying heat to different parts of the plant. After the second cracker, the process gas flows to an absorber-stripper combination where most CO_2 is removed. Then the process gas flows to a methanizer for removing any remaining CO and CO_2. Here mainly the following reactions occur:

$$\left.\begin{aligned} CO_2 + H_2 &\rightleftharpoons CO + H_2O \\ CO + 3H_2 &\rightleftharpoons CH_4 + H_2O \\ CO_2 + 4H_2 &\rightleftharpoons CH_4 + H_2O \end{aligned}\right\} \tag{17.11}$$

Carbon monoxide and dioxide, which may be present in the gas flow, are converted into methane, which remains in the gas as an inert component. The last step is the synthesis of ammonia from nitrogen and hydrogen.

Control and Protection Devices

The main controlled and correcting variables will be discussed per process unit; attention will also be paid to the consideration leading to a certain selection. The control scheme and process flow diagram can be further extended, leading to an extension of the number of control loops. However, it is not our intention to present a production proof control scheme; only the general approach and way of thinking is important. An average ammonia plant already contains 150 to 200 control loops, and it is impossible to go into enough detail to discuss them all.

The control scheme of the process is shown in Figure 17.30. The solenoid valves are denoted by SV and the control valves by CV.

First Cracking Unit

The first cracker is a tubular furnace with burners mounted in the top (Figure 17.30a). The tubes are filled with catalyst. An interruption in the

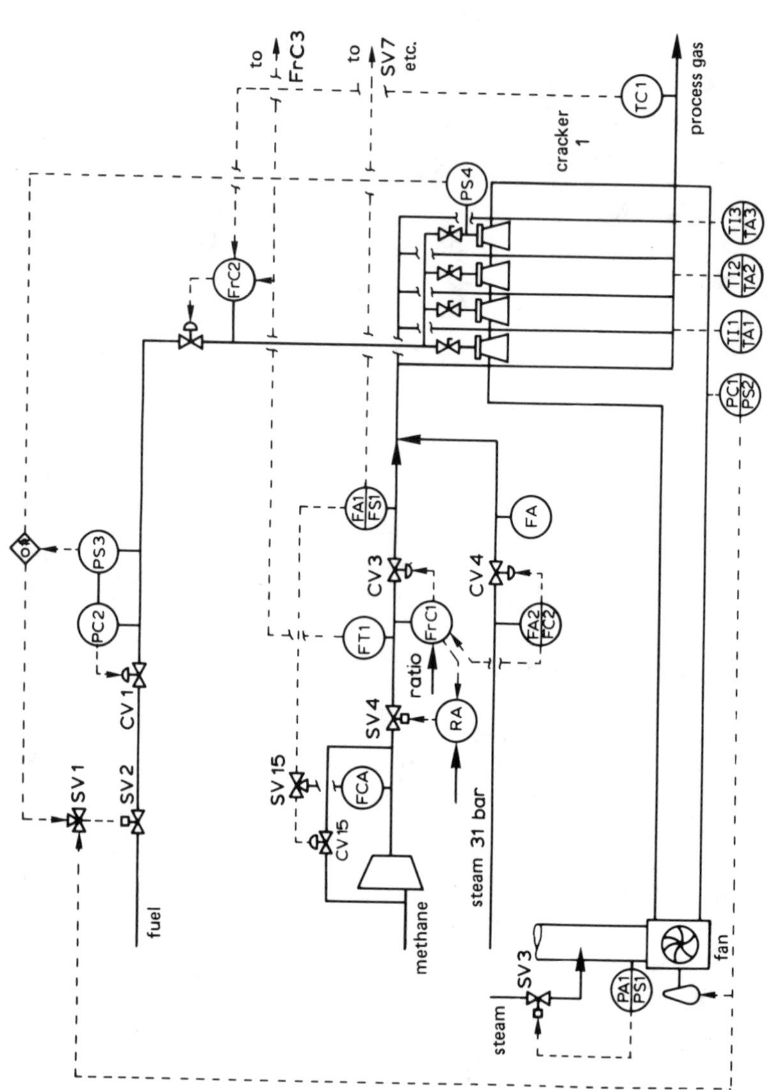

Figure 17.30a Diagram of ammonia plant: control of the first cracking unit.

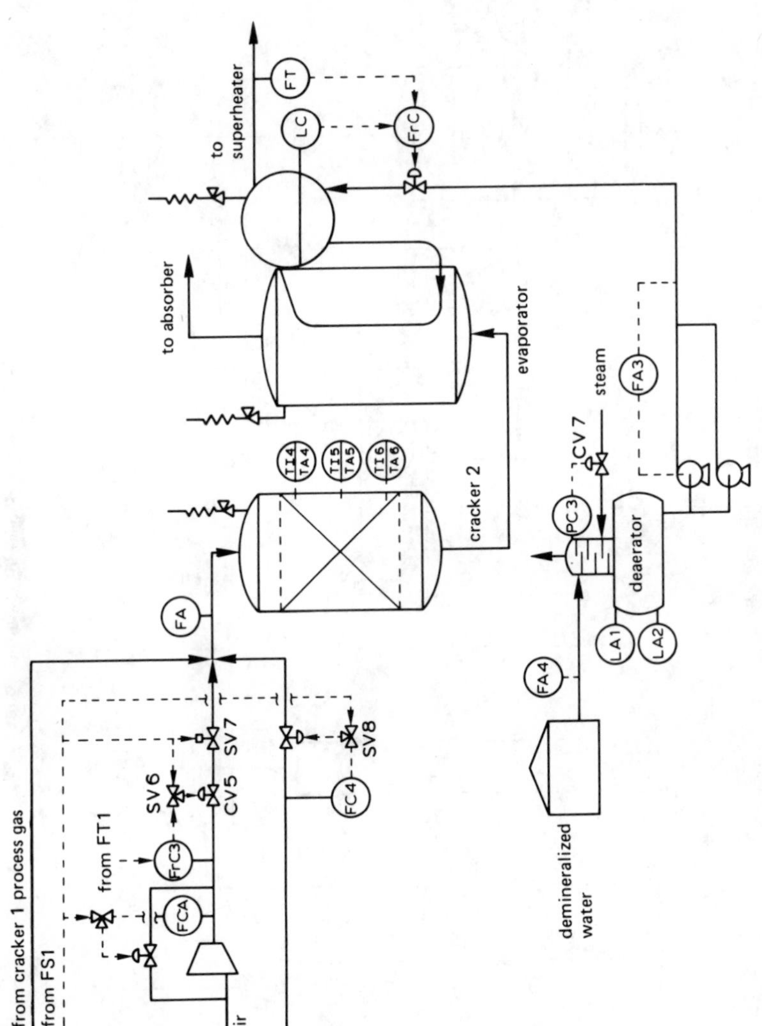

Figure 17.30b Diagram of ammonia plant: second cracking unit and evaporator.

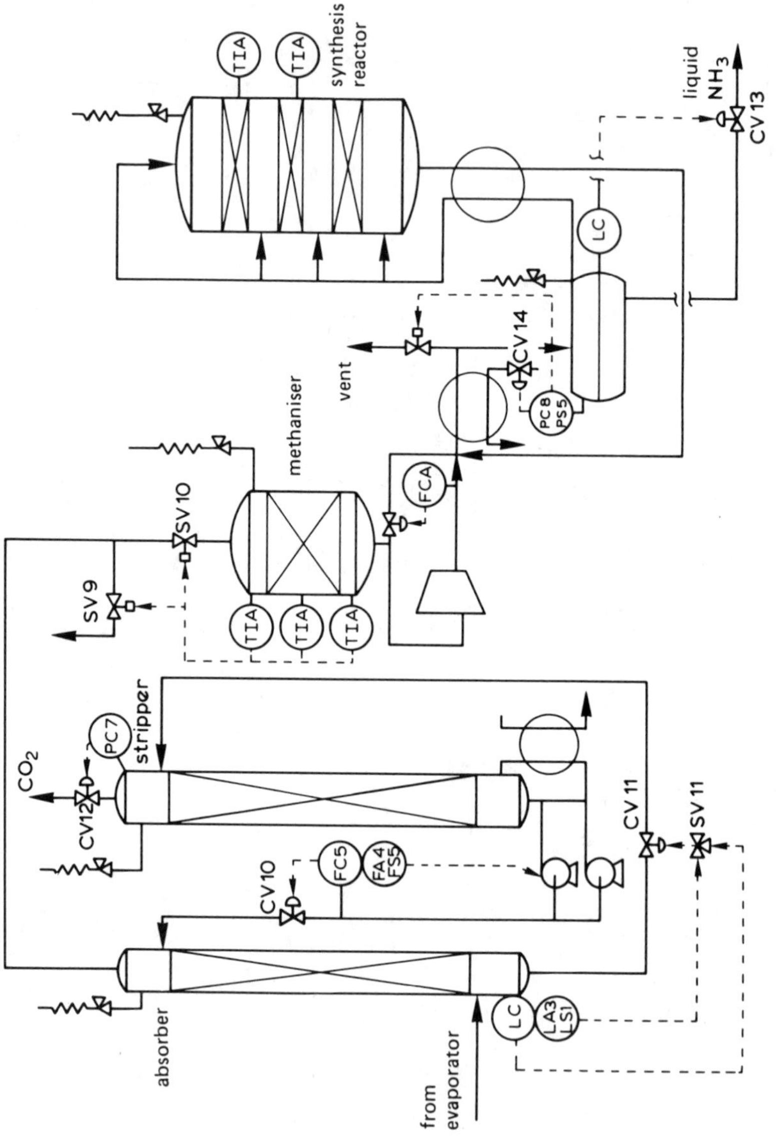

Figure 17.30c Diagram of ammonia plant: control of absorber, stripper, methanizer.

steam flow will lead to coke deposition on the catalyst. The cost of replacement for an average plant are about $150,000, and a one-week loss of production.

The steam flow is controlled (FC2) with the methane flow in ratio (FrC1). When too much methane and too little steam are mixed, a ratio alarm (RA) will close the methane supply to prevent coke deposition on the catalyst. An alarm/switch FA1/FS1 operates several solenoid valves in bypasses and steam supplies. The pressure in the furnace chamber is controlled by the ventilator speed (PC1). When the pressure is too low, startup of the furnace is blocked by SV1, as this valve blocks the full supply. Moreover, switch PS1 takes care of sufficient underpressure in the chimney by steam injection. Fuel is supplied in ratio to methane (FrC2) with temperature controller TC1 acting as a master controller.

Second Cracking Unit

In the second cracker (Figure 17.30b), the catalyst can be damaged by superheating. Therefore, temperature meters and alarms are installed. In case of alarm, the operator can make his or her own decision. When the methane supply to the first cracker is interrupted, control valve CV6 is opened by means of SV8, and steam is introduced into the second cracker. The air supply is stopped and a bypass over the compressor is opened.

The airflow is controlled to obtain a nitrogen:hydrogen ratio of 1 to 3. Often the air is preheated to about 500°C. For a good ratio control, the air density is important. The latter can be computed from a temperature measurement and introduced into the ratio controller FrC3 after multiplication with the air flow.

Waste Heat Boiler

In the evaporator (Figure 17.30b), steam is produced at about 100 bar. A ratio controller determines the ratio between steam and water flow. The ratio setting is adjusted by a level controller. When the water supply to the evaporator fails, a standby pump can be switched on by switch FA3.

Absorber-Stripper and Methanizer

When the bottom level in the absorber is too low, process gas can be blown into the stripper, damaging column and trays. When too little solution is pumped into the absorber, the absorption of carbon dioxide may be insufficient, and carbon dioxide will proceed to the methanizer. This will lead to an increase in temperature and damage to the catalyst. The bottom level in the absorber is critical, hence valve CV11 can be closed by alarm/switch

LA3-LS1 (Figure 17.30c). When the liquid supply to the absorber fails, a standby pump can be switched on. However, if there were insufficient absorption of carbon dioxide, the temperature in the methanizer will rise, and a temperature alarm will close valve SV10 and open SV9. This system can be extended with cold gas injection. Finally, the control scheme can be further extended with a reboiler control on the stripper.

Synthesis Reactor and Separator

The temperature in the synthesis reactor (Figure 17.30c) is controlled by pressure controller PC8. When the pressure becomes too high, pressure switch PS5 will open a vent. The liquid level in the separator is controlled via CV13. All compressors are provided with a bypass with control valve, which only opens when the flowrate in the main pipe line is too small. In the case of the methane and air compressor, this control valve can also be operated by a solenoid valve driven by a switch/alarm.

EXAMPLE

Selection of a control scheme for the preparation of formaldehyde from methanol.

Process Description

The process flow diagram is shown in Figure 17.31. Methanol is converted into formaldehyde with the aid of air over a nickel catalyst at 1000°K. The main reactions are:

$$\left.\begin{array}{ll} CH_3OH \rightleftharpoons CH_2O + H_2 & \text{(endothermic)} \\ H_2 + \frac{1}{2}O_2 \rightarrow H_2O & \text{(exothermic)} \end{array}\right\} \qquad (17.12)$$

The following exothermic secondary reactions also occur:

$$\left.\begin{array}{l} CH_3OH + O_2 \rightarrow CO + 2H_2O \\ CO + \frac{1}{2}O_2 \rightarrow CO_2 \end{array}\right\} \qquad (17.13)$$

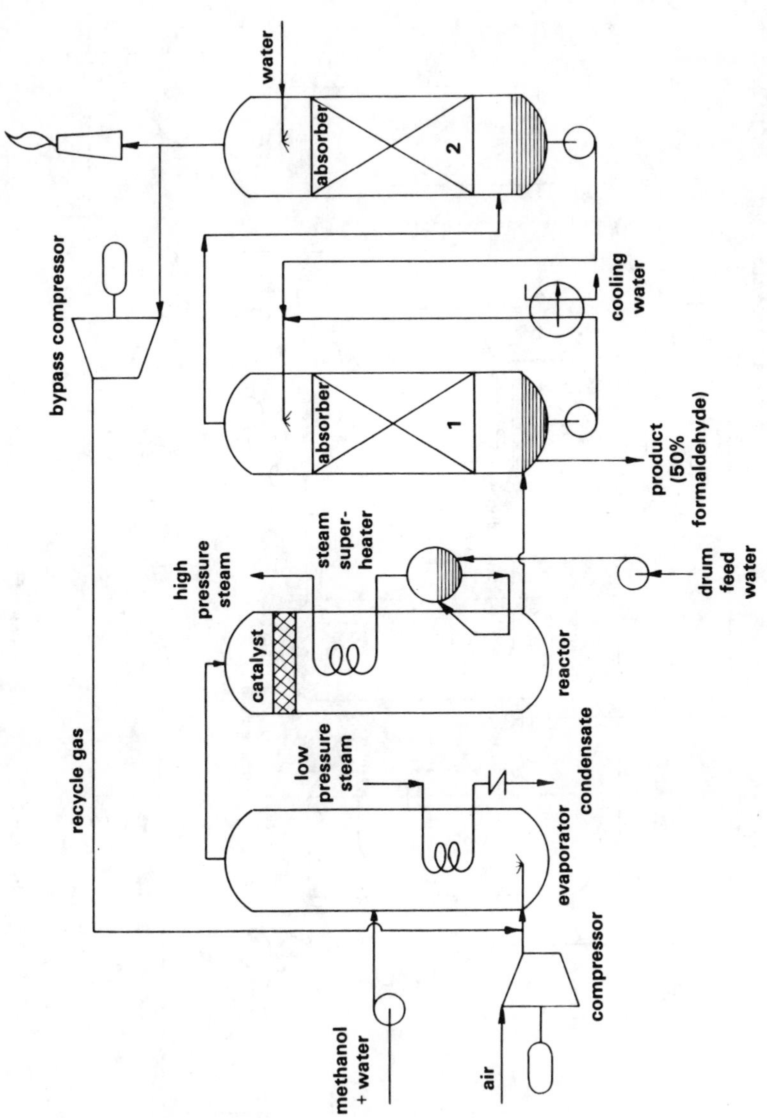

Figure 17.31 Formaldehyde from methanol.

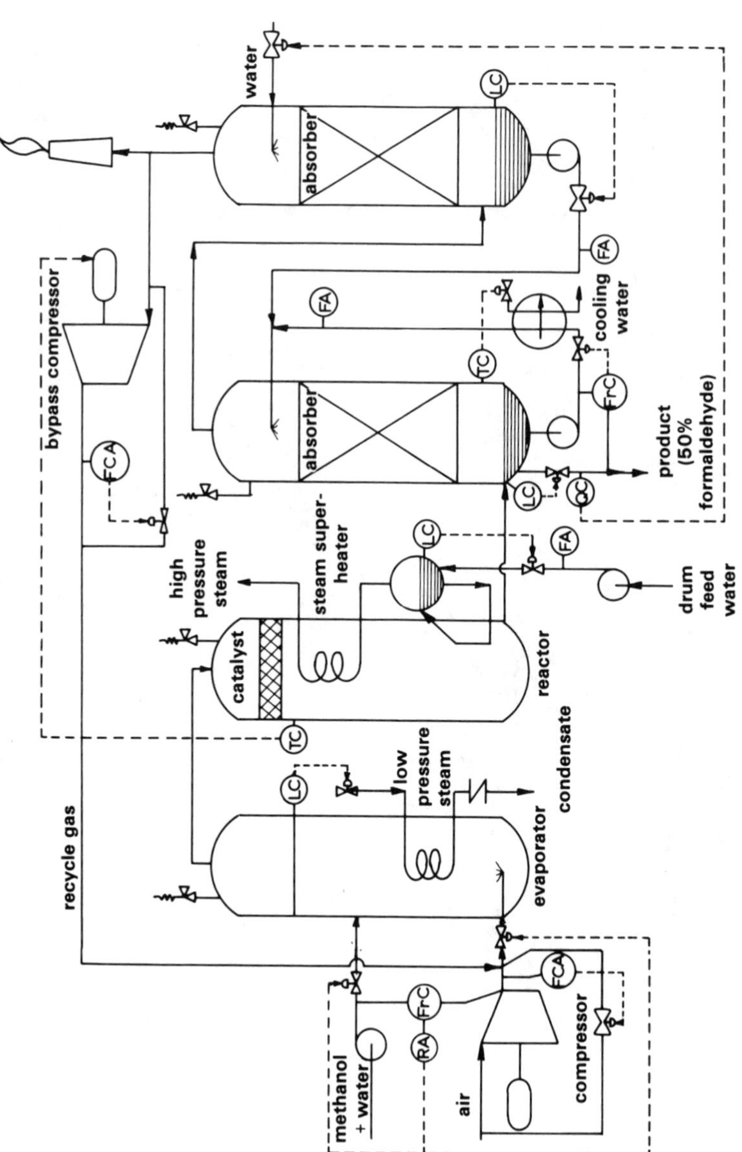

Figure 17.32 Control scheme for the preparation of formaldehyde.

The amount of oxygen needed for the reaction is chosen in such a way that there is a large margin to the explosion limit. The feed (60 wt% methanol, 40 wt% water) is evaporated with low-pressure steam. The water is used for absorbing a part of the heat of reaction. The recirculation of part of the waste gas serves the same purpose. The hot gases in the reactor are cooled and steam is generated. The gases flow to an absorption column in which a methanol solution circulates. A water cooler absorbs the heat of condensation and absorption. The product (50 wt% formaldehyde in 50 wt% water) is drawn off from the bottom. The gases flow to a second absorption column to which water is supplied. The waste gas partially goes to the evaporator; the remainder goes to a flare where hydrogen, carbon monoxide and some formaldehyde are burned.

Control Scheme (Figure 17.32)

From the process description it is clear that the methanol and water flow should be in ratio with the air supply. A ratio alarm can stop both flows. A flow control alarm (FCA) is installed on the air and bypass compressors. Normally the bypasses over these compressors are closed; only when the pumping limit is approached the bypass is opened.

The level in the evaporator is controlled by manipulation of the heat flow. The temperature in the reactor can be controlled by the bypass compressor as this more or less dilutes the reactants. In the absorption columns, both levels have to be controlled. Concentration can be controlled by adding more or less water. The temperature in the first absorber can be controlled by the cooling water flow. Further extension of this control scheme is possible; for example, a flow ratio controller (FrC) can be installed on the steam generating system which controls the steam and water flow in ratio. The level controller can then be used to determine the ratio setting. The pressure in the second absorber could be controlled by controlling the waste gas flow. The absorber can also be controlled in another way. The bottom levels may be controlled by the liquid supplies, and quality control can then be obtained by manipulation of the methanol circulation. Thus it can be seen that a control scheme is not unique; there are usually several solutions. However level control by discharge flow manipulation has some dynamical advantages over manipulation of the inlet flow.

CHAPTER 18

CONTROL OF DISTILLATION COLUMNS

Distillation is one of the most frequently applied separation processes in the chemical industry; this process often occurs in continuous flow columns (Figure 18.1) with one feed (F), one top product (D) and one bottom product (B). The separation originates from countercurrent contact between vapor and liquid (V' and L' in the stripping section; V'' and L'' in the rectifying section) on the trays. The vapor flow is generated in the reboiler, which is heated by steam (H) or a hot stream (waste heat). At the feed entrance, the vapor is changed by the feed:

$$V'' = V' + F_v \tag{18.1}$$

where F_v is the vaporized part of the feed. F_v is negative when the feed is introduced below boiling point. The vapor flow from the top of the column is condensed in a condenser by cooling water or air (C). The heat of condensation can also be transferred to another process.

The distillate goes to an accumulator, from which a part of the liquid flows back to the column as reflux (R). The other part is top product (D), which is withdrawn from the accumulator by a pump. At the feed tray, the liquid flow in the column (L'') is enlarged by the liquid part of the feed:

$$L' = L'' + F_\ell \tag{18.2}$$

In the bottom, the liquid flow is partially evaporated in the reboiler, the remainder is bottom product (B). For this distillation column the following subjects will be discussed:

1. objectives for the operation;
2. selection of the correcting conditions;

3. selection of the controlled variables;
4. speed and power of control of combinations between correcting and controlled variables;
5. selection of the basic control scheme; and
6. extension of the basic control scheme.

For more detailed information, reference is made to the literature [16,44].

OBJECTIVES FOR PROCESS OPERATION

The main objective of the distillation process is to obtain a separation between the more and less volatile components in the feed. In many cases, there is a distinct specification for the purity of one or both products. If the purity is below its specification, the economic value is much less than if the

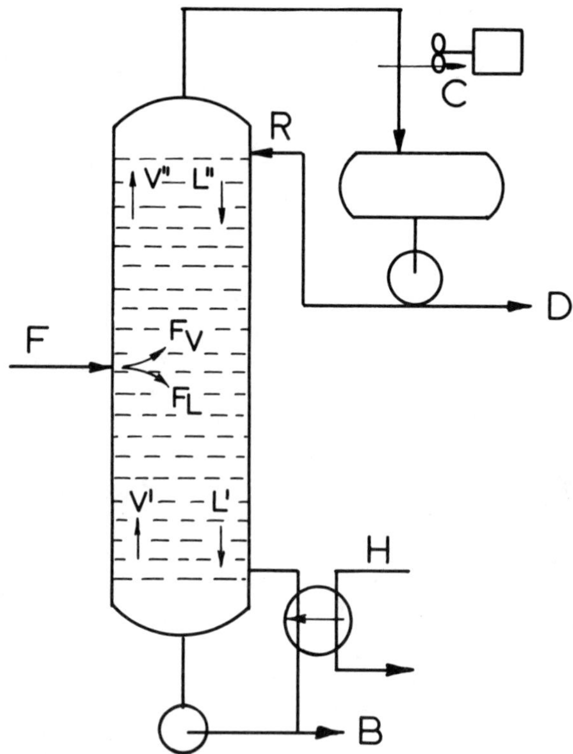

Figure 18.1 Diagram of a simple distillation column.

specification would have been satisfied. The purity is sometimes expressed in the concentration of the main component, e.g., for technical solvents.

Some components are more volatile and others less volatile. In an ordinary distillation column the so-called "light-ends" always go to the top and the "heavy ends" to the bottom. Variations in the amount of light ends in the top product or the amount of heavy ends in the bottom product must therefore be handled in other columns or in a pasteurizing section (Figure 18.2).

In other cases the purity is expressed in the concentration of the key components: the heavy key component in the top product and the light key component in the bottom product. Sometimes other variables may be important for the quality, for example sulfur content (in the preparation of gas oil) or viscosity (lubricating oil).

It is also possible that when there is no distinct quality requirement, the economic value of the product is a continuous function of certain properties, e.g., octane number. Let us assume that there is a distinct quality requirement only for the top product, which can be met by an adjustment of column operation. There are no quality specifications for the bottom product; of course a certain amount of potential top product disappears with the bottom product. The yield of the top product can be weighted against the costs of the distillation in the economical objective of the process:

$$J = c_D D + c_B B - c_F F - c_H H - I \tag{18.3}$$

where c_B = the specific value of the bottom product (\$/kg)
c_D = the specific value of the top product (\$/kg)
c_F = the specific costs of the feed (\$/kg)
c_H = the specific costs of evaporation (\$/J)

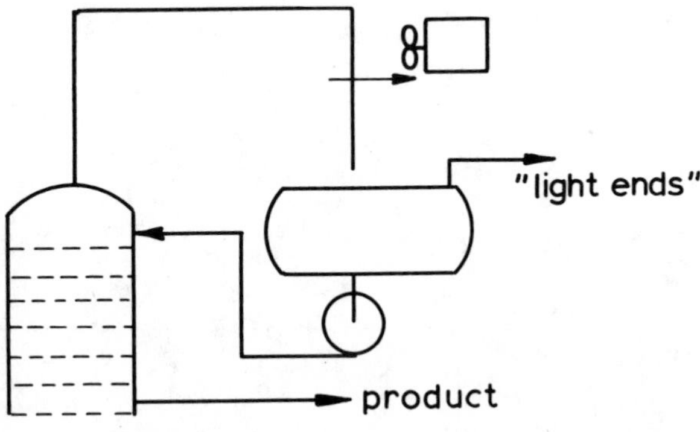

Figure 18.2 Pasteurizing section.

For reasons of simplicity, the variable costs of condensation have been included in the costs of evaporation. If I = fixed costs (write-off, personnel, stores, etc.), with the total mass balance:

$$F = D + B \tag{18.4}$$

Equation 18.3 can be written as:

$$J = (c_D - c_B)D - c_H H - (c_F - c_B)F - I \tag{18.5}$$

If the feed flow varies independently, only the first two terms of the right side of Equation 18.5 can be influenced. They can be written as follows:

$$J^* = D - \frac{c_H}{c_D - c_B} H \tag{18.6}$$

Figure 18.3 shows both terms as a function of the heat flow in the evaporator. As H increases, the yield of valuable product D goes asymptotically to a value which is smaller than the amount of top product which is potentially present in the feed:

$$D_F = \frac{F x_{1k,F}}{x_{1k,D}} \tag{18.7}$$

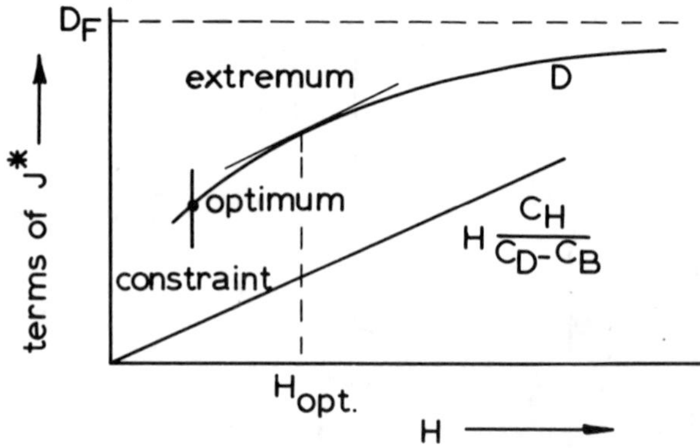

Figure 18.3 Optimal reboiling rate.

where $x_{lk,F}$ = the concentration of light key component in the feed
$x_{lk,D}$ = the concentration of light key component in the top product

The maximal value of J* can be found by drawing a tangent to the curve for D parallel to the straight line for $Hc_H/(c_D - c_B)$, as the distance between the curve and the straight line is then maximal.

The extreme point can lie within the possible operation area, but for higher loads (higher values of F) some constraint makes this point unreachable. The optimum then lies on the critical constraint (Figure 18.3).

In some distillation processes, there is a distinct requirement for both products. This, however, does not mean that both products must always be maintained on the specification limit. It may be advantageous to reduce the loss of a more valuable product in the less valuable product so much that the latter is above the specification limit. This may be clear from the static optimization [45]; to the optimal point (Figure 18.3) belongs a certain purity of the less valuable product. At higher load (higher F) the specification constraint may be reached again; thus, a second quality control becomes important. In practice one will not easily use entirely automated quality control on both products. Automation for one product gives a large improvement for the other one, which makes the benefit of the additional instrumentation rather questionable.

SELECTION OF CORRECTING CONDITIONS

The separation in a column section is determined by two degrees of freedom: vapor and liquid flow. The pressure may also have an influence [46]. If the feed is supplied directly from another process without intermediate storage and heat exchange, vapor flows in both column sections cannot be selected independently. The same holds for the liquid flows. The minimal number of correcting conditions is then two: H or C and R (Figure 18.1).

On the vapor side, an extra degree of freedom (C or H respectively) can be used, if this one serves pressure control. On the liquid side the tray holdups are determined by inherent regulation; hence, no extra degree of freedom is available. From the static point of view there are no more extra independent adjustable degrees of freedom at the top and bottom. Dynamically there is a need for level controls in the top and bottom accumulator, thus B and D can be used as extra correcting conditions.

Summarizing, there are now five correcting conditions (H, C, R, D, B) and three controlled conditions: the pressure (P), the level in the top accumulator (L_D) and the level in the bottom accumulator (L_B). Figure 18.4 shows the result for the case when the speed of rotation of the pumps are adjusted. Principally, the set value of the pressure controller can be considered as an additional degree of freedom [46].

SELECTION OF CONTROLLED CONDITIONS

In the previous section three controlled conditions have been selected: the pressure and both levels. As there are five correcting conditions in total, another two can be controlled automatically.

For good operation of the column it is very important that the distribution of the feed over the top and the bottom is correctly adjusted. This holds particularly for sharp separations which can be illustrated with the following example: a feed, consisting of 40 ton/day "light" component and 60 ton/day "heavy" component has to be separated in products of 99% purity. If one should select a top product flow of 41 ton/day, this flow will contain at least 1 ton/day of heavy component, as only 40 ton/day of light component enters the column. Hence, the purity will now be less than 97.6%. Similar

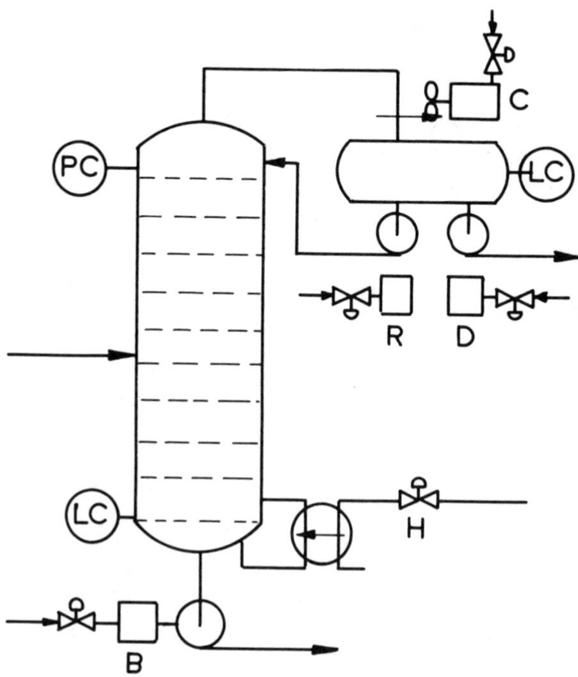

Figure 18.4 Correcting conditions of a distillation column with an independent feed.

reasoning holds for the bottom product if the top product flow is, for example, equal to 39 ton/day.

The adjustment of D/B (or D/F or B/F or another ratio between the same variables) is therefore very critical and must be adapted continuously under fluctuating conditions (e.g., varying feed composition). A good result is possible with automatic control for the most valuable product. As the quality controller maintains a certain degree of purity, the comparison of the other product cannot vary very much.

The fifth correcting condition can, if required, be used to meet remaining variations by manual control or to realize optimal process operation. In practice, a temperature control is often used as a simple form of quality control. For binary mixtures, according to the phase rule of Gibbs, the composition is fixed, if temperature and pressure are kept constant. Evidently, this does not hold for multicomponent mixtures, but one can hope that variations are reduced. One can also try to estimate the product quality from a number of tray temperatures and other easily measurable process conditions by using a model. The microcomputer offers the possibility for an economic solution (model building is the most expensive part). In the following section we shall restrict ourselves to two cases:

1. the concentration of the heavy key component in the top ($x_{hk,c}$); and
2. the concentration of the light key component in the bottom ($x_{lk,h}$).

For a good correlation with the product composition the measurement point must be installed at the condenser and before the reboiler. Attention is paid below to control of the corresponding temperatures.

POWER AND SPEED OF CONTROL

For every pair of correcting and controlled conditions the power of control will be examined first and then the speed of control. The result is a table, which can be an aid in the selection of the basic control scheme.

Pressure Control

B and D are unsuitable for pressure control due to a lack of power of control. R cannot be used either, unless the degree of undercooling is very large. C and H have a large power of control, because they directly determine the incoming and outgoing vapor flow. The dynamic response of the pressure to variations in C and H can be analyzed with the aid of Figure 18.5, where several parts and aspects of the column are shown in an information flow diagram. The response of the top vapor flow (V_c) on the cooling of the condenser (C) is fast; the time constant is determined by the heat capacity

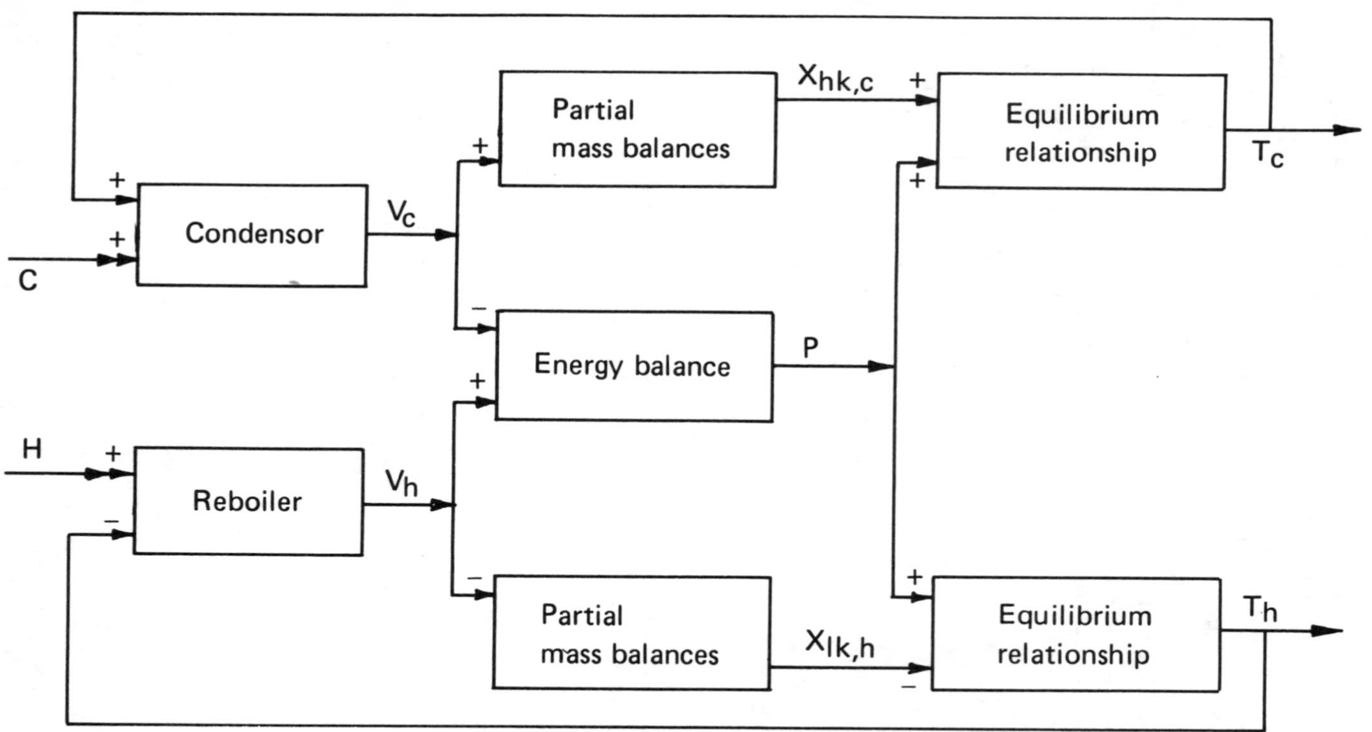

Figure 18.5 Information flow diagram with responses on C and H.

of the (finned) pipes, divided by the sum of the heat transfer rates per degree Kelvin inside and outside the pipes.

The reboiler will also generally show a fast dynamic response. The difference between the vapor flows from the reboiler and to the condenser gives the response of the pressure via an integration. (For simplicity the pressure difference is ignored.) The rate of pressure change is limited by the total heat capacity of the trays, bottom and top. The static value of the pressure response is determined by the inherent regulation in reboiler and condenser (Figure 18.5). A higher pressure results in higher top and bottom temperatures, which improve the heat transfer in the condenser and deteriorate the heat transfer in the reboiler. Finally the equilibrium between V_0 and V_h is restored. The inherent regulation, however, is also influenced by composition responses (the inherent composition regulation). A step increase in H results (via higher pressure and top temperature) in an increase in the vapor flow in the top. Consequently, the concentration of less volatile components in the top ($x_{hk,c}$) will increase, resulting in a higher top temperature and consequently a higher vapor flow in the top. This positive feedback can sometimes be so strong that the pressure finally arrives at a lower value (Figure 18.6).

Evidently such a nonminimum-phase behavior makes H unsuitable for pressure control. Inherent regulation via the composition response in the bottom (the concentration of the light key component $x_{lk,h}$) gives, however, a negative feedback; thus, the pressure response on C is not influenced in an unfavorable way. The conclusion is that C and H can both be used as correcting conditions for pressure control (the speed of control is determined by smaller time constants) unless in the case of H, the inherent composition regulation in the condenser is too strong.

Top Level Control

B is unsuitable, as the power of control is nil. D is less suitable when the reflux ratio (R/D) is large. C and R, however, have usually a large power of

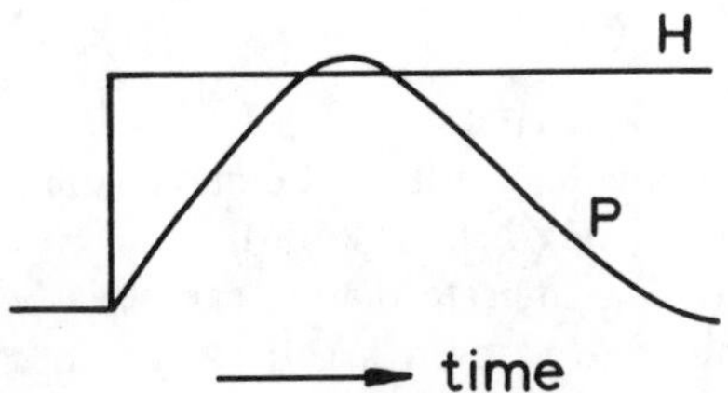

Figure 18.6 Pressure response to a step in H.

control. The responses of the level to variations in D, R and C are favorable for control: the most important dynamic element is an integration. The response to variations in H is not favorable: the condensate flow from the condenser only increases when the pressure increases. The response can therefore be approximated by the cascade of the pressure dynamics (large time constant) and the level dynamics (integration). The conclusion is that C, R and D can be used as correcting conditions for top level control, unless, in the case of D, the reflux ratio is large.

Bottom Level Control

D seems to be unsuitable, since the power of control is nil. B is less suited if the reboiling ratio (V_h/B) is large. H, C and R, however, usually have a large power of control. The response of the liquid flow from the first tray (L_1) to variations in R consists of a cascade of first-order responses:

$$\frac{\delta L_1}{\delta R} = \frac{1}{(1 + s\tau_\ell)^N} \qquad (18.8)$$

where τ_ℓ = hydraulic time constant per tray (the downcomer included)
N = number of trays

Even when τ_ℓ is small, 2 sec, for example, a large number of trays (for example, 40) will give a large effective dead time, which, together with the hydraulic dynamics of the reboiler and integration in the bottom level, leads to slow level control (oscillation period larger than 6 times 80 sec). The response to H is dynamically unfavorable if the sensitivity of the liquid flow in the column for variations in the vapor flow has an unfavorable value. This is the case when the parameter

$$\lambda_1 = \left(\frac{\partial L}{\partial V}\right)_{M_\ell} \qquad (18.9)$$

is greater than 0.5 (see explanation below). In this expression, L is the liquid flow, V the vapor flow and M_ℓ the liquid holdup of the tray (the downcomer included). Unfortunately, very little is known about λ_1 as a function of tray load. The few data seem to indicate that λ_1 can be large at low tray load, at higher tray load small, zero or even negative. The conclusion is that B is a favorable correcting condition, unless V_h/B is large and that H is favorable unless $\lambda_1 > 0.5$.

Quality Control

The response of the key components can be approximated by the response of the algebraic sum of the local relative flow variations, followed by a first order with a very large time constant [47] (see also the explanation below):

$$\delta x_i = \left(\frac{\delta L_{i+1}}{\overline{L}_{i+1}} - \frac{\delta \overline{V}_{i-1}}{V_{i-1}}\right) \frac{K_x}{1 + s\tau_x} \tag{18.10}$$

where $\overline{L}_{i+1}$ = liquid flow to tray i
$\overline{V}_{i-1}$ = vapor flow to tray i
τ_x = a large time constant, approximately proportional to the square of the number of trays; it can be in the order of many hours for long columns [48]

The response of the bottom quality to the reflux is therefore unfavorable for automatic control, for the same reason that holds for the bottom level. This does not hold for the top quality or for the composition on a tray near the top; the intermediate number of trays is then zero or small respectively. Similarly, the vapor flow (C or H with pressure control on H or C respectively) is less suitable for bottom quality control if the tray parameter

$$\hat{\lambda}_1 = \frac{V}{L}\left(\frac{\partial L}{\partial V}\right)_{M_\ell} \tag{18.11}$$

is larger than 0.5. The difference with Equation 18.9 is related to the appearance of relative flow variations in Equation 18.10. B and D have no direct influence on the product qualities, they could have an indirect influence via a level controller.

Summarizing, we can say that R, C and H are suitable for control of the top quality. C and H are suitable for control of the bottom quality, unless $\hat{\lambda}_1$ is less than 0.5. Then there is, however, hardly any alternative.

SELECTION OF THE BASIC CONTROL SCHEME

Selection Table

With the aid of Table 18.1, which summarizes the conclusions of the previous sections, different control schemes can be found. An X indicates an undesired combination, and a blank a favorable combination.

The control schemes must be able to cope with variations in feed quantity and composition. For example, more feed at constant composition means more top and bottom product. A more volatile feed, at constant flow, must find expression in more top product and less product, etc. D and B may therefore not be omitted from the control scheme. From the table it is clear that they both have to be used for level control. The problem of the limited power of control at high reflux and/or reboiling ratio is, however, not solved. This will be discussed in the following section. For quality and pressure control, H, C and R are now available. This leads to the alternative shown in Table 18.2.

Table 18.1 Possibilities for the Control Scheme[a]

	D	B	H	C	R
$x_{hk,D}$	x	x			
$x_{lk,B}$	x	x	unless $\hat{\lambda} > 0.5$	unless $\hat{\lambda}_1 > 0.5$	x
P	x	x	b		x
L_D	unless $R/D \gg 1$	x	x		
L_B	x	unless $V_h/B \gg 1$	unless $\lambda_1 > 0.5$	x	x

[a]x = undesired combination; blank = favorable combination.
[b]Unless the inherent composition regulation in the condenser is very strong.

Table 18.2 Alternatives for Quality and Pressure Control

Controlled Condition	Correcting Conditions			
	Quality Control at the Top			
$x_{hk,D}$	H	C	R	R
P	C	H	C	H
L_D	D	D	D	D
L_B	B	B	B	B
	Quality Control at the Bottom			
$x_{lk,B}$	H	C		
P	C	H		
L_D	D	D		
L_B	B	B		

Interaction

When selecting these alternatives, one should first pay attention to the interaction between control loops. Interaction means a mutual influence of control loops; a one-sided influence may be annoying but does not change the dynamics and stability of the system. An undesired effect of interaction can be that the dynamic behavior of one control loop is strongly affected by the other one. Sometimes also a strong effect in the other direction is present.

A very extreme case of influencing is a sign change: negative feedback turns to positive feedback and the relating control loop becomes monotonically unstable. We will analyze this for the scheme CHDB: without pressure control more C will result in a larger vapor flow (Figure 18.5) and consequently more top and less bottom product (the reflux remains constant). As a result the concentration of the heavy key component in the top increases and that of the light key component in the bottom decreases.

With pressure control, more C will give more H, and consequently the vapor flow is increased more. Thus the sign of the response does not change, apparently the interaction is not very unfavorable. If, however, the temperature were used as a measure for the composition, the situation is entirely different: without pressure control more C gives less volatile mixtures in the column and a lower pressure. The net effect, at least in the short term, is lower temperatures. With pressure control, more C gives less volatile mixtures. As the pressure is now constant, the temperatures will increase. We see that the sign of the temperature response depends on the effectiveness of the pressure control, which is undesirable for control. The scheme CHDB can therefore not be recommended when the temperature is used as a measure for product quality.

Other Arguments for the Selection of the Basic Control Scheme

If many disturbances enter with the cooling of the condenser, pressure control on C can be effective as an environmental control. The same principle holds for the heating of the reboiler. If C, H or R offer a critical constraint for optimal operation, the relating condition should not be used for control, but must be made maximal.

EXTENSION OF THE CONTROL SCHEME

Ratio Control

From Equation 18.10 it can be seen that keeping L/V constant in a column section reduces variations in the concentration. In fact, only a slow and smoothed propagation from tray to tray remains, which is hardly a problem for quality control. Figure 18.7 shows a ratio control at the top of the column. The reflux flow is adjusted in ratio to the top vapor flow. The quality controller adjusts the ratio. An extra advantage of this ratio control is the power of level control: an increase in V_c is immediately compensated by a proportional increase of R, resulting in a decreased influence on the level. The level controller can now operate well also at large reflux ratios. Figure 18.8 shows a ratio control at the bottom. The heat flow in the reboiler is calculated from the temperature difference and flow of the heating medium (here no phase change is assumed). The level controller acts on a slave flow controller on the bottom discharge flow. If $\lambda_1 > 0.5$, this scheme is not satisfactory, because the bottom level controller also adjusts the heating of the reboiler by means of the ratio controller. One could remove this by applying a feedforward ratio control. Figure 18.9 shows this for a steam heated reboiler, from which the condensate flow is adjusted to maintain a certain condensate level in the reboiler, making part of the heat transfer area ineffective for heat transfer from the steam to the tubes. From F and R the liquid flow to the reboiler is estimated by dynamic compensation devices, and thus the steam flow is subsequently adjusted.

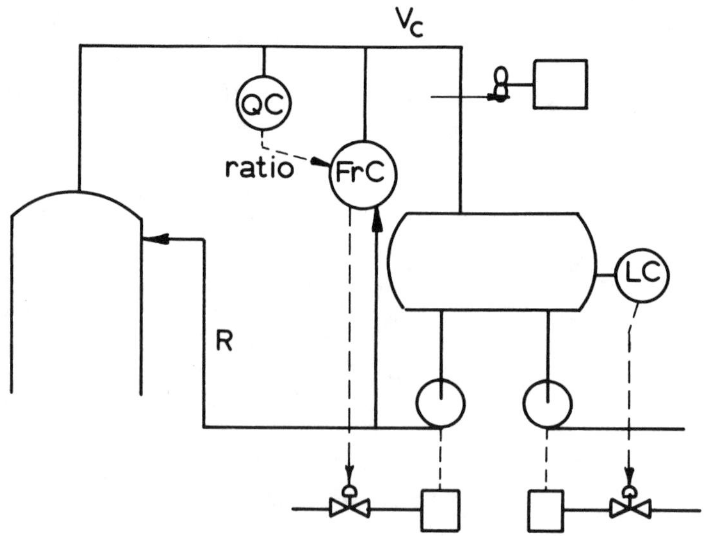

Figure 18.7 Ratio control at the top.

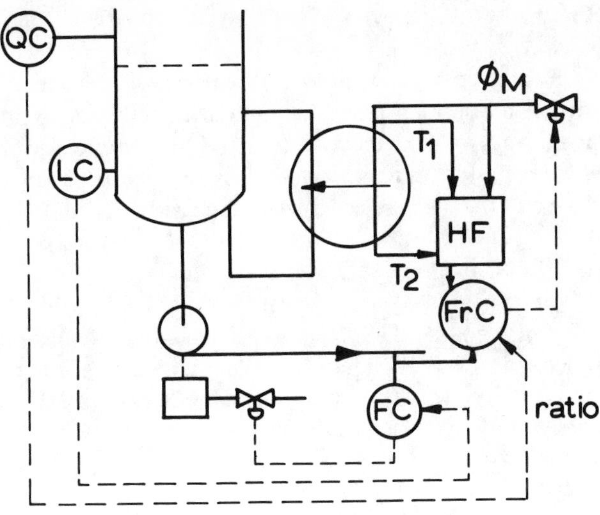

Figure 18.8 Ratio control at the bottom.

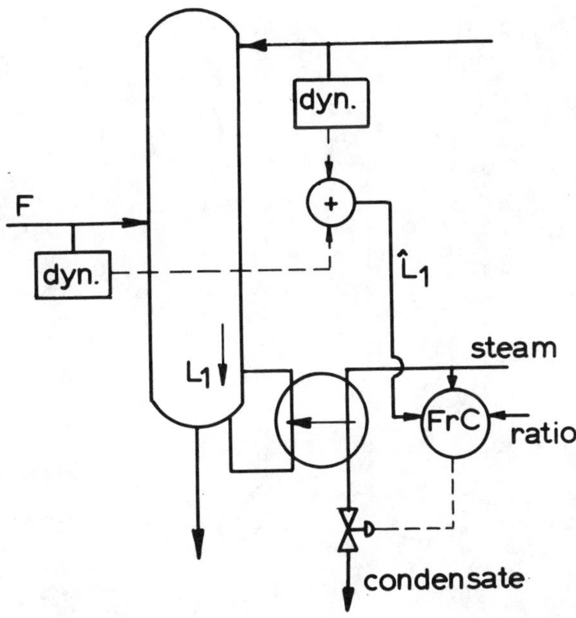

Figure 18.9 Feedforward ratio control.

Tray Load Control

If the tray loading forms a critical constraint, overloading can be avoided by control of the pressure difference. Figure 18.10 shows an example where three groups of trays, which can potentially become overloaded, act on the reboiler heating via a low value selector. Separate attention must be paid to the interaction with the pressure control (in this case on C).

Internal Reflux Control

In the case of air cooling, the liquid can become rather strongly subcooled. Then vapor condenses on the liquid in the column until the liquid is at its boiling point again. One could apply an internal reflux control (Figure 18.11) which operates according to the equation:

$$L'' = R\left[1 + \frac{C_R}{H_e}(T_T - T_R)\right] \tag{18.12}$$

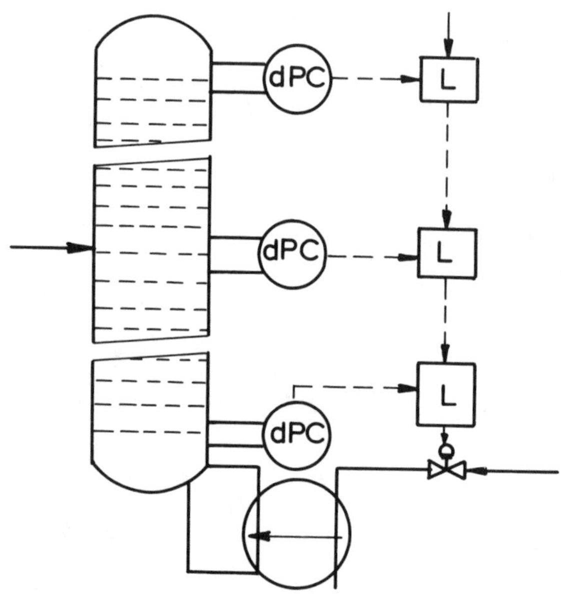

Figure 18.10 Tray load control.

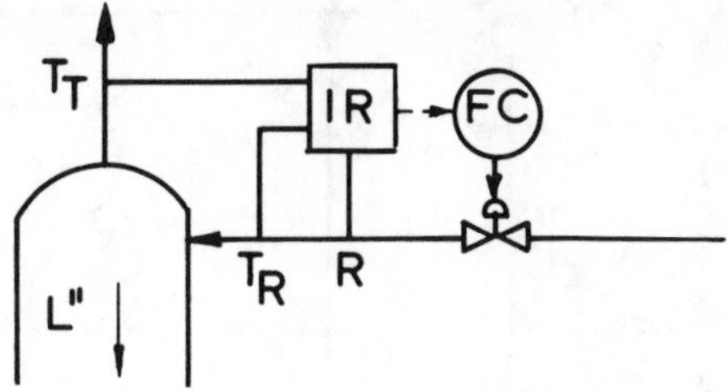

Figure 18.11 Internal reflux control.

where C_R = specific heat of the reflux
H_e = heat of condensation of the vapor
T_R = reflux temperature
T_T = top temperature (as a measure for the boiling point of the reflux)

Feedforward Optimizing Control

As long as there are no critical constraints, one can try to maintain the extremal point (Figure 18.3) by a feedforward control. Figure 18.12 gives an example where the feed composition is estimated from the feed and bottom flows. This estimate, where the dynamic behavior of the column is compensated for, together with the dynamically compensated feed flow via the geometric position of the optima, yields an adjustment for the steam flow to the reboiler.

FINAL REMARKS

In the previous section only some examples of column controls have been shown. The total number of possibilities is only restricted by the ingenuity of the control engineer and the specific circumstances in practice.

DETAILED ANALYSIS OF HYDRAULIC COLUMN DYNAMICS AND CONTROL

Influence of Vapor Flow Variations on the Liquid Flow

The liquid flow from tray i to tray i − 1 is usually a function of the liquid content on that tray and the vapor flow to the tray:

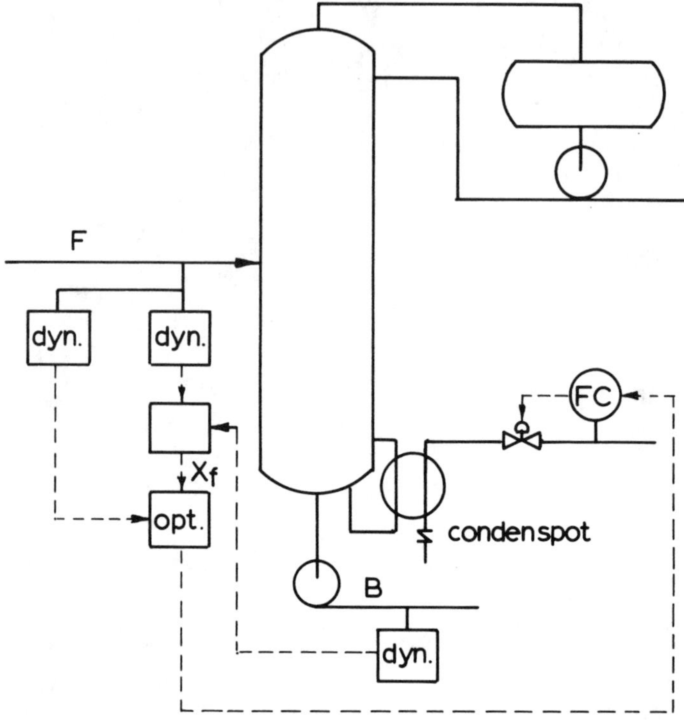

Figure 18.12 Feedforward optimizing control.

$$L_i = L_i(M_{\ell,i}, V_{i-1}) \tag{18.13}$$

For small variations:

$$\delta L_i = \frac{1}{\tau_\ell} \delta M_{\ell,i} + \lambda_1 \delta V_{i-1} \tag{18.14}$$

where τ_ℓ is the hydraulic time constant per tray:

$$\tau_\ell = \left(\frac{\partial M_\ell}{\partial L}\right)_V \tag{18.15}$$

and λ_1 is defined by Equation 18.9. An increase of the vapor flow from the reboiler will propagate relatively rapidly through the column, especially when the pressure control (in this case via the condenser) operates well. The same holds for an increase of the vapor flow to the condenser, with pressure control on H. Therefore, Equation 18.14 can be written as:

$$\delta L_i = \frac{1}{\tau_\ell} \delta M_{\ell,i} + \lambda_1 \delta V \tag{18.16}$$

Neighboring trays will approximately have the same parameter values. The term $\lambda_1 \delta V$ is then the same everywhere, thus from Equation 18.16 we have at first instance:

$$\ldots = \delta L_{i-1} = \delta L_i = \delta L_{i+1} = \ldots \tag{18.17}$$

The mass balance of the trays remain, therefore, in equilibrium:

$$s\delta M_{\ell,i} = \delta L_{i+1} - \delta L_i + \delta V - \delta V = 0 \tag{18.18}$$

resulting in no change of liquid contents on the trays. This, however, does not hold for the top tray, where the external reflux still enters with the same flowrate. Therefore, the original liquid flowrate will be restored via a first-order lag with time constant τ_ℓ. This is clear from the substitution of Equation 18.16 into 18.18 for i = n with $\delta L_{n+1} = 0$:

$$s\delta M_{\ell,n} = -\frac{1}{\tau_\ell} \delta M_{\ell,n} - \lambda_1 \delta V$$

or

$$\delta M_{\ell,n} = -\frac{\tau_\ell \lambda_1 \delta V}{1 + s\tau_\ell} \tag{18.19}$$

followed by a substitution in Equation 18.16:

$$\delta L_n = \lambda_1 \left[1 - \frac{1}{(1 + s\tau_\ell)}\right] \delta V \tag{18.20}$$

This process propagates to the bottom of the column, thus for tray i it is found that:

$$\delta L_i = \lambda_1 \left[1 - \frac{1}{(1 + s\tau_\ell)^{n-i+1}}\right] \delta V \tag{18.21}$$

and for the bottom tray:

$$\delta L_1 = \lambda_1 \left[1 - \frac{1}{(1 + s\tau_\ell)^{n}}\right] \delta V \tag{18.22}$$

Similarly, it can be found that the liquid content is (see Equation 18.19):

$$\delta M_{\ell,i} = -\frac{\tau_\ell \lambda_1 \delta V}{(1 + s\tau_\ell)^{n-i+1}} \tag{18.23}$$

and

$$\delta M_{\ell,1} = -\frac{\tau_\ell \lambda_1 \delta V}{(1 + s\tau_\ell)^{n}} \tag{18.24}$$

The total amount of liquid that finally disappears from the trays is given by:

$$\lim_{s \to 0} \sum_{i=1}^{n} \delta M_{\ell,i} = -\tau_1 \lambda_1 n \tag{18.25}$$

With Equations 18.9 and 18.15 and a well known formula from differential calculus, this can be written as:

$$\lim_{s \to 0} \sum_{i=1}^{n} \delta M_{\ell,i} = n \left(\frac{\partial M_\ell}{\partial V}\right)_L \delta V \tag{18.26}$$

which is in accordance with expectations. Bottom level control on H is strongly delayed by the λ_1 effect, if $\lambda_1 > 0.5$. This is clear from the following approximation. The closed-loop transfer is here with P control:

$$\phi_H \cdot \frac{K_L}{s} \cdot \left[1 - \lambda_1 + \frac{\lambda_1}{(1 + s\tau_\ell)^n}\right] \tag{18.27}$$

where the first s represents the integration of the level, the "one" between brackets represents the "normal" level response on δV, ϕ_H contains the reboiler dynamics and the last terms within the brackets represents Equation 18.22.

After approximation of the last term by a dead time $n\tau_\ell$, the oscillation period and closed-loop gain can be determined. For simplicity, ϕ_H is set equal to one. Then for $\lambda_1 = 0.5$:

$$P_u = 2n\tau_\ell \text{ and } K_{L,u} = \infty \tag{18.28}$$

For $\tau_\ell = 1$, Equation 18.27 reduces to the equation for bottom level control by the reflux. Then:

$$P_u = 4n\tau_\ell \;;\quad K_{L,u} = \frac{\pi}{2} \cdot \frac{1}{n\tau_\ell} \tag{18.29}$$

Ratio Control of Liquid and Vapor Flow in the Column

For each component, the partial mass balance on a tray can be written as:

$$\frac{d}{dt}(M_{\ell,i}x_i) = L_{i+1}x_{i+1} - L_i x_i + V_{i-1}y_{i-1} - V_i y_i \tag{18.30}$$

where x = concentration in the liquid phase
y = concentration in the vapor phase (the vapor content is ignored).

The total mass balance is:

$$\frac{dM_{\ell,i}}{dt} = L_{i+1} - L_i + V_{i-1} - V_i \tag{18.31}$$

The total mass balance is multiplied by x_i and subtracted from Equation 18.30. The result divided by L_{i+1} is:

$$\frac{M_{\ell,i}}{L_{i+1}}\frac{dx_i}{dt} = (x_{i+1} - x_i) + \frac{V_{i-1}}{L_{i+1}}(y_{i+1} - y_i) + \frac{V_{i-1} - V_i}{L_{i+1}}(y_i - x_i) \tag{18.32}$$

The last term can be ignored because $V_{i-1} \approx V_i$. When originally the column is in equilibrium, $dx_i/dt = 0$. If V/L_{i+1} is kept constant, Equation 18.32 has become a relationship between the concentrations on successive trays. The only way the concentrations can change is by propagation of the concentrations from tray to tray, a slow and smooth process. Evidently, V/L_{i+1} cannot be kept constant at all trays at the same time; it is, however, possible to do this near one of the ends of the column, for example by controlling the ratio of the external flows.

CHAPTER 19

PROTECTION

Nowadays more attention is being paid to safety and availability aspects in the process industries. This stems from greater public awareness of risks, more stringent legislation and increasing complexity of processing plants. When plant operation runs into troubles, there may be serious consequences for the life and health of people working in and living around the plant site. Hence, the best possible measures have to be taken to reduce the size and the probability of calamities. On the next level of priority, immediate damage to plant equipment should be avoided as much as possible.

In practice, a number of barriers are put up to avoid dangerous conditions. The first one is realized by normal control, which tries to keep the process away from critical limits. The second barrier is a standby control which only operates when a critical limit is approached, for example the antisurge control of a compressor. The third barrier is the alarm system, which warns the operator that action must be taken. The fourth barrier is an automatic protection system, which can shut down the process or bring it to a safe condition (standby mode). The fifth barrier is the final opportunity for the operators to take action in case the automatic protection system fails.

The dynamics from safe to unsafe operation and the seriousness of the consequences will determine how many of the abovementioned barriers will be selected to prevent the unsafe and undesired situation. If the dynamics are slow and if the consequences are not very serious, a normal controller and an alarm are sufficient. For slow as well as for fast dynamics with serious consequences, a combination of normal control, alarm and shut down device will often be applied. Alarms usually are obtained from one or more switches. In the case of thermal protection, one can use, for example, a bimetallic strip, or a gas-filled tube with pressure transmitter. It is evident that shutdown has immediate consequences for production, while in the case of an alarm the operator can still try to keep the process going. The safety and availability must in first instance be improved by a further development of the basic scheme; a shutdown device must only be used as a penultimate barrier to prevent undesired situations.

CONDITIONS FOR A GOOD PROTECTION SYSTEM

Figure 19.1 gives a simple scheme of a protection device. The measured value of a process variable is compared to a limit value. When the limit value is exceeded, a switch activates a relay, which operates one or more valves, thus putting a process section or an entire process out of operation. In case of such an automatic protection, some points are of great importance:

1. reliability of the system;
2. time lag between the moment at which the process variable has reached a critical value and the moment at which all of the safety valves are in the desired position;
3. characteristics of the safety valves; and
4. capacity of the safety valves.

A reliable system should fulfill two conditions: (1) It must not fail when it has to take action (error of omission); and (2) it must take action only when necessary (error of commission).

It will be evident that a single protection will have a higher probability of failing than a double one. To increase the reliability of protection, it is desirable that a critical process variable is measured in at least two different ways with the aid of two different and independent measurement methods. Sometimes this last condition is utopic, as there are not always two measurement methods that can be used. Moreover, transmission lines for the measurement signal have to be separated, as well as the different elements of the protection device. To compare a number of protection systems, it is useful to go into somewhat more detail on terms such as reliability, failure and probability.

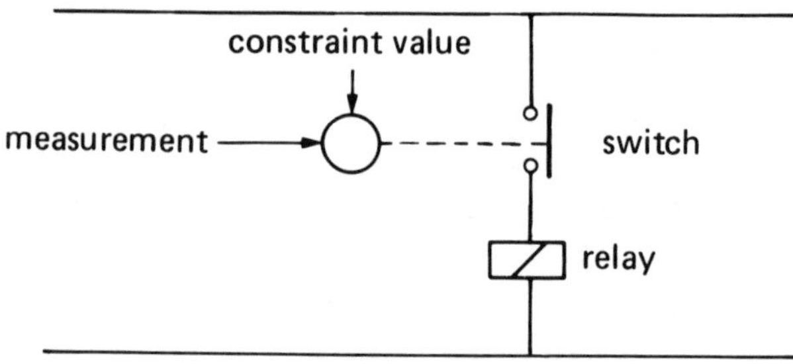

Figure 19.1 Simple protection device.

ANALYSIS OF A SINGLE PROTECTION SYSTEM

Let us assume that the occurrence of a functional failure, which is a real failure of the protection device, is an entirely random process; in other words every event is in itself, not related to other events and occurs at unpredictable moments.

Suppose that the average failure frequency of the protection device is constant and equal to μ (the main failure rate). The probability that the failure will not occur in a time interval 0,t corresponds to a Poisson probability distribution [49].

$$R(t) = e^{-\mu t} \tag{19.1}$$

where R is called the reliability. The probability the failure will occur at least once in the interval 0,t is then:

$$P(t) = 1 - R(t) = 1 - e^{-\mu t} \tag{19.2}$$

where P(t) is called the failure probability. The failure probability density function is given by:

$$f(t) = \frac{dP(t)}{dt} = \mu e^{-\mu t} \tag{19.3}$$

The expected value of the time interval between two failures (MTBF, mean time between failures) is given by:

$$E(t) = MTBF = \int_0^\infty tf(t)dt = \frac{1}{\mu} \tag{19.4}$$

For the determination of the reliability of a system, it is important to know how the components are linked. Elementary links are series and parallel links. In the case of the series link, failure of one component causes failure of the whole system. For mutual independent components, the product rule for the reliability holds:

$$R = \prod_{i=1}^{n} R_i \tag{19.5}$$

where R is the total reliability of the system and R_i the reliability of component i. From Equations 19.1 and 19.5, the failure frequency can be calculated:

$$\mu = \sum_{i=1}^{n} \mu_i \tag{19.6}$$

in other words, the total failure frequency is equal to the sum of the failure frequencies of the separate components. In the case of a parallel link between the components, the system only fails if both links fail. For mutually independent components, the product rule for unreliability now holds:

$$P = \prod_{i=1}^{n} P_i \tag{19.7}$$

and the total reliability becomes:

$$R = 1 - \prod_{i=1}^{n} (1 - R_i) \tag{19.8}$$

Another term which is often used is the unavailability U of a system. The way of calculation of U is different for systems with revealed and unrevealed failures. When the motor in a car fails, it will be immediately discovered, and this failure is called a revealed failure. However, in the case of a protection device, the situation is entirely different. The failure has an unrevealed character; it is only discovered when the protected process variable has already passed a limit value. To restrict the influence of unrevealed failures, regular inspection is desired. To calculate U for systems with revealing failures, a time interval is considered during which the system was inactive a couple of times. The period of inactivity (down time DT, time to repair TTR) is called τ_{ri}, the time of operation (time before failure) is called τ_{bi} (Figure 19.2). Usually a long period is considered in which inactivity occurred a number of times. Then expected values can be defined:

$$\tau_r \cong \frac{1}{n} \sum_{i=1}^{n} \tau_{ri} \tag{19.9}$$

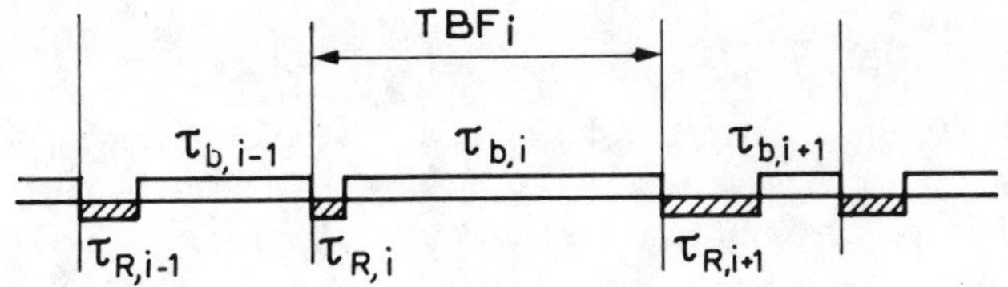

Figure 19.2 Operation of a system with revealing failures.

$$\tau_b \simeq \frac{1}{n}\sum_{i=1}^{n} \tau_{bi} \tag{19.10}$$

where n is the number of times that inactivity occurred during the considered period. The sum $\tau_r + \tau_b$ is equal to the average time between two failures (mean time between failures MTBF):

$$\text{MTBF} = \frac{1}{\mu} = \tau_r + \tau_b \tag{19.11}$$

The availability A of a system is defined as the probability to find the system in good order:

$$A = \frac{\tau_b}{\tau_r + \tau_b} \tag{19.12}$$

and the unavailability U, (fractional dead time FDT) as the probability to find the system deficient:

$$U = \frac{\tau_r}{\tau_r + \tau_b} \tag{19.13}$$

It is evident that

$$U + A = 1 \tag{19.14}$$

From Equations 19.11 and 19.13 the unavailability U can be written as:

$$U = \mu\tau_r \tag{19.15}$$

Example [50]

Consider a pump system as shown in Figure 19.3. For simplicity, the cabling, fuses, etc. are omitted. When making a detailed reliability analysis these components have also to be taken into consideration. The following data are assumed:

- Starter: average revealing failure frequency once per 50 years: $\mu_1 = 0.02$ yr^{-1}. The average time to remedy the deficiency is equal to 2 hr: $\tau_{r1} = 2.27 \times 10^{-4}$ yr.
- Motor: the average revealing failure frequency is once per 20 years: $\mu_2 = 0.05$ yr^{-1}. In case of a deficiency, the motor can be replaced within 36 hr: $\tau_{r2} = 3.12 \times 10^{-3}$ yr.
- Pump: the pump fails approximately once per 10 years: $\mu_3 = 0.1$ yr^{-1}. To install and operate a new pump will cost about 4 hr: $\tau_{r3} = 4.5 \times 10^{-4}$ yr.

The failure frequency for the system is, according to Equation 19.6

$$\mu = \mu_1 + \mu_2 + \mu_3 = 0.02 + 0.05 + 0.1 = 0.17\ yr^{-1}$$

The probability that the system will fail during the coming two years is, according to Equation 19.2:

$$P(t) = 1 - e^{-\mu t} = 1 - 0.712 = 0.288$$

The reliability is R(t) = 1 − P(t) = 0.712; in other words, there is a probability of 71.2% that the system will not fail during the coming two years. The unavailability of the system is:

$$U = \mu_1\tau_{r1} + \mu_2\tau_{r2} + \mu_3\tau_{r3}$$
$$= 0.02 \times 2.27 \times 10^{-4} + 0.05 \times 4.12 \times 10^{-3} + 0.1 \times 4.5 \times 10^{-4} = 2.6 \times 10^{-4}$$

The availability is therefore A = 1 − U = 0.99974; in other words it can be expected that the system will be in operation 99.97% of the time as far as

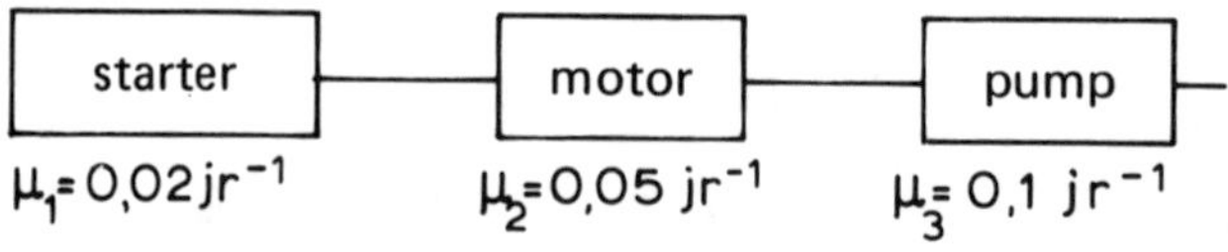

Figure 19.3 Pump system.

random deficiencies are concerned. When we have to deal with unrevealed failures, regular inspection will be necessary to guarantee good operation. The unavailability is now defined as:

$$U = \frac{\tau_0}{\tau_i} \tag{19.16}$$

where τ_0 is the average period of unavailability during the inspection interval and τ_i is the inspection interval. During the inspection interval the average period of inactivity is:

$$\tau_0 = \int_0^{\tau_i} P(t)dt \tag{19.17}$$

resulting in:

$$U = \frac{1}{\tau_i}\int_0^{\tau_i} P(t)dt \tag{19.18}$$

When $\mu t \ll 1$, Equation 19.2 can be simplified to:

$$P(t) = \mu t \tag{19.19}$$

Combination with Equation 19.18 gives:

$$U = \frac{1}{2}\mu\tau_i \tag{19.20}$$

It can be seen that when the inspection interval becomes smaller, the availability increases. It should be noted that the time to repair is ignored, which is usually permitted.

Example

In the previous chapter the protection of the first cracker was discussed. The situation is again shown in Figure 19.4. When the underpressure in the furnace is too low, a pressure switch PS operates a relay. Consequently a solenoid valve is operated, after which a valve in the fuel supply is closed.

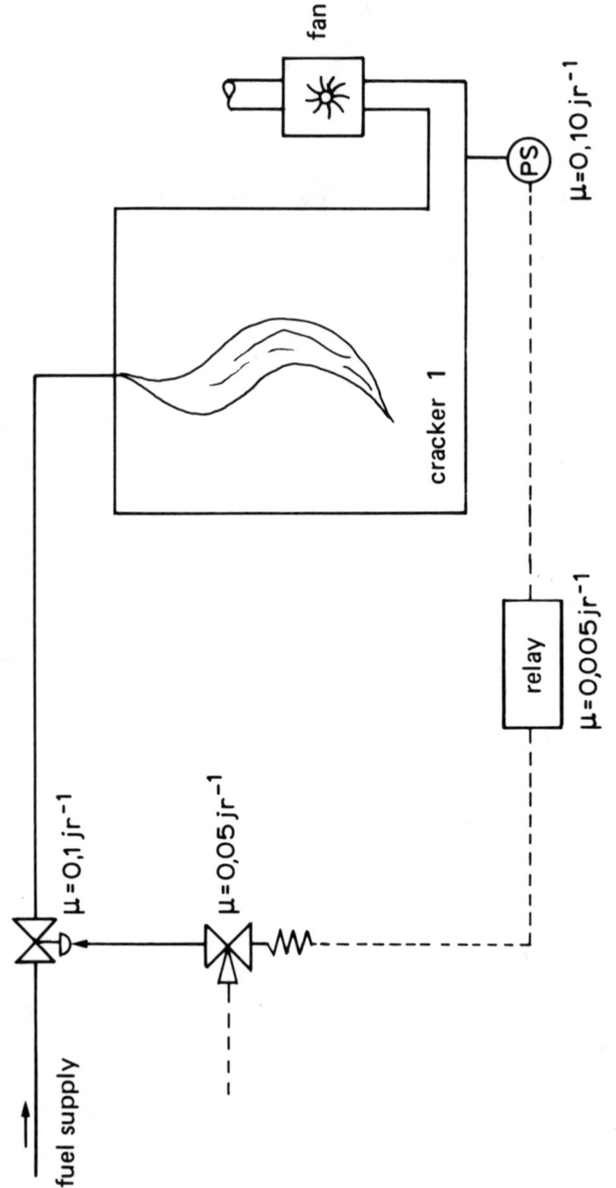

Figure 19.4 Combustion chamber protection.

Suppose the average nonrevealing failure frequencies of the components to be

- pressure switch = 0.1 yr^{-1} (1 failure in 10 years)
- relay = 0.005 yr^{-1}
- solenoid valve = 0.05 yr^{-1}
- safety valve = 0.1 yr^{-1}

Inspection is carried out once a month, thus τ_i = 1/12 year. With the aid of Equations 19.6 and 19.20 it can be found that:

$$U = \frac{1}{2}(\mu_1 + \mu_2 + \mu_3 + \mu_4)\tau_i$$

$$= \frac{1}{2}(0.1 + 0.005 + 0.05 + 0.1) \cdot \frac{1}{12} = 0.0106$$

in other words, the expected value of a system failure is once per 94 years. In the literature [51] failure frequencies for different instruments are given. Table 19.1 gives some examples. The values are averaged for three different types of chemical plants.

PROBABILITY OF COINCIDENCE

A functional failure, which is a real failure of the protection, does not always result in a failure with serious effects. Only with failure and passing of the constraint coincidence can a dangerous situation exist.

Table 19.1 Failure Frequency for Instruments in the Chemical Process Industry [51]

Instrument	Failure Frequency (number of times per year)
Control Valve	0.60
Valve Positioner	0.44
Current/Pressure Converter	0.49
Pressure Measurement	1.41
Flow Measurement	
Orifice with Pressure Transmitter	1.73
Magnetic Flowmeter	2.18
Temperature Measurement	
Thermocouple	0.52
Mercury in Steel Bulb	0.027
Controller	0.29
Pressure Switch	0.34
Gas-Liquid Chromatograph	30.6

Suppose a good operating protection device was activated p times during a time interval T. The average frequency λ of passing the constraint by the process variable is then:

$$\lambda = \frac{p}{T} \tag{19.21}$$

When the functional unavailability of the protection device is equal to U; in other words, when the probability of finding the protection device deficient is equal to U, $p \cdot U$ constraint passings from a total of p will lead to dangerous coincidences [52]. The average frequency r_c of dangerous coincidences is then:

$$r_c = \frac{pU}{T} \tag{19.22}$$

With the aid of Equations 19.20 and 19.21, r_c becomes:

$$r_c = \frac{1}{2} \lambda \mu \tau_i \tag{19.23}$$

The expected value of a risk (mean time between coincidence MTBC) is now:

$$\text{MTBC} = \frac{1}{r_c} = \frac{2}{\lambda \mu \tau_i} \tag{19.24}$$

Example

A protected process variable runs into troubles once per 16 months on the average, hence $\lambda = {}^{12}\!/_{16} = 0.75\ \text{yr}^{-1}$. The protection device which is applied fails once per 25 years ($\mu = 0.04\ \text{yr}^{-1}$). After a failure, the time to repair the protection device can be ignored. Inspection takes place every month ($\tau_i = 0.083$). For a single protected process variable, the following data can be calculated:

$$U = \frac{1}{2} \mu \tau_i = \frac{1}{2} \times 0.04 \times 0.083 = 1.66 \times 10^{-3}$$

$$r_c = \lambda U = 0.75 \times 1.66 \times 10^{-3} = 1.25 \times 10^{-3}\ \text{yr}^{-1}$$

$$\text{MTBC} = r_c^{-1} = 1/1.25 \times 10^{-3} \simeq 800 \text{ years}$$

in other words, the expected value of a dangerous situation is once per 800 years.

DOUBLE PROTECTION (1–OUT OF–2 SYSTEM)

A considerable increase in the availability can be obtained by doubling the protection. The protection devices are installed in such a way that activation of one of the two protection devices or both results in an action to prevent an unsafe situation. When one protection fails, the other one is able to take over. When we have to deal with two independent identical systems, each with a total average nonrevealing failure frequency μ, and both an inspection time τ_i, the probability U_2 to find both systems inactive is:

$$U_2 = U_1 U_2 = U^2 = \frac{1}{4}\mu^2\tau_i^2 \tag{19.25}$$

As there is one process variable with a constraint passing frequency of λ, the average frequency of dangerous coincidences is:

$$r_{c2} = \frac{1}{4}\lambda\mu^2\tau_i^2 \tag{19.26}$$

Example

When a double protection is applied in the previous example, the values are:

$$U_2 = 2.76 \times 10^{-6}$$

$$r_{c2} = \lambda U_2 = 2.07 \times 10^{-6}\ \text{yr}^{-1}$$

$$\text{MTBC} = \frac{1}{r_{c2}} = 4.8 \times 10^5\ \text{yr}$$

When inspection takes place every two months, the expected value of a dangerous situation becomes 1.2×10^5 yr.

OPERATIONAL FAILURES

When a 1–out of–1 system is replaced by a 1–out of–2 system, the expected value of dangerous situations decreases, as shown before. Apart from a functional failure, this is a real failure of the protection (sometimes also called open-mode failure), there can also be an operational failure (sometimes called short-mode failure). In the latter case, there is action, although there is no passing of the constraint of the process variable (spurious trip).

After one or both channels of the protection device show an operational failure, the protection system will be inactive. Suppose that the average operational failure frequency of a single channel is equal to m, in other words: the protection device is active m times per time unit without any need. When both channels are identical and when the operational failures are random events, the average operational failure frequency of two channels is two times the failure frequency of one channel. The operational failure frequency μ_s from a 1–out of–2 system is therefore:

$$\mu_s(1\text{–out of–}2) = 2\text{ m} \tag{19.27}$$

THE 2–OUT OF–3 PROTECTION DEVICE

When we compare a 1–out of–1 and a 1–out of–2 system, it can be seen that by doubling the protection the functional availability A increases, but the probability of an operational failure also increases; thus, the operational reliability decreases. This dilemma can be avoided by using more complicated systems. The most frequently applied one is a 2–out of–3 system: a protection operates when two or more channels are activated. For the calculation of the availability U of a 2–out of–3 system, we start with the probability to find one channel deficient: U_A, U_B, U_C. The protection system will fail, if two or three channels are deficient. The probabilities to meet with these situations are:

$$U_AU_B(1 - U_C)\ ,\quad U_AU_C(1 - U_B)\ ,\quad U_BU_C(1 - U_A)\ ,\quad U_AU_BU_C$$

The probability to find the protection system in a deficient state is equal to the sum of all possible combinations:

$$\begin{aligned} U(2\text{–out of–}3) &= U_AU_B(1 - U_C) + U_AU_C(1 - U_B) + U_BU_C(1 - U_A) + U_AU_BU_C \\ &= U_AU_B + U_BU_C + U_AU_C - 2U_AU_BU_C \end{aligned} \tag{19.28}$$

If

$$U_A = U_B = U_C = \frac{1}{2}\mu\tau_i \tag{19.29}$$

then Equation 19.28 can be written as:

$$U(2\text{-out of-}3) = \frac{3}{4}\mu^2\tau_i^2 - \frac{1}{4}\mu^3\tau_i^3 \tag{19.30}$$

Usually $\mu\tau_i$ is much smaller than one, thus Equation 19.30 can be further simplified to:

$$U(2\text{-out of-}3) \simeq \frac{3}{4}\mu^2\tau_i^2 \tag{19.31}$$

As we have to deal with one process variable with a constraint passing frequency of λ, the average frequency of dangerous coincidences r_c is equal to:

$$r_c(2\text{-out of-}3) = \lambda U(2\text{-out of-}3) = \frac{3}{4}\lambda\mu^2\tau_i^2 \tag{19.32}$$

and the expected value of a risk, or the average time between two dangerous situations:

$$\text{MTBC}\ (2\text{-out of-}3) = \frac{4}{3\lambda\mu^2\tau_i^2} \tag{19.33}$$

For the determination of the average time between two operational failures of a 2-out of-3 system we proceed as follows. Suppose the channels are identical with an average failure frequency of m for each channel. The average operational failure frequency of one channel from three (for example A) is equal to 3m. The system becomes inactive if one of the two remaining channels shows an operational failure. The probability that channel B fails is equal to $U_B = \frac{1}{2} m\tau_i$ (analogous to Equation 19.20). The probability that channel C fails is $U_C = \frac{1}{2} m\tau_i$. The frequency of coincidence of A and B is $3m.\ \frac{1}{2} m\tau_i = 3m^2\tau_i/2$. In a similar way the frequency of coincidence of a failure of A and C is equal to $3m^2\tau_i/2$. The average frequency for the 2-out of-3 system is equal to the sum of the frequencies with which the combinations AB and AC can occur:

$$\mu_s = 3m\left(\frac{1}{2}m\tau_i + \frac{1}{2}m\tau_i\right) = 3m^2\tau_i \tag{19.34}$$

SURVEY OF REDUNDANT PROTECTION SYSTEMS

The general formulas are given below for the unavailability, and the functional and operational failure frequencies of M–out of–N systems. The characteristic of these systems is that action is taken when M of the N channels are activated. For the derivation of these equations, one is referred to the literature [49,52]. Table 19.2 summarizes the characteristics of the most commonly used systems; for more complicated systems reference is made to the literature [49].

μ_f indicates that the protection system operates without constraint passing of the process variable (spurious trip). The more complicated protection systems can be made with the aid of AND and OR gates. In Figure 19.5, the characteristics are given. Using these gates one can easily make a 1–out of–2 or 2–out of–3 system as shown in Figure 19.6. In this figure,

Table 19.2 Survey of Some Protection Systems

System	Functional Unavailability U	Functional Failure Frequency μ_f	Operational Failure Frequency μ_s
1–out of–1	$\frac{1}{2}\mu\tau_i$	μ	m
1–out of–2	$\frac{1}{4}\mu^2\tau_i^2$	$\mu^2\tau_i$	2m
2–out of–3	$\frac{3}{4}\mu^2\tau_i^2$	$3\mu^2\tau_i$	$3m^2\tau_i$

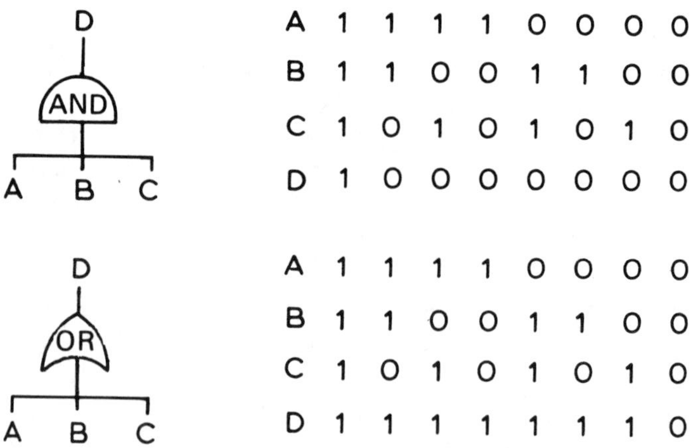

Figure 19.5 Truth table for AND and OR gates.

C is a transmitter, which may be a pressure transmitter, temperature transmitter, gas chromatograph, etc. A qualitative picture of the unavailability of different protection systems is shown in Figure 19.7. Depending on the consequences of constraint passing, one can select a certain system. It is evident that when the protection system has a higher availability, it will be more complicated and more expensive. As can be seen from the figure, the smallest unavailability is obtained when using two separated systems. For such a system, different methods are used for measurement and logics. Each of these different measurement methods can exist of instruments which are redundantly connected (for example 2–out of–3).

COMPARISON OF THREE PROTECTION SYSTEMS

To illustrate the previously treated matter, the protection of a compressor will be analyzed. A compressor fails once per two years due to insufficient cooling. The related damage is $250,000. The protection system which can be installed fails once per 15 years. Once per ten years the protection

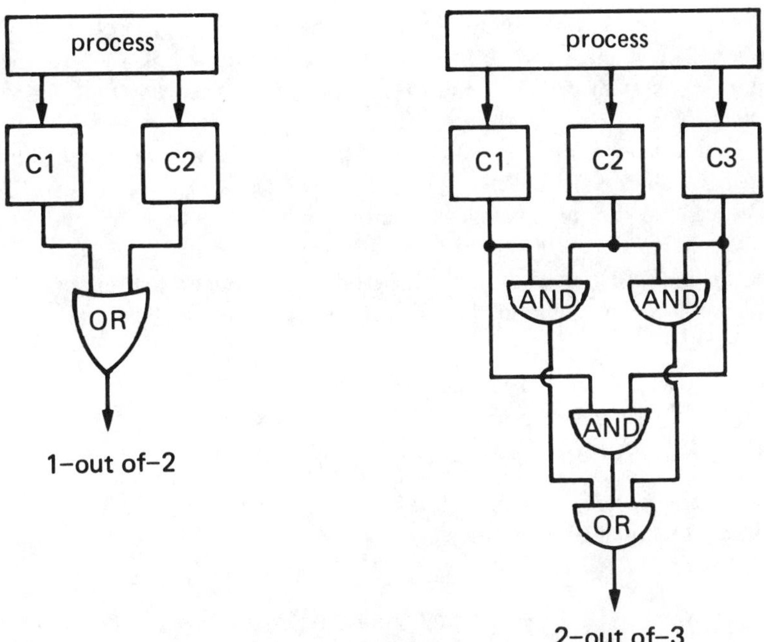

Figure 19.6 1–out of–2 and 2–out of–3 systems.

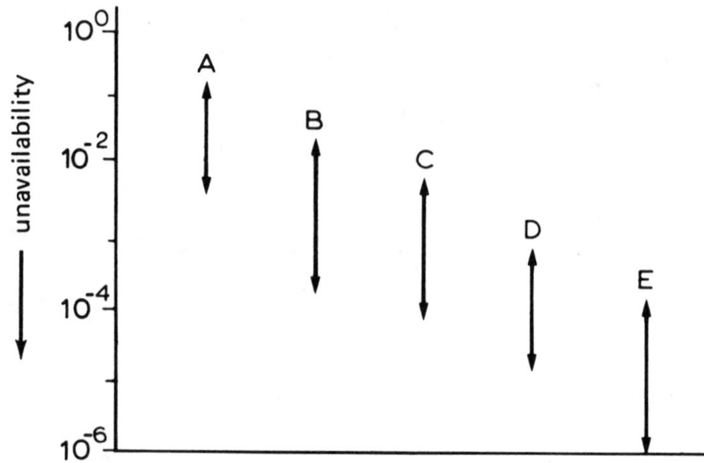

Figure 19.7 Unavailability of some protection systems at monthly inspection. A = single protection; B = redundant system; C = system with partially different instrumentation methods; D = system with entirely different instrumentation methods; E = two separated systems (entirely different).

operates, without this being necessary. The related damage is \$12,500. Investment and maintenance of a single protection system is \$200/yr; for a 1-out of-2 system, \$450/yr; and for a 2-out of-3 system, \$700/yr. Inspection takes place every month.

For the calculation of the expected value of the damage, it is assumed that the unavailability of the AND and OR gates does not play any role. This is the case if the unavailability of these gates is in the order of 10^{-7} or smaller. When using less reliable logic components, the calculations will give too optimistic results, and it is better to incorporate the unavailability of the logic components in the calculation for that case. From the data it follows that:

- frequency of constraint passing $\lambda = 0.5\ \text{yr}^{-1}$
- functional failure frequency $\mu = 0.0667\ \text{yr}^{-1}$
- operational failure frequency $m = 0.1\ \text{yr}^{-1}$
- inspection interval $\tau_i = 0.083$ yr.

When there is no protection, the potential damage is \$125,000/yr. For a single protection device it holds that:

$$U = \frac{1}{2}\mu\tau_i = \frac{1}{2} \times 0.0667 \times 0.083 = 2.77 \times 10^{-3}$$

$$r_c = \lambda U = 0.5 \times 2.77 \times 10^{-3} = 1.38 \times 10^{-3}\ \text{yr}^{-1}$$

$$\text{MTBC} = r_c^{-1} = 723\ \text{yr.}$$

The expected value of the damage is then:

$$\frac{\$250{,}000}{723} = \$346/\text{yr}$$

For a 1-out of-2 protection the following data may be calculated:

$$U = \frac{1}{4}\mu^2\tau_i^2 = 7.67 \times 10^{-6}$$

$$r_c = \lambda U = 0.5 \times 7.67 \times 10^{-6} = 3.84 \times 10^{-6}\ \text{yr}^{-1}$$

$$\text{MTBC} = r_c^{-1} = 2.61 \times 10^5\ \text{yr}$$

The expected value of the damage is now:

$$\frac{\$250{,}000}{260{,}661} = \$0.96/\text{yr}$$

In a similar way, by using Table 19.2, the expected value of the damage can be calculated for a 2-out of-3 system. This damage is equal to \$2.88/yr. Due to spurious trips, there is also a certain potential damage. For a single protection device this is equal to 0.1 × \$12,500 = \$1250. For a 1-out of-2 system, the operational failure frequency is equal to 2m, hence the expected value of the damage is \$2500; in a similar way for a 2-out of-3 system \$31.13. Table 19.3 summarizes the results.

Table 19.3 Compressor Protection

Protection	Costs of Protection Plus Maintenance ($)	Expected Loss due to: Constraint Passing ($)	Expected Loss due to: Spurious Trip ($)	Expected Value of the Costs per Year ($)
None				125,000
1-out of-1	200	346	1,250	1,796
1-out of-2	450	0.96	2,500	2,950.96
2-out of-3	700	2.88	31.13	732

SELECTION OF A PROTECTION DEVICE

The selection of a certain type of protection will not only depend on the expected value of the costs per year. It is clear that when one improves the protection, the expected value of the costs per year will approach to the costs of the protection device and maintenance. When the protection system becomes more complicated, these costs will increase. Based on the data used for Table 19.3, one can easily see that for a M–out of–N system with N equal to four, the costs of protection and maintenance will lie in the order of $950/yr. This amount is already higher than the expected value of the costs per year for a 2–out of–3 system. Also, the availability of financial means will play a role in the selection. A nice example of the selection of a certain protection device is given in the literature [53].

In this case, the selection is not based on the expected value of the costs per year but on the average risk frequency r_c. For a complete chemical plant a value is used equal to $r_c = 3 \times 10^{-5}\ yr^{-1}$, hence the probability of a dangerous situation is once per 33,000 years. Starting from this value, the protection of a reactor is calculated, and a 2–out of–3 system is selected. The expectancy of a certain risk has already been the subject of many interesting discussions, and the discussion is still going on [54,55]. For the calculation of the average risk frequency r_c of a complex system, fault tree analysis is often used, in which the consequences, causes and combination of events that may lead to a dangerous situation are represented. For more information see Lees [56].

REDUNDANCY

When we have N identical independent protection channels, each with an unavailability U, and where action is taken when M of the N channels are activated, we call the system a M–out of–N system, for which the total unavailability is [49]:

$$U(M\text{–out of–}N) = \sum_{k=N-M+1}^{N} \binom{N}{k} U^k (1 - U)^{N-k} \qquad (19.35)$$

If U is much smaller than one, powers of a higher order than N – M + 1 can be ignored; thus Equation 19.35 becomes:

$$U(M\text{–out of–}N) \simeq \frac{N!}{(N - M + 1)!(M - 1)!} U^{N-M+1} \qquad (19.36)$$

For the average functional failure frequency μ_f (without confrontation with constraint passing), it can be derived that:

$$\mu_f(\text{M-out of-N}) = \mu N \sum_{k=N-M}^{N-1} \binom{N-1}{k} U^k (1-U)^{N-k-1} \qquad (19.37)$$

where μ is the functional failure frequency of a single protection. One can easily see that for a single protection, μ_f(1–out of–1), can be written as:

$$\mu_f(\text{1-out of-1}) = \mu \qquad (19.38)$$

Besides functional (real) failures, there can also be an operational failure. When the average operational failure frequency of a protection channel is equal to m, it can be derived that the average operational failure frequency μ_s of a M–out of–N system is equal to:

$$\mu_s(\text{M-out of-N}) = m \binom{N-1}{M-1} (U^*)^{M-1} (1-U^*)^{N-M} \qquad (19.39)$$

where U* is defined according to Equation 19.16, but where τ_0 now is (analogous to Equations 19.17 and 19.19):

$$\tau_0 = \int_0^{\tau_i} m t dt = \frac{1}{2} m \tau_i^2 \qquad (19.40)$$

thus:

$$U^* = \frac{1}{2} m \tau_i \qquad (19.41)$$

If U* is much less than one, Equation 19.39 can be simplified to:

$$\mu_s(\text{M-out of-N}) = \frac{N!}{(N-M)!\,M!} M \cdot m \cdot (U^*)^{M-1} \qquad (19.42)$$

APPENDIX

LAPLACE TRANSFORMS

The Laplace transform is a method for converting linear differential equations with constant coefficients into algebraic equations. The algebraic equation is then solved explicitly for the Laplace transformation of the dependent variable, and finally inverse transformation is applied, which results in the actual solution in the time domain. The transformation technique can also be used to reduce linear partial differential equations to ordinary differential equations, for which solutions may be found more easily. Instead of the time, Laplace transform introduces a new variable, the Laplace operator s. For a function f(t) the one-sided Laplace transform is defined as:

$$f(s) \triangleq \pounds[f(t)] \triangleq \int_0^\infty f(t)e^{-st}dt \ , \quad t > 0 \tag{1}$$

where it is assumed that for $t < 0$ the function $f(t) = 0$. The operator s is a number that usually is complex (in contradiction to the operational notation). In any case s must have a real part, which is larger than possible exponential factors in f(t), because the integral must converge.

LAPLACE TRANSFORM OF SIMPLE FUNCTIONS

The Laplace transform on step at $t = 0$ is:

$$\pounds[Au(t)] = \int_0^\infty Au(t)e^{-st}dt = -\frac{Ae^{-st}}{s}\Big/_0^\infty = \frac{A}{s} \tag{2}$$

where u(t) is the unit step function. For a sine function, starting at t = 0:

$$\pounds[u(t)\sin\omega t] = \int_0^\infty u(t)\sin\omega t\, e^{-st}dt = \frac{\omega}{s^2+\omega^2} \tag{3}$$

For an exponential function e^{at} starting at t = 0:

$$\pounds[u(t)e^{at}] = \int_0^\infty u(t)e^{at-st}dt = -\frac{e^{-(s-a)t}}{s-a}\Bigg/_0^\infty = \frac{1}{s-a} \tag{4}$$

In the literature [22,26] tables are given of the transformation of simple functions. The transformation of the derivative of a function can be determined as follows:

$$\pounds\ \frac{df(t)}{dt} = \int_0^\infty \frac{df(t)}{dt}e^{-st}dt = \int_0^\infty e^{-st}df(t)$$

$$= e^{-st}f(t)\Big/_0^\infty + s\int_0^\infty f(t)e^{-st}dt = -f(0^+) + sf(s) \tag{5}$$

where $f(0^+)$ indicates the limit value of f(t), if t approaches zero from the positive side (f(t) can have a discontinuity at t = 0).

Equation 5 shows that taking the derivative in the time domain corresponds to multiplication with s in the Laplace domain. In addition, there is a term which incorporates the influence of the initial condition. It can also be seen that when this initial condition is equal to zero, the Laplace transform is similar to the operational notation where the operator s = d/dt was introduced.

INITIAL VALUE THEOREM

A theorem that can be used to check the Laplace-transformed expressions for errors is the value at the time $t = 0^+$:

$$f(0^+) = \lim_{s\to\infty}[sf(s)] \tag{6}$$

This can be seen from the following formula:

$$sf(s) \equiv s\int_0^\infty e^{-st}f(t)dt \tag{7}$$

When s approaches infinity, the term e^{-st} will go to zero for every positive value of t. The value of f(t) is then irrelevant (it is assumed that f(t) is finite). Therefore f(t) in Equation 7 may be replaced by $f(0^+)$, resulting in:

$$\lim_{s\to\infty} [sf(s)] = s\int_0^\infty e^{-st} f(0^+)dt = f(0^+) \tag{8}$$

FINAL VALUE THEOREM

Another important theorem is the final value theorem.

$$\lim_{t\to\infty} [f(t)] = \lim_{s\to\infty} [sf(s)] \tag{9}$$

This can be made plausible as follows:

$$\begin{aligned} sf(s) &= s\int_0^\infty e^{-st} f(t)dt = -\int_0^\infty f(t)d(e^{-st}) \\ &= -e^{-st}f(t)\Big/_0^\infty + \int_0^\infty e^{-st}df'(t) \end{aligned} \tag{10}$$

in which:

$$f'(t) = \frac{df(t)}{dt} \tag{11}$$

In the limit case:

$$\begin{aligned} \lim_{s\to 0} [sf(s)] &= f(0^+) + \int_0^\infty df'(t) \\ &= f(0^+) + f(t)\Big/_0^\infty = f(\infty) \end{aligned} \tag{12}$$

EXAMPLE

The Laplace transform of a step function is A/s, the Laplace transform of a first order system $1/1 + \tau s$; thus the expression for the Laplace transform of the step response of the first-order system becomes

$$y(s) = \frac{A}{s} \cdot \frac{1}{1 + \tau s} \tag{13}$$

Transformation to the time proceeds as follows. Write Equation 13 as:

$$y(s) = \frac{A}{s} - \frac{A}{s + \tau^{-1}} \tag{14}$$

With the aid of a table of Laplace transform functions the transformation becomes:

$$y(t) = Au(t) - Au(t)e^{-t/\tau} \tag{15}$$

The final value of the step response is:

$$\lim_{t \to \infty} y(t) = \lim_{s \to 0} [sy(s)] = \lim_{s \to 0} \left[\frac{A}{\tau s + 1}\right] = A \tag{16}$$

REFERENCES

1. Amundson, N. R. *Mathematical Methods in Chemical Engineering* (Englewood Cliffs, NJ: Prentice-Hall, Inc., 1969).
2. Pontryagin, L. S., V. G. Boltyanski, R. V. Gramkrelidze and E. F. Mischenko. *The Mathematical Theory of Optimal Processes* (New York: John Wiley & Sons, Inc., 1962).
3. Hafez, M. M., and A. Prochazka. "The Dynamic Effects in Vibrating Plate and Pulsed Extractors," *Chem. Eng. Sci.* 29:1745-1762 (1974).
4. Larsen, J., and M. Kümmel. "Hydrodynamic Model for Controlled Cycling in Tray Columns," *Chem. Eng. Sci.* 34(4):455-462 (1979).
5. Robertson, D. C., and A. J. Engel. "Particle Separation by Controlled Cycling," *Ind. Eng. Chem. PDD* 6(1):2-6 (1967).
6. Eckmann, D. P. *Automatic Process Control* (New York: John Wiley & Sons, Inc., 1958).
7. Bird, R. B., W. E. Stewart and E. N. Lightfoot. *Transport Phenomena* (New York: John Wiley & Sons, Inc., 1960).
8. Staroselsky, N., and L. Ladin. "Improved Surge Control for Centrifugal Compressors," *Chem. Eng.* (May 21, 1979), pp. 175-184.
9. Himmelblau, D. M., and K. B. Bischoff. *Process Analysis and Simulation* (New York: John Wiley & Sons, Inc., 1968).
10. Roffel, B., and J. E. Rijnsdorp. "Dynamics and Control of a Gas-Fired Furnace," *Chem. Eng. Sci.* 29:2083-2092 (1974).
11. Monod, J. "The Growth of Bacterial Cultures," *Ann. Rev. Microbiol.* 3:371-394 (1949).
12. Lawrence, A. W., and P. L. McCarty. "Unified Basis for Biological Treatment Design and Operation," *J. San. Eng. Div., ASCE* 96(SA3):757-778 (1970).
13. Sewards, G. J., and G. A. Holder. "Disposal of Dairy Whey by Two Stage Biological Treatment," *Water Res.* 9:409-416 (1975).
14. Westberg, N. "An Introductory Study of Regulation in the Activated Sludge Process," *Water Res.* 3:613-621 (1969).
15. Cool, J. C., F. J. Schijff and T. J. Viersma. *Regeltechniek* (in Dutch) (Amsterdam, The Netherlands: Agon/Elsevier, 1969).
16. Rademaker, O., J. E. Rijnsdorp and A. Maarleveld. *Dynamics and Control of Continuous Distillation Units* (New York: Elsevier North-Holland, Inc., (1975).
17. Rijnsdorp, J. E. "Some Recent Trends in the Dynamics and Control of Continuous Distillation Units," paper presented at the Meeting of the

Working Party on Distillation, Absorption and Extraction, The Hague, The Netherlands, 1974.

18. Perry, J. H. *Chemical Engineers Handbook* (New York: McGraw-Hill Book Company, 1968).
19. Roffel, B. "Constraint Control of a Gas-Liquid Contacting Plant," *Chem. Eng. Sci.* 31:751-757 (1976).
20. Roffel, B., and H. J. Fontein. "Constraint Control of Distillation Processes," *Chem. Eng. Sci.* 34:1007-1018 (1979).
21. Douglas, J. M. *Process Dynamics and Control, Vol. 1, Analysis of Dynamic Systems* (Englewood Cliffs, NJ: Prentice-Hall Inc., 1972).
22. Le Page, W. R. *Complex Variables and Laplace Transforms for Engineers* (New York: McGraw-Hill Book Company, 1961).
23. Harriot, P. *Process Control* (New York: McGraw-Hill Book Company, 1964).
24. Coulson, J. M., and J. F. Richardson. *Chemical Engineering* (Elmsford, NY: Pergamon Press, Inc., 1965).
25. Thé, G. "Parameter Identification in a Model for the Conductivity of a River Based on Noisy Measurements at Two Locations," *Proceedings of the IFIP Conference* (New York: Elsevier North-Holland, Inc., 1977).
26. *Handbook of Chemistry and Physics*, 53rd ed. (Cleveland, OH: CRC Press, Inc., 1973).
27. van der Laan, E. T. "Notes in the Diffusion Type Model for the Longitudinal Mixing in Flow," *Chem. Eng. Sci.* 7(3):187-191.
28. Crider, J. E., and A. S. Foss. "An Analytic Solution for the Dynamics of a Packed Adiabatic Chemical Reactor," *Am. Inst. Chem. Eng. J.* 14(1):77-84 (1968).
29. Douglas, J. M., and L. C. Eagleton. "Analytic Solutions for Some Adiabatic Reactor Problems," *Ind. Eng. Chem. Fund.* 1(2):116-119 (1962).
30. Bischoff, K. B. "A Note on Boundary Conditions for Flow Reactors," *Chem. Eng. Sci.* 16(1,2):131-133 (1961).
31. Fan, L.-T., and Y.-K. Ahn. "Critical Evaluation of Boundary Conditions for Tubular Flow Reactors," *Ind. Eng. Chem. PDD* 1(3):190-195 (1962).
32. Froment, G. F., and K. B. Bischoff. "Non Steady State Behavior of Fixed Bed Catalytic Reactors due to Catalyst Fouling," *Chem. Eng. Sci.* 16:189-201 (1961).
33. Aris, R., and N. R. Amundson. *Mathematical Methods in Chemical Engineering* (Englewood Cliffs, NJ: Prentice-Hall, Inc., 1973).
34. Friedly, J. C. *Dynamic Behavior of Processes* (Englewood Cliffs, NJ: Prentice-Hall, Inc., 1972).
35. Gould, L. A. *Chemical Process Control, Theory and Applications* (Reading, MA: Addison-Wesley Publishing Co., Inc., 1969).
36. Du Fort, E. C., and S. P. Frankel. "Stability Conditions in the Numerical Treatment of Parabolic Differential Equations," *Math. Tabl. Aids. Comp.* 7:135-152 (1953).
37. Carnahan, B., H. A. Luther and J. O. Wilkes. *Applied Numerical Methods* (New York: John Wiley & Sons, Inc., 1969).
38. Saul'yev, V. K. *Integration of Equations of Parabolic Type by the Method of Nets* (New York, McMillan Publishing Co., Inc., 1964).
39. Jansen, J. M. L. "Control System Behavior Expressed as a Deviation Ratio," *Trans. Am. Soc. Mech. Eng.* 76:1303-1312 (1954).

40. Ziegler, J. H., and B. Nichols. "Optimum Settings for Automatic Controllers," *Trans. Am. Soc. Mech. Eng.* 64:759-768 (1942).
41. Gilliland, E. R., and C. E. Reed. "Degrees of Freedom in Multicomponent Absorption and Rectification Columns," *Ind. Eng. Chem.* 34:551 (1942).
42. Bertrand, L., and J. B. Jones. "Controlling Distillation Columns," *Chem. Eng.* 68(4):139-144 (1961).
43. Buckly, P. S. *Techniques of Process Control* (New York: John Wiley & Sons, Inc., 1964).
44. Shinskey, F. G. *Distillation Control* (New York: McGraw-Hill Book Company, 1977).
45. Johnson, M. L., D. E. Lupfer, J. R. Parsons and D. N. Pierson. "Distillation Columns Control," *Control Eng.* 11(8):68-73 (1964).
46. Maarleveld, A., and J. E. Rijnsdorp. "Constraint Control on Distillation Columns," *Automatica* 6:51-58 (1970).
47. Rijnsdorp, J. E., and A. Maarleveld. "Use of Electrical Analogues in the Study of the Dynamic Behaviour and Control of Distillation Columns," *Inst. Symp. Instrum. Comput., Inst. Chem. Eng.* (1959), pp. 135-144.
48. Wahl, E. F., and E. P. Harriot. "Understanding and Prediction of the Dynamic Behaviour of Distillation Columns," *Ind. Eng. Chem. PDD* 9(3):396-406 (1970).
49. de Heer, H. J. "A Basic Theory on the Probability of Failure of Safeguarding Systems," Proceedings of the First Symposium on Loss Prevention and Safety Promotion in the Chemical Industry, The Hague, May 28-30, 1974.
50. Burgess, L. H. "System Reliability," *Hosp. Eng.* 28:3-11 (1974).
51. Anyahora, S. N., G. F. M. Engel and F. P. Lees. "Some Data on the Reliability of Instruments in the Chemical Plant Environment," *Chem. Eng.* (November 1971), pp. 396-402.
52. de Heer, H. J. "A Simple Theory on the Reliability of Automatic Protective Systems," paper presented at the 10th International TNO Conference on Risk Analysis in Industry, Government and Society, February 24-25, 1977, Rotterdam, The Netherlands.
53. Stewart, R. M., and G. Hensley. "High Integrity Protective Systems on Hazardous Chemical Plants," *Inst. Chem. Eng. Symp. Ser. No. 34, Chem. Eng.* London (1971).
54. Ravetz, J. "The Political Economy of Risk," *New Scientist* (September 1977), pp. 751-757.
55. Vlek, C. A. J., and P. J. M. Stallen. "Judgment of Riskful Activities," (in Dutch) *De Ingenieur* (48):842-848 (1979).
56. Lees, F. P. *Loss Prevention in the Process Industries* (London: Butterworths, 1980).

NOMENCLATURE

A, A_{sub}	constant or parameter defined by corresponding equation; also used for area (m^2) and availability
A_i	constants in the solution of a differential equation ($i = 1 \ldots 3$)
a	parameter defined in the corresponding equation
a_i	constant or parameter defined by the corresponding equation ($i = 0 \ldots 5$); also used as constants in a transfer function or difference equation
a_T	parameter defined in Equation 6.84
a_-, a_+	parameters defined in Equation 14.91
a_0, a^*	parameters defined in Equation 14.98
a_{ij}	coefficients in differential equation ($i = 1 \ldots n; j = 1 \ldots n$)
B	constant or parameter defined by corresponding equation; also used for bottom flow distillation column (kg/sec)
b	dimensionless parameter according to Equation 12.53
b_i	constants in a transfer function ($i = 0 \ldots 2$)
b_{ij}	coefficients in differential equation ($i = 1 \ldots n; j = 1 \ldots n$)
C	capacity; also used for concentration (kg/m^3, mol/m^3) and conversion
c, c_{sub}	constant or parameter defined by corresponding equation; also used for specific heat (J/kg-°K)
c_b, c_s	microorganism and substrate concentrations, respectively (kg/m^3)
c_c	coke concentration (kg/kg catalyst)
C_D, C_B, C_F, C_H	cost coefficient, value per unit mass or energy (\$/kg, \$/J)
c_s	sound velocity (m/sec)
c_1^*, c_2^*	parameters defined in Equations 6.101 and 6.103

NOTE: *sub* indicates a subscript, either numeric or alphanumeric.

D	diffusion coefficient (m/sec^2); also used for distillate flow (kg/sec)
d	diameter (m)
E	activation energy (J/mol)
e	parameter defined by Equation 12.55
F	feed flow (kg/sec); also used for area (m^2)
f	parameter defined in Equation 12.50; also used for friction factor
G(s)	transfer function
g	parameter defined in Figure 7.4; also used for gravity acceleration (m/sec^2)
$g_i(s)$	transfer function (i = 1 . . . 3)
H_{sub}	enthalpy (J/kg)
ΔH_R	heat of reaction (J/kg)
h	step length in integration; also used for height (m)
h_w	weir height (m)
h_{sub}	heat transfer coefficient (W/m^2-°K)
I	intensity delta function (kg/m^2); also used for fixed costs ($)
i	current (A)
$J_{1,2}$	roots of characteristic equation
k	mass transfer coefficient (m^2/sec)
k, k_1, k'	reaction rate constant (sec^{-1})
K_c	controller gain
K_θ, K_p, K_i	process gain (i = 1 . . . 5)
K_s, K_e, K_0	parameters defined by corresponding equation
k_0	preexponential factor (sec^{-1})
k_d, k_g	attrition, growth rate constant (sec^{-1})
k_{g0}	maximum specific growth rate (sec^{-1})
k_t	heat conduction coefficient (W/m-°K)
L	level, length (m); also coefficient of self-induction
L_i	liquid flow (kg/sec)
Δl, l	length (m)
M, M_{sub}	mass (kg)
M_f	mass per unit length (kg/m)
m	molecular weight; also used for distribution coefficient
Δm	orifice output signal
N	speed of rotation (sec^{-1}); also used for number of segments, i.e., reactors
n	section number
P, P_{sub}	pressure (N/m^2); P also used for probability
$\Delta p, \Delta P$	pressure drop (N/m^2)
P_0	oscillation period (sec) for $\zeta = 0.25$
p	parameter (m^{-1}) defined in Equation 9.56

Pe*	Péclet number according to Equation 11.59
Pe	Péclet number according to Equation 11.62
Q	volumetric flow (m^3/sec)
Q_{sub}	supplied heat (W)
R, R_{sub}	resistance coefficient; R also used for reflux flow (kg/sec) and reliability
R_g	gas constant (J/mol-°K)
R_v	valve rangeability
R(z,t)	rate of reaction (kg/m^3-sec)
r_A	rate of reaction for component A (mol/m^3-sec)
r_{s1}	gross rate of substrate consumption (kg/m^3-sec)
r_1, r_2	parameters defined in Equation 6.75
r_c	frequency of dangerous coincidences (sec^{-1})
S	extract flow (kg/sec)
s	Laplace operator
s*	parameter defined in Equation 11.56
T	temperature (°K)
T_f, T_{ff}, T_{ws}, T_{wf}	time constants (sec) according to Equations 10.3, 10.4, 10.5 and 10.14
t	time (sec)
U	overall heat transfer coefficient (W/m^2-°K); also used for voltage and unavailability
u	process input variable; signal to valve shaft
u_r	relative valve input signal
V, V_{sub}	volume (m^3); V also used for vapor flow (kg/sec)
v	velocity (m/sec)
W	amount of pollutant per unit cross-sectional area (kg/m^2)
w	disturbance variable
x	process state variable; also used for mole fraction liquid
x_f	feed compound mole fraction
x_i	parameters defined in Equations 12.51 and 12.55 (i = 1 . . . 3)
Y	conversion factor according to Equation 7.3
y	process output variable; also used for mole fraction gas
z	geometric coordinate (m); also used for mole fraction in the feed
z_i	compressibility

GREEK SYMBOLS

α, α_{sub}	constant or parameter, defined by the corresponding equation; also used for heat transfer coefficient (W/m^2-°K)
β	constant or parameter defined by the corresponding equation

γ, γ_{sub} constant or parameter defined by the corresponding equation; also used for specific heat (J/kg-°K)

δ variation around the equilibrium state

ϵ deviation between measured and desired value; also used for gas fraction

φ phase shift (rad)

ϕ, ϕ_{sub} constant or parameter defined by the corresponding equation; also used for flow (kg/s, m^3/sec)

κ ratio of specific heats; also used for conductivity (mmho)

$\lambda, \lambda_{1,2}$ constant or parameter, defined by the corresponding equation

μ dynamic viscosity (kg/m-sec); also used for mean future rate (sec^{-1})

μ_i i^{th} moment

μ_f, μ_s functional, i.e., operational, failure frequency (sec^{-1})

ν correlation coefficient

ω, ω_{sub} oscillation frequency (rad/sec)

π pi, 3.1415

θ_{sub} temperature (°K)

ρ, ρ_{sub} density (kg/m^3)

ψ constant or parameter defined by corresponding equation

$\psi_{a,b}$ dimensionless concentration

τ, τ_{sub} time constant, defined by corresponding equation

σ_z variance

ζ damping coefficient; also used for dimensionless geometric coordinate

INDEX